A TEXTBOOK OF
MACHINE DRAWING

This Page has been intentionally left blank

A TEXTBOOK OF MACHINE DRAWING

[ENGINEERING DRAWING-2]

[For Degree, Diploma and A.M.I.E. Students]

By

Prof. P. J. SHAH, M.E. (Mech.)

Formerly Professor of Mechanical Engineering,
L. D. College of Engineering,
Ahmedabad - 380 015

Assisted by

(1) Shri Parag P. Shah [B. E. Mechanical]
(2) Shri Keyur P. Shah [B. E. Mechanical]
(3) Shri Umang P. Shah [Graphic Designer]

[With more than 550 Diagrams, 250 Solved Problems and 200 Exercises]

S. CHAND
PUBLISHING

S Chand And Company Limited

(ISO 9001 Certified Company)

S Chand And Company Limited

(ISO 9001 Certified Company)

S. CHAND
PUBLISHING

Head Office: Block B-1, House No. D-1, Ground Floor, Mohan Co-operative Industrial Estate, New Delhi – 110 044 | Phone: 011-66672000

Registered Office: A-27, 2nd Floor, Mohan Co-operative Industrial Estate, New Delhi – 110 044
Phone: 011-49731800

www.**schandpublishing.com**; e-mail: **info@schandpublishing.com**

Branches

Chennai	:	Ph: 23632120; chennai@schandpublishing.com
Guwahati	:	Ph: 2738811, 2735640; guwahati@schandpublishing.com
Hyderabad	:	Ph: 40186018; hyderabad@schandpublishing.com
Jalandhar	:	Ph: 4645630; jalandhar@schandpublishing.com
Kolkata	:	Ph: 23357458, 23353914; kolkata@schandpublishing.com
Lucknow	:	Ph: 4003633; lucknow@schandpublishing.com
Mumbai	:	Ph: 25000297; mumbai@schandpublishing.com
Patna	:	Ph: 2260011; patna@schandpublishing.com

First Edition 1988
Second Revised Edition 1990-91
Third Revised & enlarged Edition 1996-97 (In 1st Angle Projection System)
Fourth Edition 1999, 2000
Fifth Improved & Enlarged Edition 2005-2006
S. Chand Edition 2008
Reprints 2009, 2016

ISBN : 978-81-219-2966-0 **Product Code:** H3MDR62EDES10ENAE16R

PRINTED IN INDIA

By Vikas Publishing House Private Limited, Plot 20/4, Site-IV, Industrial Area Sahibabad, Ghaziabad – 201 010 and Published by S Chand And Company Limited, A-27, 2nd Floor, Mohan Co-operative Industrial Estate, New Delhi – 110 044.

PREFACE TO THE THIRD REVISED & ENLARGED EDITION

New illustrative problems with solution are added to strengthen the subject matter. Diagrammes are made more illustrative and most of them in The 1st Angle System of Projection.

Corrections, suggestions, modifications, etc., to improve the book will be gratefully received by me.

Ahmedabad-9. **Prof. P. J. Shah**

PREFACE TO THE SECOND REVISED EDITION

The overwhelming response to the First Edition has necessitated the bringing out of its Second Revised Edition within a short span of time.

I am pleased to add chapters on (1) Couplings (2) Bearings and (3) Detail and Assembly Drawings in this volume.

Many Exercises are given at the end to each chapter with solutions given at the back of the book. Each chapter contains an adequate number of solved and unsolved problems to imbue self-confidence in the reader. More problems with solutions on Isometric and Ortho-Reading are added at the back of the book.

Corrections, suggestions, modifications etc. to improve the book will be gratefully received by me.

Ahmedabad-15.

Prof. P. J. Shah

CONTENTS

		Page
Ch. 15	Orthographic Projections	15.01-15.28
Ch. 16	Sectional Orthographic Projections	16.01-16.44
Ch. 17	Orthographic Reading	17.01-17.37
Ch. 18	Isometric (Projection/View/Drawing) (Axonometric Projection)	18.01-18.58
Ch. 19	Detail and Assembly Drawings	19.01-19.08
Ch. 20	Dimensioning	20.01-20.16
Ch. 21	Fasteners	21.01-21.26
Ch. 22	Couplings	22.01-22.07
Ch. 23	Bearings	23.01-23.09
	Solutions of exercises	Ex-Sol. 01-09

CHAPTER 15
ORTHOGRAPHIC PROJECTIONS

1. Introduction :

Three kinds of projections are used by engineers to describe the objects.

1. Orthographic ; 2. Oblique ; 3. Perspective.

They differ from each other in the relationship of the point of sight and the rays of projections to the plane of projection. [See Fig.15.1 (a) (b) (c)].

In orthographic projection, it is assumed that observer is at infinite distance and rays or projection lines are parallel and perpendicular (i.e. ortho.) to the plane of projection. It is extensively used by engineers for the purpose of communication because these projections reveal eact shape and size of the object.

In oblique projection, it is assumed that observer is at infinite distance and rays or projection lines are parallel and not perpendicular (i.e. oblique) to the plane of projection. Oblique projection is a three dimensional projection. Principal face of the object is kept parallel to the plane of projection.

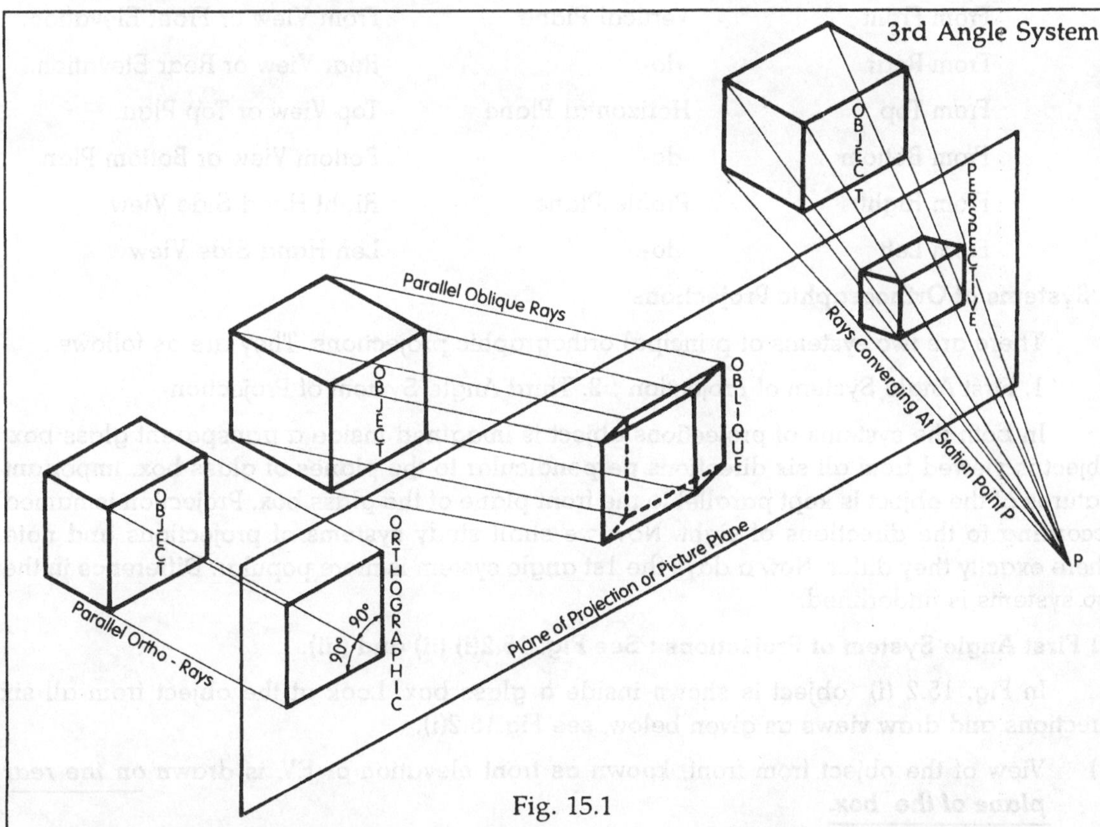

Fig. 15.1

Principles of (a) Orthographic (b) Oblique (c) Perspective projections

In perspective projection, it is assumed that observer is at finite distance and rays or projection lines are converging at observer's eye known as '**station point**'. Civil engineers and architects use perspective projection to show the apperance of the buildings and large complexes. It does not reveal exact size and shape of the object and hence is not used as a working drawing.

2. Orthographic Projection :

Orthographic projections are named according to the direction of sight or according to the picture plane on which projections are taken and drawn. They are as follows :

1. Principal orthographic projections; 2. Auxiliary orthographic projections;

3. Axonometric projections.

In this chapter, study of only principal orthographic projections is covered. Study of Auxiliary projections and Axonometric projections will be done separately in chapter No.20 and No.18 respectively.

There are six directions of sight, six principal planes perpendicular to these six directions of sight and correspondingly there are six principal projections. They are given in the table below :-

Sr. No.	Direction of Sight	Plane of Projection	Name of Projection
1.	From Front	Vertical Plane	Front View or Front Elevation.
2.	From Rear	-do-	Rear View or Rear Elevation.
3.	From Top	Horizontal Plane	Top View or Top Plan.
4.	From Bottom	-do-	Bottom View or Bottom Plan
5.	From Right	Profile Plane	Right Hand Side View
6.	From Left	-do-	Left Hand Side View

3. Systems of Orthographic Projections :

There are two systems of principal orthographic projections. They are as follows :

1. First Angle System of Projection ; 2. Third Angle System of Projection

In both the systems of projections object is imagined inside a transparent glass box. Object is viewed from all six directions perpendicular to the planes of glass box. Important features of the object is kept parallel to the front plane of the glass box. Projection is named according to the directions of sight. Now we shall study systems of projections and note where exactly they differ. Now a days the 1st angle system is more popular. Difference in the two systems is underlined.

3.1 First Angle System of Projections : See Fig. 15.2(i) (ii) and (iii).

In Fig. 15.2 (i) object is shown inside a glass box. Look at the object from all six directions and draw views as given below, see Fig.15.2(i).

(1) View of the object from front, known as front elevation or F.V. is drawn ***on the rear plane of the box.***

(2) View of the object from rear, known as rear elevation or R.V. is drawn ***on the front plane of the box.***

1st Angle System

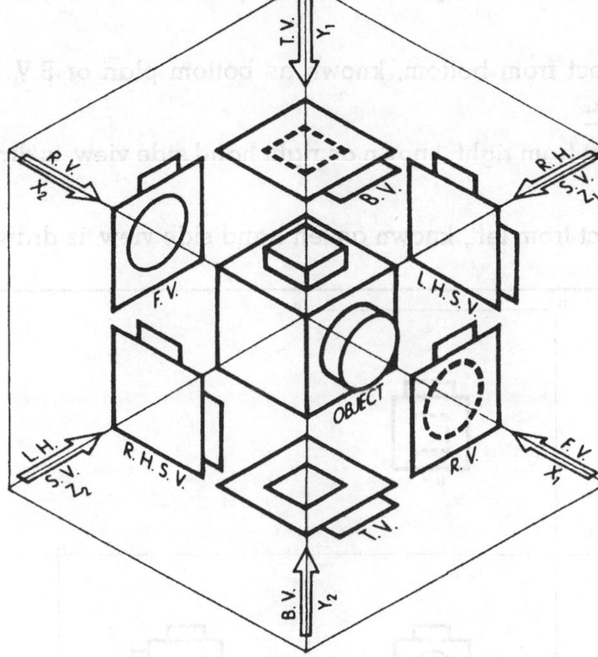

Fig. 15.2 (i)

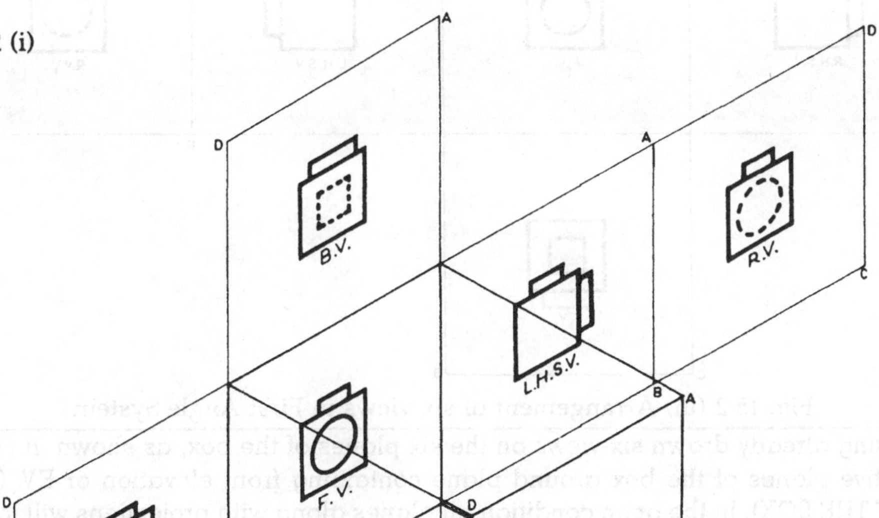

Fig. 15.2 (ii)

(3) View of the object from the top, known as top plan or T.V. is drawn ***on the bottom plane of the box.***

(4) View of the object from bottom, known as bottom plan or B.V. is drawn ***on the top plane of the box.***

(5) View of the object from right, known as right hand side view, is drawn ***on the left plane of the box.***

(6) View of the object from left, known as left hand side view, is drawn ***on the right plane of the box.***

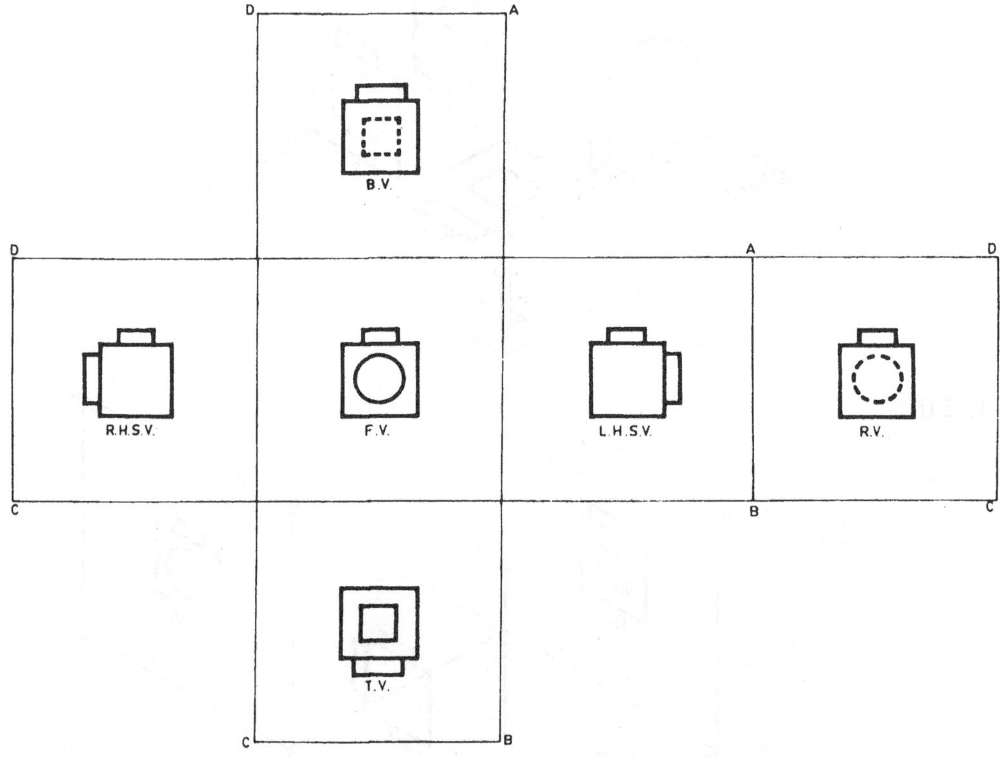

Fig. 15.2 (iii) Arrangement of six views in First Angle System

Having already drawn six views on the six planes of the box, as shown in Fig. 15.2(i), open out five planes of the box around plane containing front elevation or F.V. (i.e. REAR PLANE OF THE BOX). In the open condition all planes along with projections will look like as shown in Fig. 15.2(ii) . Views of Fig.15.2(ii) are finally drawn as shown in Fig.15.2 (iii).

Observe these views to understand following conclusions for **1st Angle System of Projection.**

(1) Front Elevation or F.V. remains in the centre of all views.

(2) Top Plan or T.V. is vertically ***below*** F.V.

(3) Bottom Plan or B.V. is vertically ***above*** F.V.

(4) R.H.E.V. or R.H.S.V. is horizontally on ***left*** of F.V.

(5) L.H.E.V. or L.H.S.V. is horizontally on ***right*** of F.V.

(6) Rear Elevation or Rear View is horizontally on extreme right of F.V.

(7) Front portion of the object , in all views surrounding F.V., ***faces away*** from F.V.

(8) Rear portion of the object, in all views surrounding F.V., ***faces*** F.V.

(9) Length of F.V. and T.V. is same.

(10) Height of the F.V. and S.V. is same.

(11) The depth of the top view is the same as the depth of the side view.

3.2. Third Angle System of Projections : See Fig. 15.3(i), (ii) & (iii)

In Fig. 15.3(i) object is shown inside a glass box. Look at the object from all six directions and draw six views, as given below, see Fig. 15.3 (i).

Third Angle System

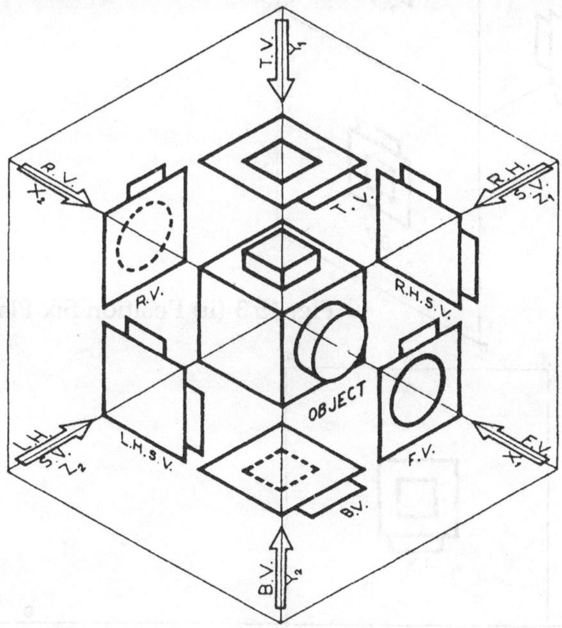

Fig. 15.3 (i) Glass Box

(1) View of the object from front, known as front elevation or F.V., is drawn on the ***Front Plane of the Box.***

(2) View of the object from rear, known as rear elevation or R.V., is drawn on the ***Rear Plane of the Box.***

(3) View of the object from top, known as top plan or T.V., is drawn on the **Top *Plane of the Box.***

(4) View of the object from bottom, known as bottom plan or B.V., is drawn on the ***Bottom Plane of the Box.***

(5) View of the object from right, known as right hand side view, is drawn on the ***Right Plane of the Box.***

(6) View of the object from left, known as left hand side view, is drawn on the ***Left Plane of the Box.***

3rd Angle System

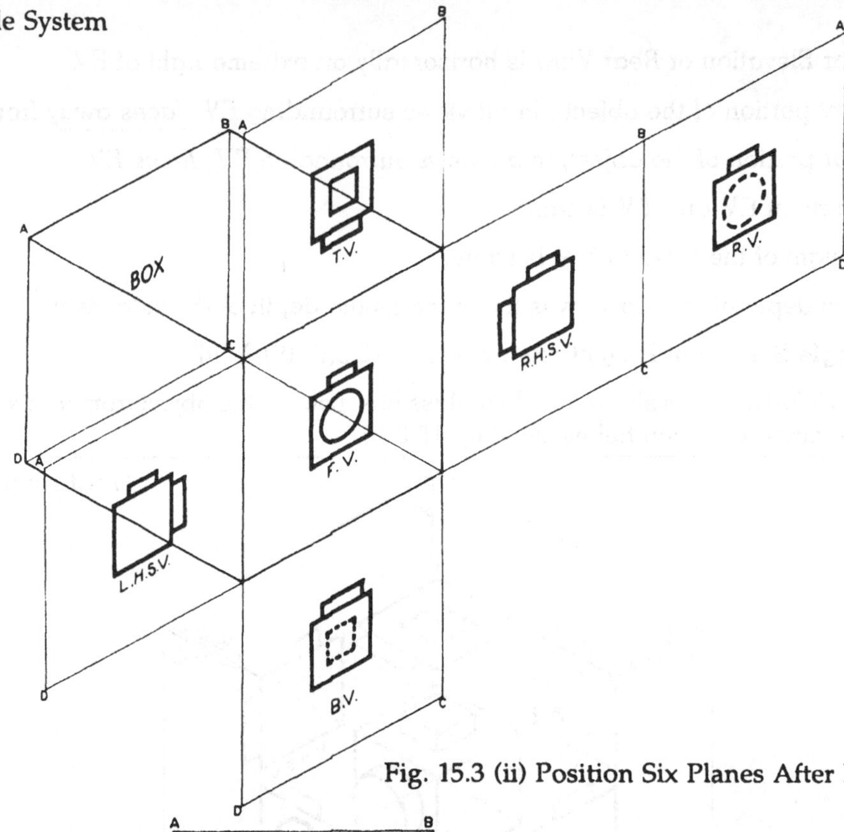

Fig. 15.3 (ii) Position Six Planes After Rotation

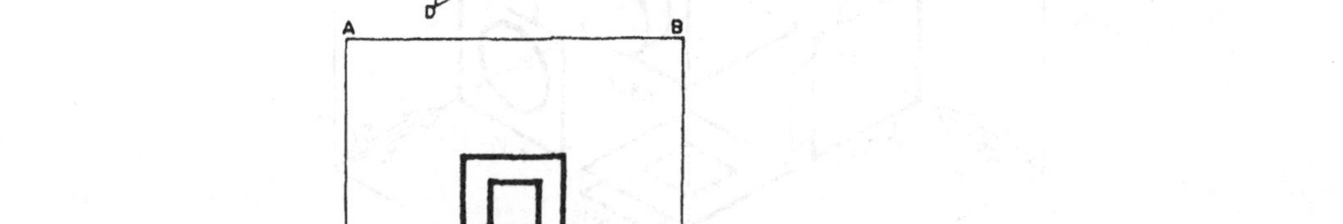

Fig. 15.3 (iii) Arrangement of Six Views in 3rd Angle System

Having already drawn six views on the six planes of the box, as shown in Fig. 15.3(i), open out 5 planes of the box around plane containing front elevation or F.V. i.e. Front Plane of the Box. In the open condition all planes along with projections will look like as shown in Fig. 15.3(ii). Views of Fig.15.3(ii) are finally drawn, as given in Fig.15.3(iii).

Observe these views to understand following conclusions for the **3rd Angle System of Projection.**

(1) Front Elevation or F.V. remains in the centre of all views.

(2) Top Plan or T.V. is vertically above F.V.

(3) Bottom Plan or B.V. is vertically below F.V.

(4) R.H.E. V. or R.H.S.V. is horizontally on right of F.V.

(5) L.H.E.V. or L.H.S.V. is horizontally on left of F.V.

(6) Rear Elevation or Rear view is horizontally on extreme right of F.V.

(7) Front portion of the object, in all views, surrounding F.V. faces the front view.

(8) Rear portion of the object in all views surrounding F.V. faces away from the F.V. Conclusions (9) (10) and (11) of the 1st angle system remain same for the 3rd angle system.

4. Important Conclusions :

Now it is advisable to go through some important conclusions regarding projections of straight lines and projections of planes.

(1) If a line is perpendicular to one of the principal planes then it will be parallel to two other principal planes.

(2) If a line is perpendicular to one of the principal planes then its projection on that principal plane will be point view and on other two principal planes it will show the true length.

(3) If a line is parallel to one of the principal planes and inclined to two other P.Ps. then its projection on parallel P.P. will be the true length showing true inclinations and on other P.Ps it will show shorter lines parallel to ground lines (Vertical or Horizontal).

(4) If a line is inclined to all P.Ps. then its all projections will be shorter and inclined to ground lines.

(5) If a surface is parallel to one of the principal planes then its projection on it will be the true shape and projections on other two principal planes will be straight lines parallel to corresponding ground lines.

(6) If a surface is perpendicular to one of the principal planes and inclined to two other, its projection on it will be a straight line and other two views will be foreshortened.

(7) If a plane is inclined to all principal planes then nowhere it will show straight line view and all projections will be foreshortened.

(8) Edge in an object is formed by intersection of two surfaces.

(9) At every corner of solid object , atleast three surfaces meet.

(10) Corner of a solid is formed by intersection of atleast three edges. Most of the

corners of solids are formed by three mutually perpendicular lines.

5. Methods of Projections

The projections of an object are obtained by one of the following methods :

(1) The glass box method, see Fig. 15.2 (i) (ii) (iii) or 15.3 (i) (ii) (iii)

(2) The natural method, see Fig. 15.4

Results, i,e. projections, will be same whether you adopt the glass box method or you adopt the natural method.

(1) Glass Box Method : See Fig. 15.2 and Fig. 15.3

We have already studied glass box method while studing the systems of projections.

(2) Natural Method : See Fig. 15.4

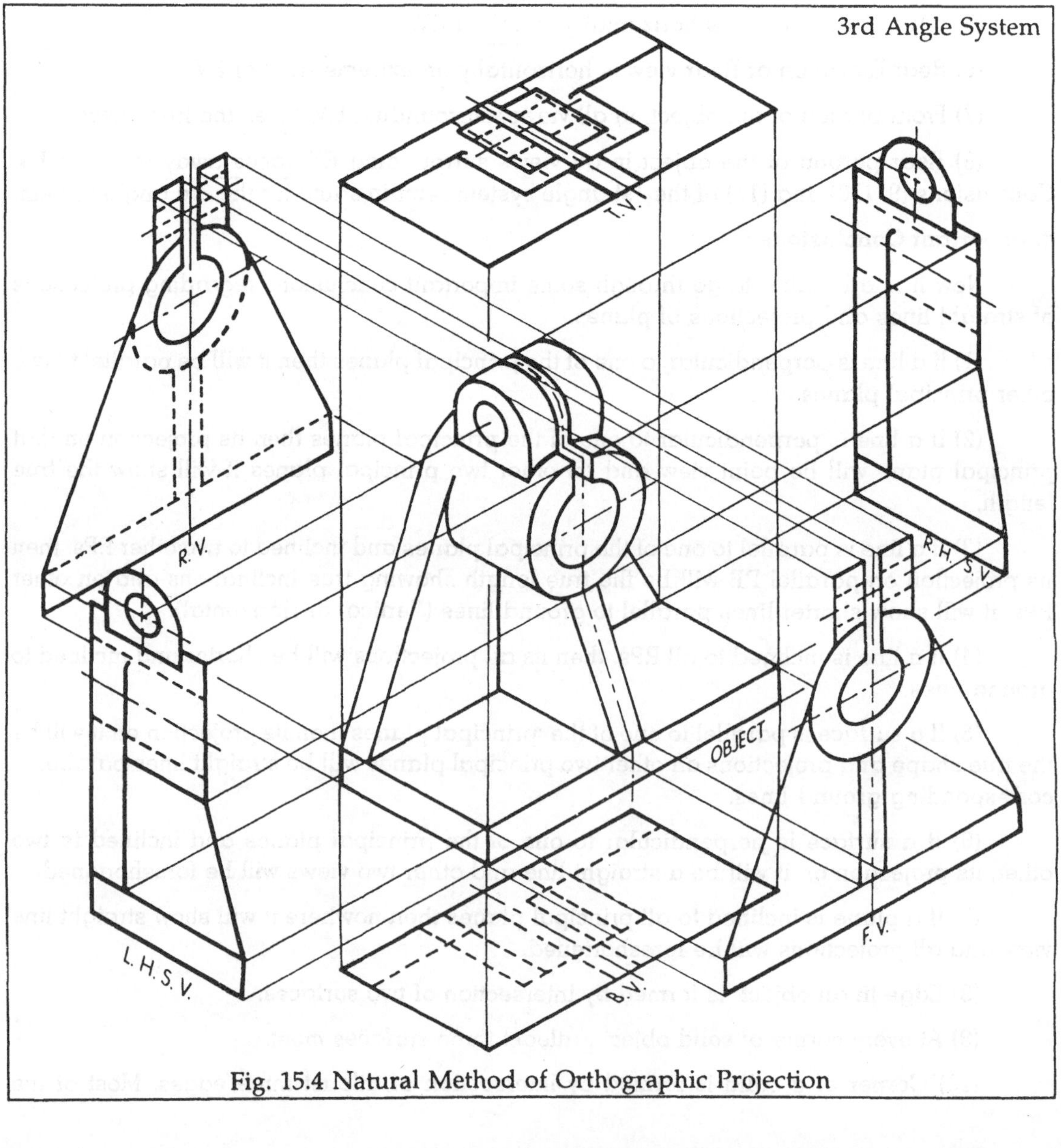

Fig. 15.4 Natural Method of Orthographic Projection

Object is kept in a position such that its important features face front side. Now look at the object from front, top, bottom, right, left and rear to get front view, top view, bottom view, right hand side view, left hand side view and rear view respectively.

In Fig. 15.4, six principal projections according to 3rd Angle System are shown. In the application of this method either observer changes his positions and observes the object from six different positions or observer remains at one position and turn the object to observe 6 faces of the object to get six different projections.

6. Selection of Views

Decision about number of views to be drawn and selection of views, should be such that no misinterpretation occures about the shape of object to the reader. Whenever dispute arises between two firms the drawing on which the contract is made, becomes the important document. Number of views and selection of views should be such that the reader of drawing can imagine one and only one object. Views should clearly and completely describe the object. Superfluous views should be avoided. Views should be selected in such a way that dotted lines will be minimum in views.In case of more complex objects, sectional views, auxiliary views, etc. are drawn to make the object more clear. Projections/views showing contours, profiles, etc. of the object must be selected. Rear and bottom views are rarely used. Mostly important features of the object are taken on the front view, known as the principal view.

7. Analysis of Points, Lines and Surfaces in Principal Views or Projections

Exercise - 1 : In Fig. 15.5(i) three dimensional view of the object is given. Draw (i) F.V./Front Elevation (ii) L.H. S.V. and (iii) T. V. Use 1st Angle Projection System.

To get the front elevation look at it in the direction of arrow X. Elevation is the king of all views and hence selection of direction X should be done in such a way as to get maximum details in elevation.

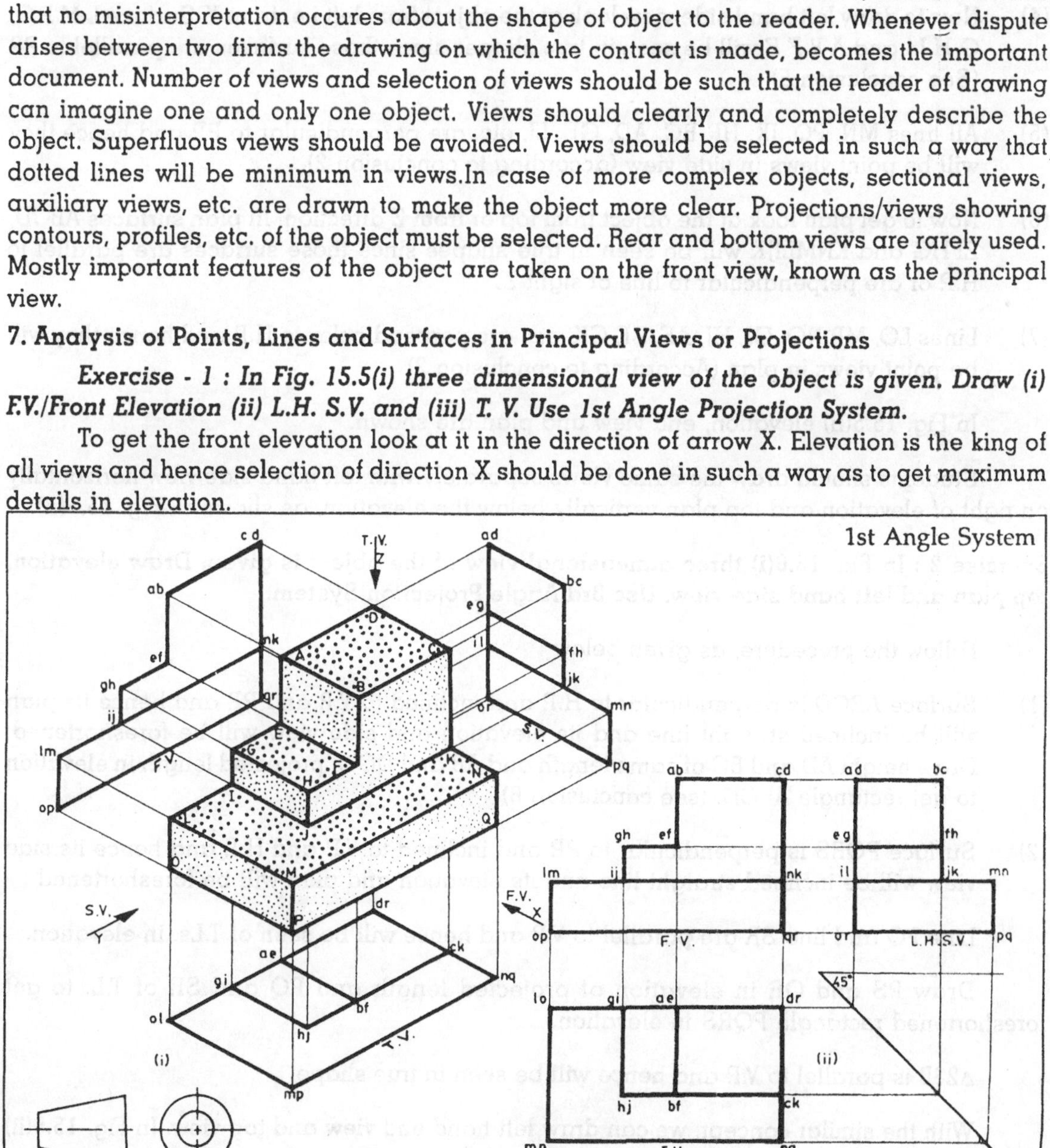

Fig. 15.5 Exercise 1

(1) Draw rectangle mpqn of the same size as rectangle MPQN of the object, since rectangle MPQN is parallel to the plane of projection (V.P.). Its elevation mpqn will be T. S. (See conclusion : 5).

(2) Surface BFHJKCB is also parallel to V.P. or perpendicular to the line of sight and hence it will be seen in true shape in elevation. Draw it (According to the same conclusion 5)

(3) All lines OP, LM, IJ, GH, EF, AB, DC, KN, etc. are perpendicular to V.P. and hence they will be point views in elevation. (According to conclusion: 2).

(4) Now to draw left hand side view look at the object from left i.e. from Y. Surface L M P O, G H J I and A B F E will be seen in true shapes since those surfaces are parallel to P.P. (See conclusion 5)

(5) All lines MN, PQ, JK, HF, BC, AD, GE, LI, etc. are perpendicular to P.P. and hence they will be point views in side view (according to conclusion 2).

(6) Now to get plan look at the object from top or from Z direction. In plan surfaces ABCD, EFHG and KNMLIJK will be seen in true shapes since those surfaces are paraliel to H.P. or are perpendicular to line of sight Z.

(7) Lines LO, MP, NQ, GI, HJ, AE, BF, CK, etc. are perpendicular to H.P. and hence they will be point views in plan (According to conclusion 2)

In Fig. 15.5(ii) elevation, end view and plan are shown.

Students should draw the same views separately with left hand side view horizontally on right of elevation and top plan vertically below the elevation, as shown in Fig. 15.5(ii).

Exercise 2 : In Fig. 15.6(i) three dimensional view of the object is given. Draw elevation, top plan and left hand side view. Use 3rd Angle Projection System.

Follow the procedure, as given below :

(1) Surface ABCD is perpendicular to H.P. and inclined to V.P. and P.P. and hence its plan will be inclined straight line and its elevation and side view will be foreshortened. Draw height AD and BC of same length and AB and DC of projected length in elevation to get rectangle ABCD. (see conclusion 6)

(2) Surface PQRS is perpendicular to P.P. and inclined to V.P. and H.P. and hence its side view will be inclined straight line and its elevation and plan will be foreshortened.

Line PQ and line SR are parallel to V.P. and hence will be seen of T.Ls. in elevation.

Draw PS and QR in elevation of projected length and PQ and SR of T.L. to get foreshortened rectangle PQRS in elevation.

Δ23B is parallel to V.P. and hence will be seen in true shape.

With the similar concept, we can draw left hand end view and top view. In Fig. 15.6(ii) elevation, left hand end view and top plan are given.

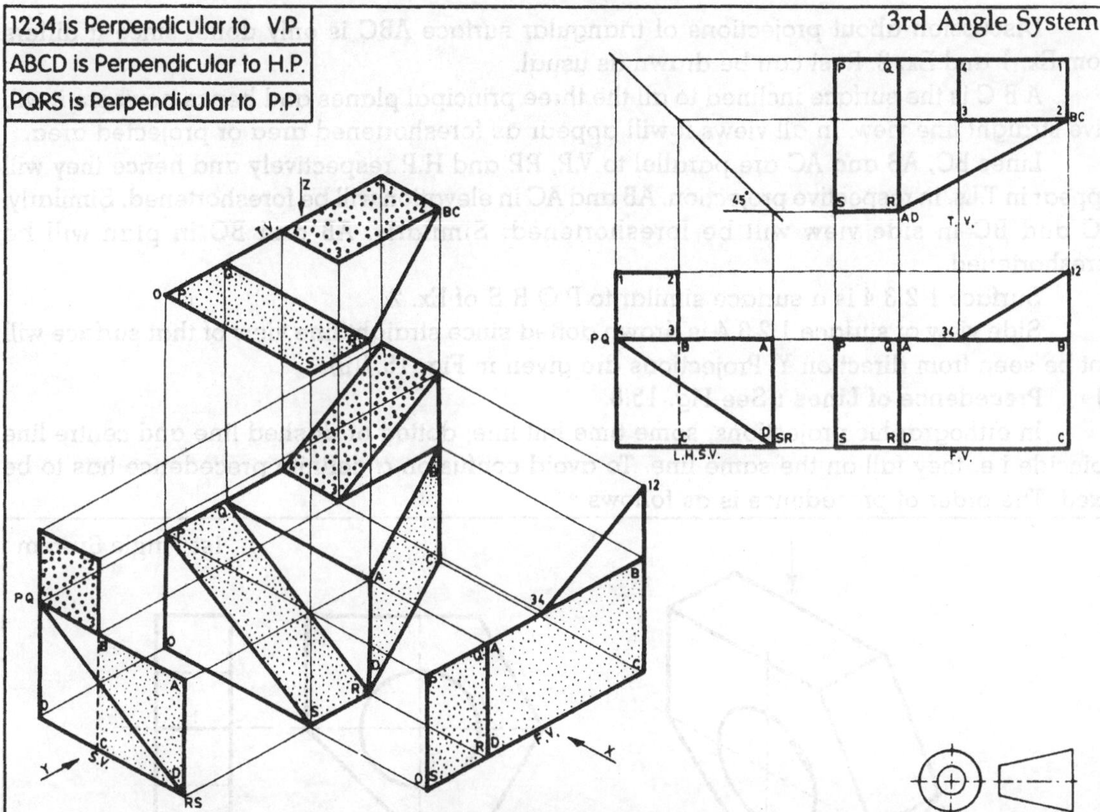

Fig. 15.6 Exercise 2

Exercise 3: In Fig. 15.7(i) three dimensional view of the object is given. Draw the elevation, left hand side view and top plan. Use 1st Angle Projection System.

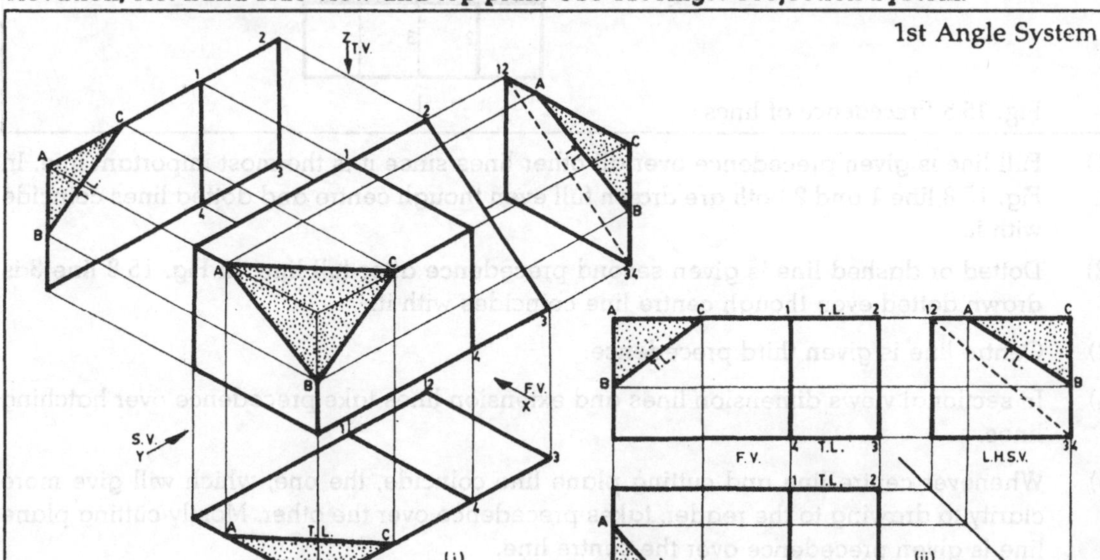

Fig. 15.7 Exercise 3

Discussion about projections of triangular surface ABC is only done, since it differs from Ex. 1 and Ex. 2. Rest can be drawn as usual.

A B C is the surface inclined to all the three principal planes and hence nowhere it will give straight line view. In all views it will appear as foreshortened area or projected area.

Lines BC, AB and AC are parallel to V.P., P.P. and H.P. respectively and hence they will appear in T.Ls. in respective projection. AB and AC in elevation will be foreshortened. Similarly, AC and BC in side view will be foreshortened. Similarly, AB and BC in plan will be foreshortened.

Surface 1 2 3 4 is a surface similar to P Q R S of Ex. 2.

Side view of surface 1 2 3 4 is drawn dotted since straight line view of that surface will not be seen from direction Y. Projections are given in Fig. 15.7(ii)

8) Precedence of Lines : See Fig. 15.8.

In orthographic projections, some time full line, dotted or dashed line and centre line coincide i.e. they fall on the same line. To avoid confusion a definite precedence has to be fixed. The order of precedence is as follows :

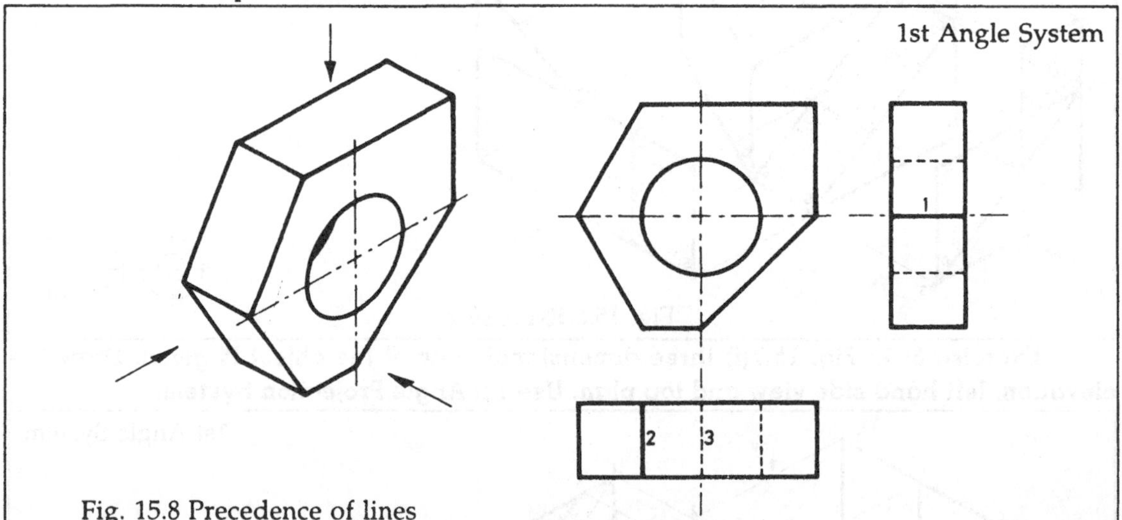

Fig. 15.8 Precedence of lines

(1) Full line is given precedence over all other lines since it is the most important line. In Fig. 15.8 line 1 and 2 both are drawn full even though centre and dotted lines coincide with it.

(2) Dotted or dashed line is given second precedence after full line. In Fig. 15.8 line 3 is drawn dotted even though centre line coincides with it.

(3) Centre line is given third precedence.

(4) In sectional views dimension lines and extension lines take precedence over hatching lines.

(5) Whenever centre line and cutting plane line coincide, the one, which will give more clarity to drawing to the reader, takes precedence over the other. Mostly cutting plane line is given precedence over the centre line.

9 Procedure for Preparing Orthographic Projections to Scale

Follow the procedure step by step as given below :

(1) First of all decide the direction of principal view i.e. F.V. for the given pictorial view or

a real object. Decide also whether pictorial drawing given is oblique drawing or isometric drawing.

(2) Do selection of views and decide about number of views to be drawn.

(3) Decide whether sectional and/or auxiliary views are required or not

(4) Prepare free hand sketches of required projections by soft pencil.

(5) Knowing overall dimensions of the views and knowing the size of the paper decide a suitable scale.

(6) Prepare border line and title block.

(7) Prepare rectangles for the views allowing sufficient space between two views and between views and the border line of the sheet. Keep in mind space required for dimensioning.

(8) Draw centre lines, if any, in all the views.

(9) Draw details simultaneously in all the views. It is a bad practice to prepare one view completely and then to start another view.

Draw details in the following order :

(i) Full, half and quarter circle.

(ii) Lines for the major shape of the object.

(iii) Lines for the minor shape of the object.

(10) After completing all the views, erase extra lines and fair required lines with 2H, 3H or 4H pencil.

(11) Give dimensions in all the views. As far as possible do not give dimensions inside the object, inside the hatched zone, etc. Give dimensions of features in appropriate views.

(12) Draw hatching lines in the sectional views.

(13) Draw cutting plane line and show arrows for direction of sight.

(14) Name the views.

(15) If views for many objects are drawn in one sheet then give numbers.

(16) Check carefully for missing dimensions and try to read the object from the views.

10. Types of Pictorial Projections

Having studied and grasped the system of projections, methods of projections, important conclusions and analysis of points, lines and planes in projections, we shall now take some most simple practice problems. Complicated problems are nothing but combination of many simple problems.

Along with the knowledge, imagination of object plays very important role. Pictorial projections given are either **Axonometric Projections** (i. e. Isometric, diametric & trimetric) or **Oblique Projections.** Isometric is very widely used projection, see Fig. 15.9. In pictorial projections mostly hidden parts are not shown even by dotted lines and hence we have to imagine the shape of hidden parts on the basis of symmetry.

To get principal orthographic projections we look at the object in the directions of axes

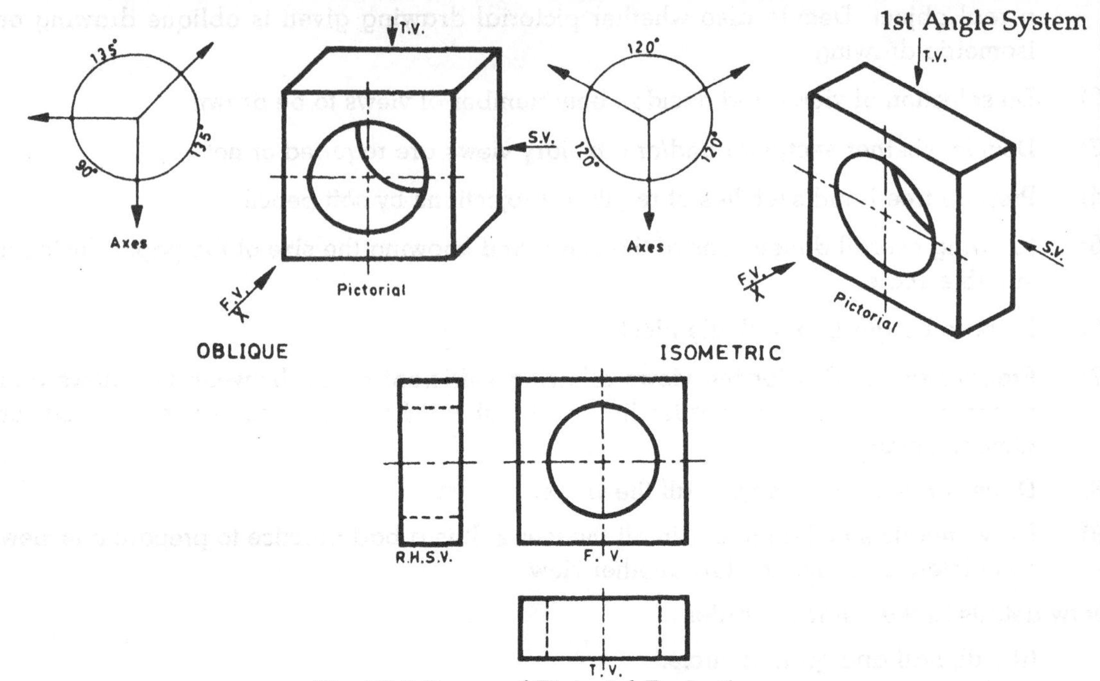

Fig. 15.9 Types of Pictorial Projections

of pictorial projection. Axes of projections for oblique as well as for isometric are shown in Fig. 15.9. along with both the type of projections. When we look at the object in the direction of any one axis, lines and dimensions of the object in that particular direction will not be seen. Dimensions and lines in the directions of two other axes will be seen in the true length, e.g. when you are on a railway track you cannot see the length of the train but you can see height and width of the guard's compartment or of the engine.

11. Practice Problems

***Problem 1* : In Fig. 15.10(1) pictorial view of a rectangular plate is given, draw its (i) F.V. (ii) T. V. and (iii) R.H.S.V. Use 1st Angle System.**

From direction X, dimension W will not be seen and dimensions L and H will be seen true and hence we get rectangle (L x H) as F. V. Similarly, R.H.S.V. will be rectangle (W x H) and plan will be rectangle (L x W). All three views are shown in Fig. 15.10(1)

***Problem 2* : In Fig. 15.10(2) pictorial projection of a plate is given, draw its (i) F.V. (ii) T.V. and (iii) R.H.S.V. Use 1st Angle System.**

From direction X, (dimension W) thickness will not be seen and shape of front plane will be seen in true shape. And hence F.V. will be as shown in the figure. R.H.S.V. and T.V. will be same as the views of problem 1.

***Problem 3* : In Fig. 15.10(3) pictorial projection of a plate is given, draw its (i) F.V. (ii) T.V. and (iii) R.H.S.V. Use 1st Angle System.**

Projections are shown in the figure. Note that R.H.S.V. and T.V. are same as for problems 1 and 2.

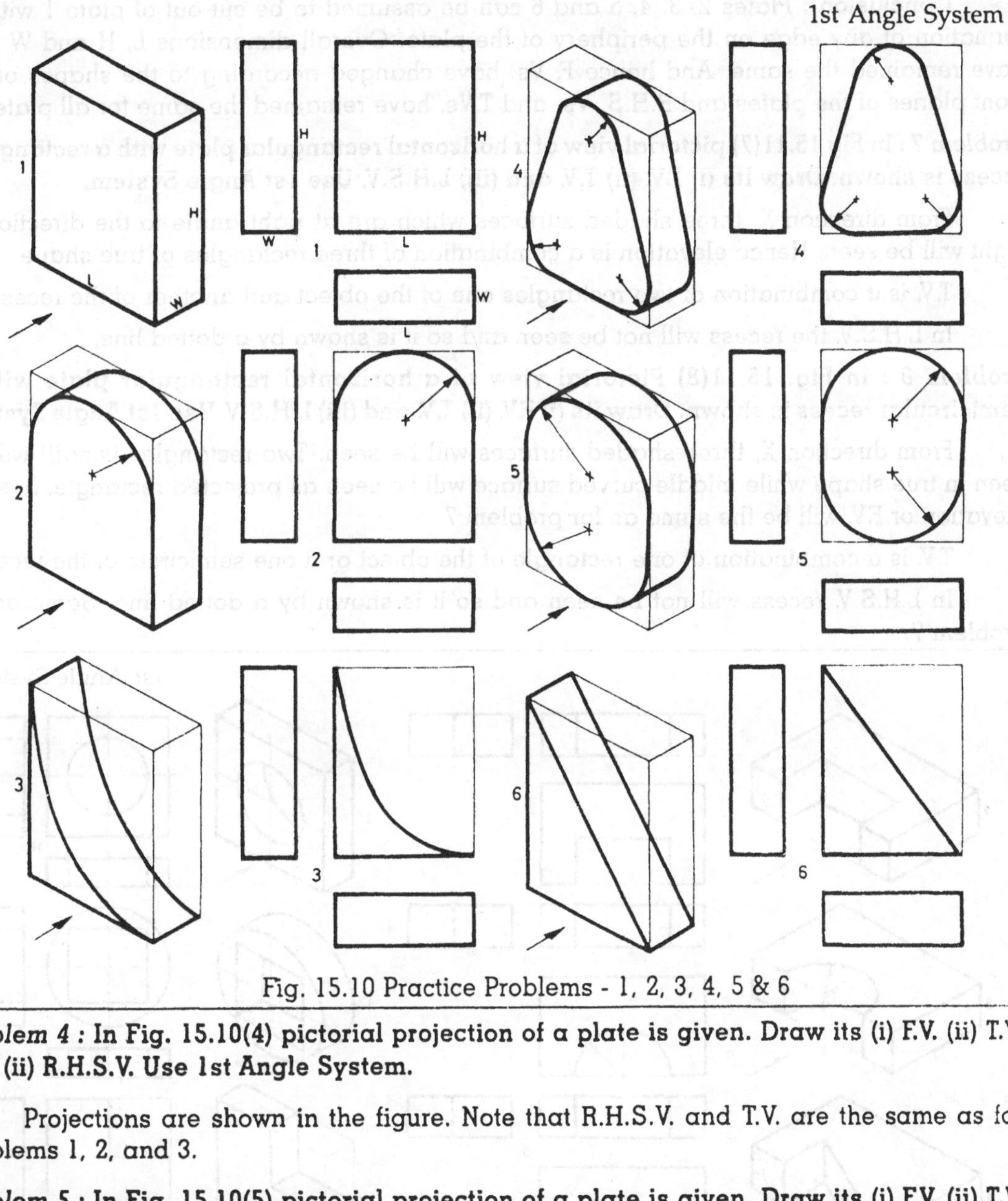

Fig. 15.10 Practice Problems - 1, 2, 3, 4, 5 & 6

***Problem 4* : In Fig. 15.10(4) pictorial projection of a plate is given. Draw its (i) F.V. (ii) T.V. and (ii) R.H.S.V. Use 1st Angle System.**

Projections are shown in the figure. Note that R.H.S.V. and T.V. are the same as for problems 1, 2, and 3.

***Problem 5* : In Fig. 15.10(5) pictorial projection of a plate is given. Draw its (i) F.V. (ii) T.V. and (iii) R.H.S.V. Use 1st Angle System.**

Projections are shown in the figure. Note that its R.H.S.V. and T.V. are the same as for problems 1, 2, 3 and 4.

***Problem 6* : In Fig. 15.10(6) pictorial projection of a plate is given. Draw its (i) F.V. (ii) T.V. and (iii) R.H.S.V. Use 1st Angle System.**

Projections are shown in the figure. Note that its R.H.S.V. and T.V. are the same as for problems 2, 3, 4 and 5.

Conclusion : Plates 2, 3, 4, 5 and 6 can be assumed to be cut out of plate 1 without formation of any edge on the periphery of the plate. Overall dimensions L, H and W also have remained the same. And hence F. Vs. have changed according to the shapes of the front planes of the plates and R.H.S. Vs. and T.Vs. have remained the same for all plates.

***Problem 7 :* In Fig 15.11(7) pictorial view of a horizontal rectangular plate with a rectangular recess is shown. Draw its (i) F.V. (ii) T.V. and (iii) L.H.S.V. Use 1st Angle System.**

From direction X, three shaded surfaces which are at right angle to the direction of sight will be seen. Hence elevation is a combination of three rectangles of true shape.

T.V. is a combination of two rectangles one of the object and another of the recess.

In L.H.S.V. the recess will not be seen and so it is shown by a dotted line.

***Problem 8 :* In Fig. 15.11(8) Pictorial view of a horizontal rectangular plate with a semicircular recess is shown. Draw its (i) F.V. (ii) T.V. and (iii) L.H.S.V. Use 1st Angle System.**

From direction X, three shaded surfaces will be seen. Two rectangles (small) will be seen in true shape while middle curved surface will be seen as projected rectangle. In short elevation or F.V. will be the same as for problem 7.

T.V. is a combination of one rectangle of the object and one semicircle of the recess.

In L.H.S.V. recess will not be seen and so it is shown by a dotted line. Same as for problem 7.

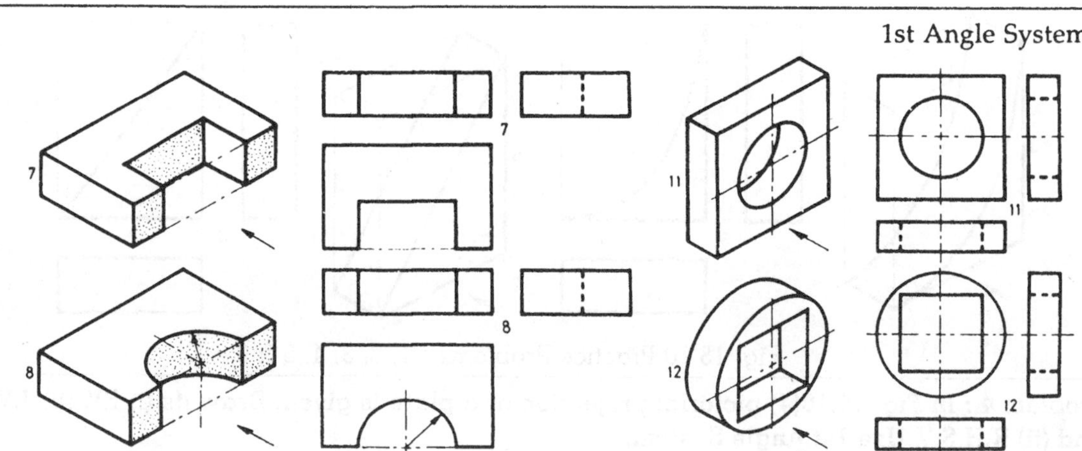

Fig. 15.11 Practice Problem - 7, 8, 9, 10, 11, 12, 13 & 14

***Problem 9* : In Fig. 15.11(9) pictorial view of a horizontal rectangular plate with a triangular recess is shown. Draw its (i) F.V. (ii) T.V. and (iii) L.H.S.V. Use 1st Angle System.**

F.V. and L.H.S.V. will be same as of problem 7 and 8. T.V. is a combination of one rectangle of the object and one triangle of the recess.

***Problem 10* : In Fig. 15.11(10) pictorial view of a horizontal rectangular plate with recess is shown. Draw its (i) F.V. (ii) T.V. and (iii) L.H.S.V. Use 1st Angle System.**

Three projections are given in the figure.

***Conclusion* :** Recesses in plates of problems 8, 9 and 10 can be assumed to be cut out of rectangular recess of plate 7 without formation of any edge on the periphery of the recess. Overall dimensions of recess have also remained the same and hence T. Vs. have changed according to shapes of top planes of the plates and F. Vs. and L.H.S.Vs. have remained same for all plates.

***Problem 11* : In Fig. 15.11(11) pictorial projection of a vertical square plate with a round hole is given. Draw its (i) F.V. (ii) T.V. and (iii) L.H. S.V. Use 1st Angle System.**

From direction X, surface having square boundary and a circular hole will be seen. In T.V. and L.H.S.V. a rectangle will be seen and the hole will give pair of dotted lines at a distance of diameter of the hole.

***Problem 12* : In Fig. 15.11(12) pictorial projection of a vertical round plate with square hole is given. Draw its (i) F.V. (ii) T.V. and (iii) L.H.S.V. Use 1st Angle System.**

From direction X, surface having round boundary and square hole will be seen. T.V. and L.H.S.V. will be the same as for problem 11.

***Problem 13/14* : In Fig. 15.11(13) (14) pictorial projection of a vertical square plate with a recess is shown. Draw its (i) F.V. (ii) T.V. and (iii) L.H.S.V. Use 1st Angle System.**

In F.V. front plane of the object is seen in true shape in both the problems.

Irrespective of shape of recess T.Vs. and S. Vs. are the same for both the problems.

In short shapes of recesses are seen only in one view. In other views overall dimensions of recesses decide views irrespective of shapes of recesses.

***Problem 15* : In Fig. 15.12 (A) pictorial projection of an object (complicated one) is given. Draw (i) F.V. (ii) T.V. and (iii) L.H.S.V. Use 1st Angle System of Projections.**

As the problem is complicated one, to make it easier and simpler, let us divide the object into 5 simple parts, as shown in the Fig. 15.12(B) and draw first the F.V., T.V. and L.H.S.V. of each part separately. Views as they will appear are shown in Fig. 15.12(B).

Now, if we combine together F. Vs. of all parts taking care about their positions and locations with little changes where they join, we can easily draw the F.V. of the complete object. F.V. of full object is shown in Fig. 15.12(A).

Similarly, if we combine together T.Vs. of all parts taking care about their positions and locations with little changes where they join, we can easily draw the T.V. of the complicated object. T.V. of full object is shown in Fig. 15.12(A).

Similarly, we can get L.H.S.V. also. It is shown in Fig. 15.12(A).

This above method is adopted in the beginning but after having sufficient practice of these types of problems you may solve the problem as a whole instead of dividing it into parts.

Shaded area indicates the area (zone) where the joint was existing before separate pieces were made. Number near shaded area indicates the part number with which the joint was existing.

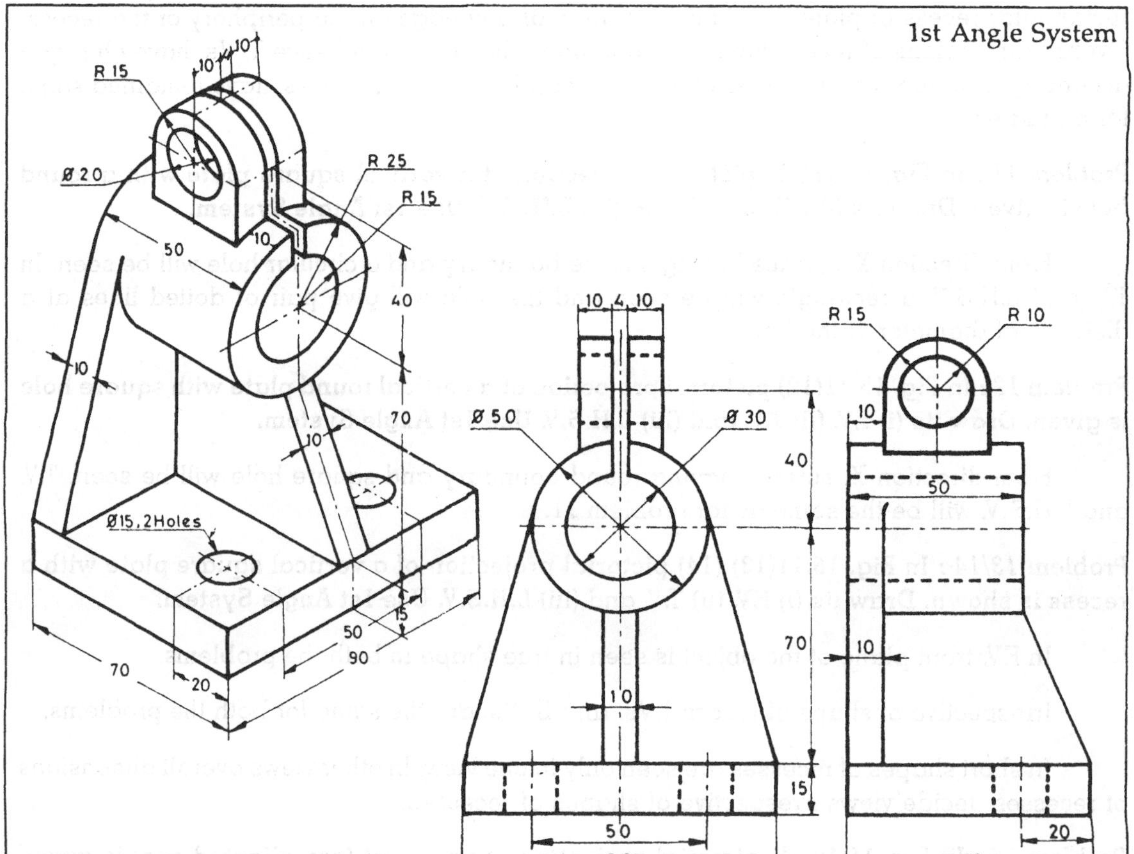

Fig. 15.12 (A) Pictorial projection - Complicated object

1st Angle System

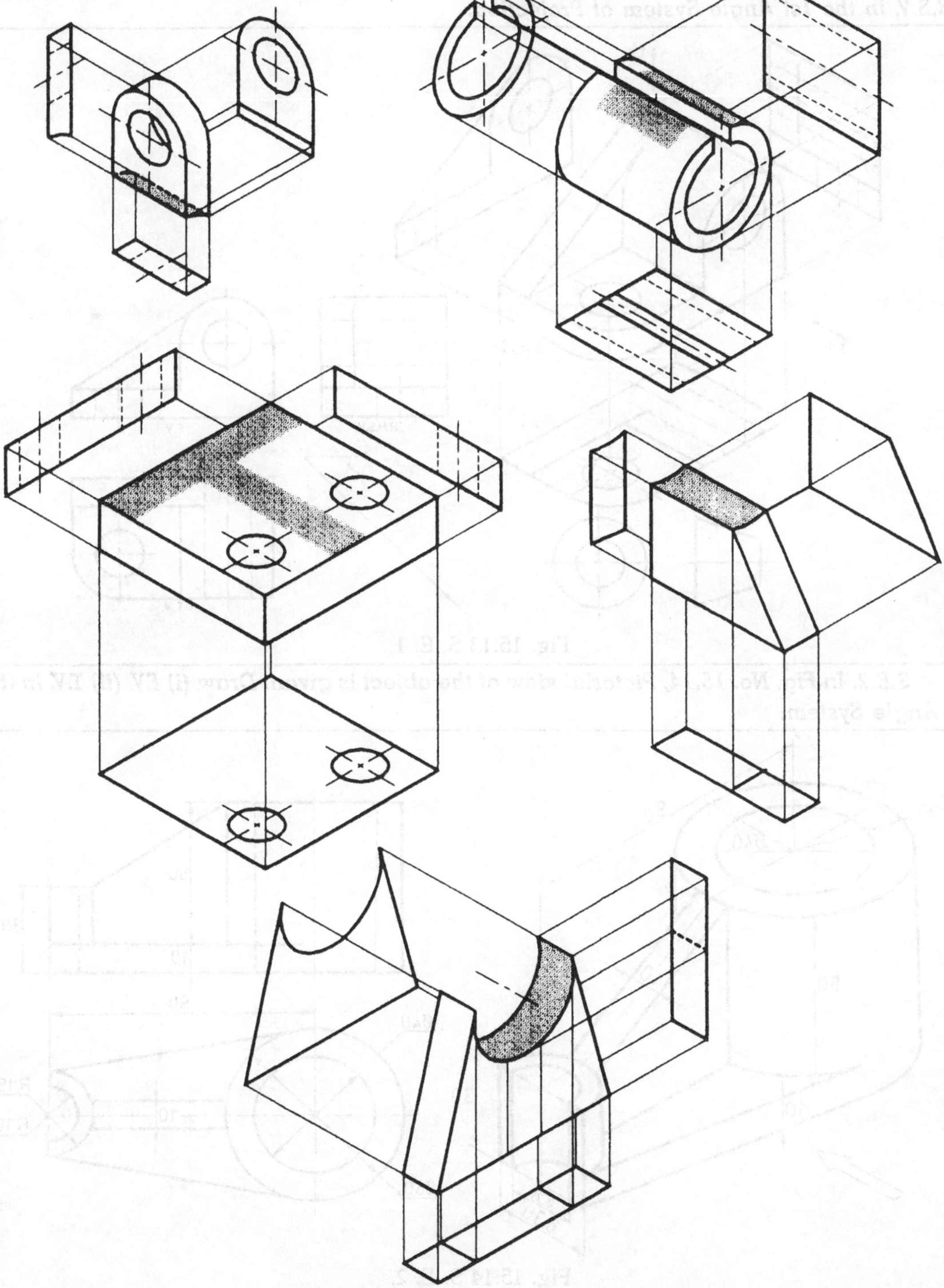

Fig. 15.12 - B. 5 Pieces of complicated object

Solved-Exercise [S. E.]

S.E.1. In Fig. No. 15.13, Pictorial view of the object is given. Draw (i) F.V. (ii) T.V. (iii) R.H.S.V. in the 1st Angle System of Projection.

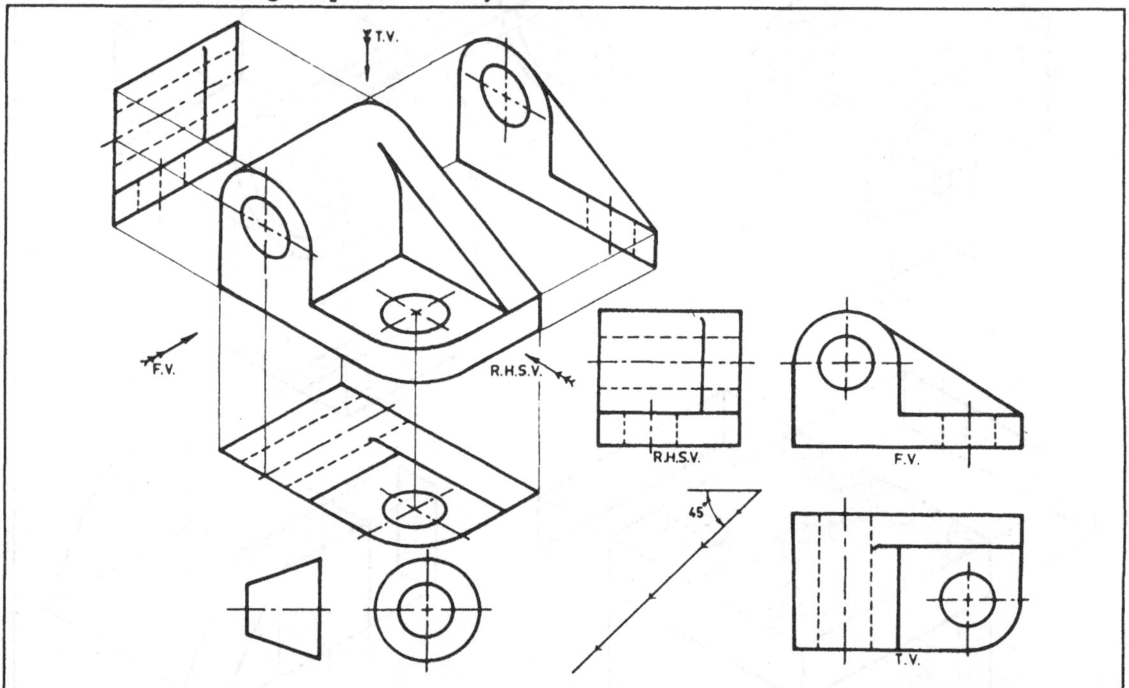

Fig. 15.13 S. E. 1.

S.E.2. In Fig. No. 15.14, Pictorial view of the object is given. Draw (i) F.V. (ii) T.V. in the 1st Angle System.

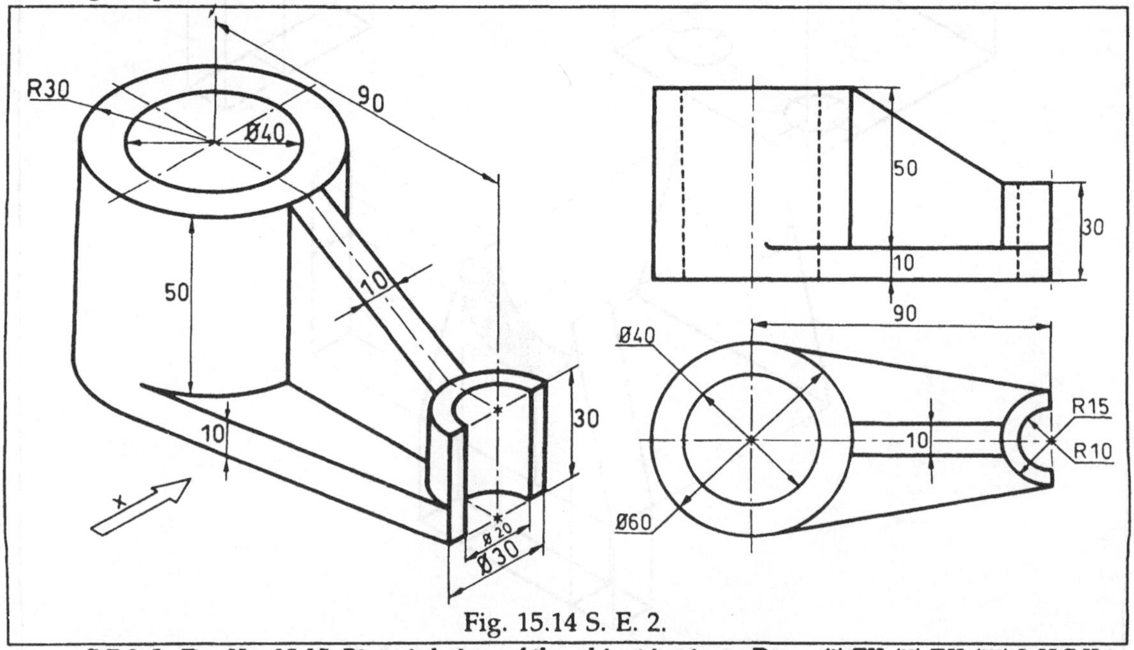

Fig. 15.14 S. E. 2.

S.E.3. In Fig. No. 15.15, Pictorial view of the object is given. Draw (i) F.V. (ii) T.V. (iii) L.H.S.V. in the 1st Angle System. Insert necessary dimensions in the Unidirectional System of Dimensioning.

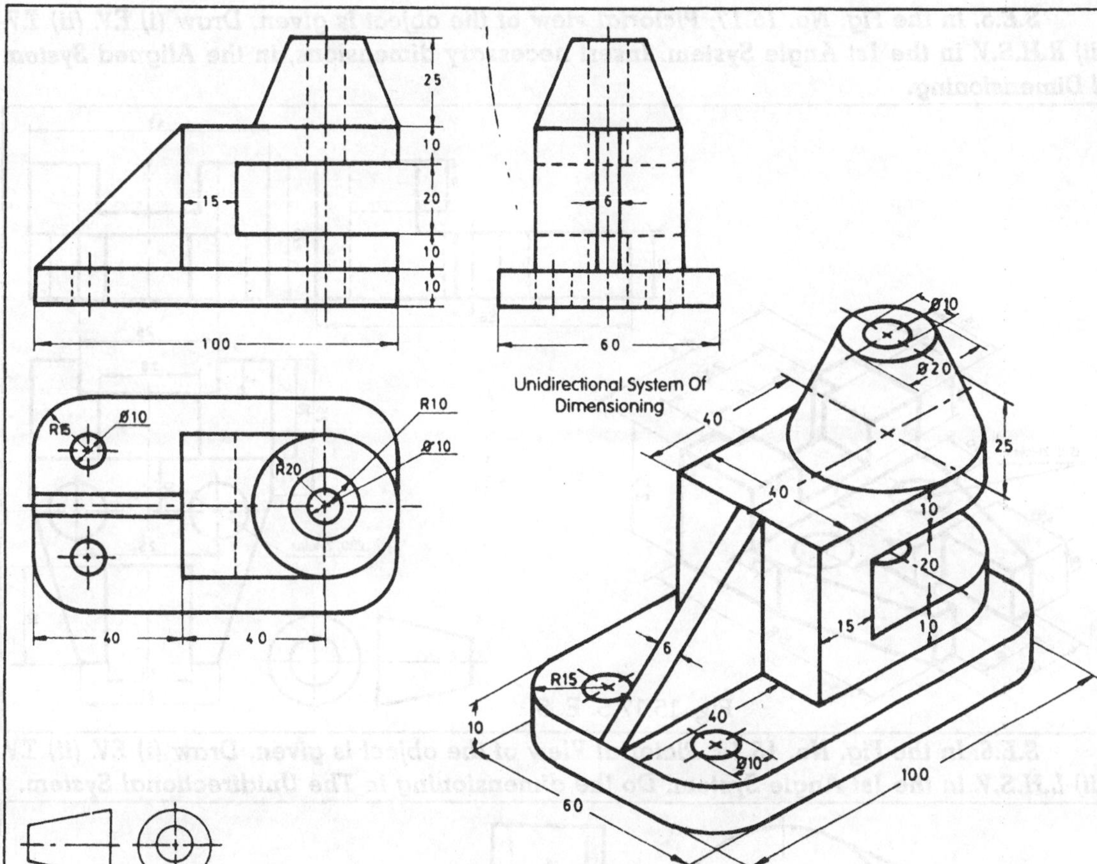

Fig. 15.15 S. E. 3.

S.E.4. In Fig. No. 15.16, Pictorial view of the object is given. Draw (i) F.V. (ii) L.H.S.V. (iii) T.V. in the 3rd Angle system. Insert necessary dimensions, in the Unidirectional system .

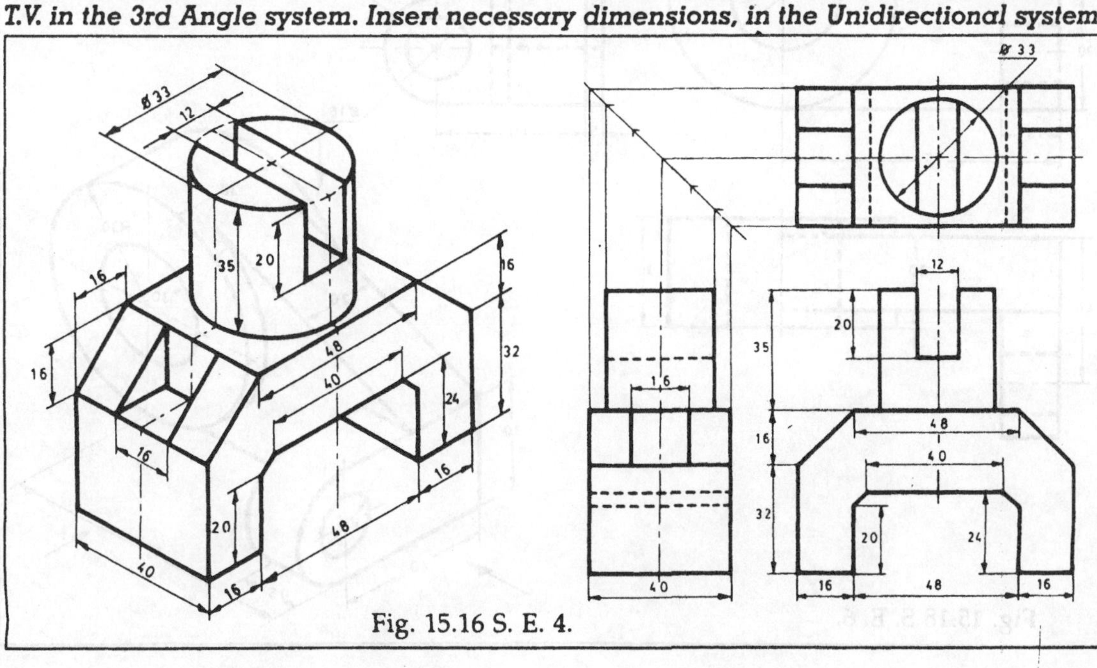

Fig. 15.16 S. E. 4.

S.E.5. In the Fig. No. 15.17, Pictorial view of the object is given. Draw (i) F.V. (ii) T.V. (iii) R.H.S.V. in the 1st Angle System. Insert necessary dimensions, in the Aligned System of Dimensioning.

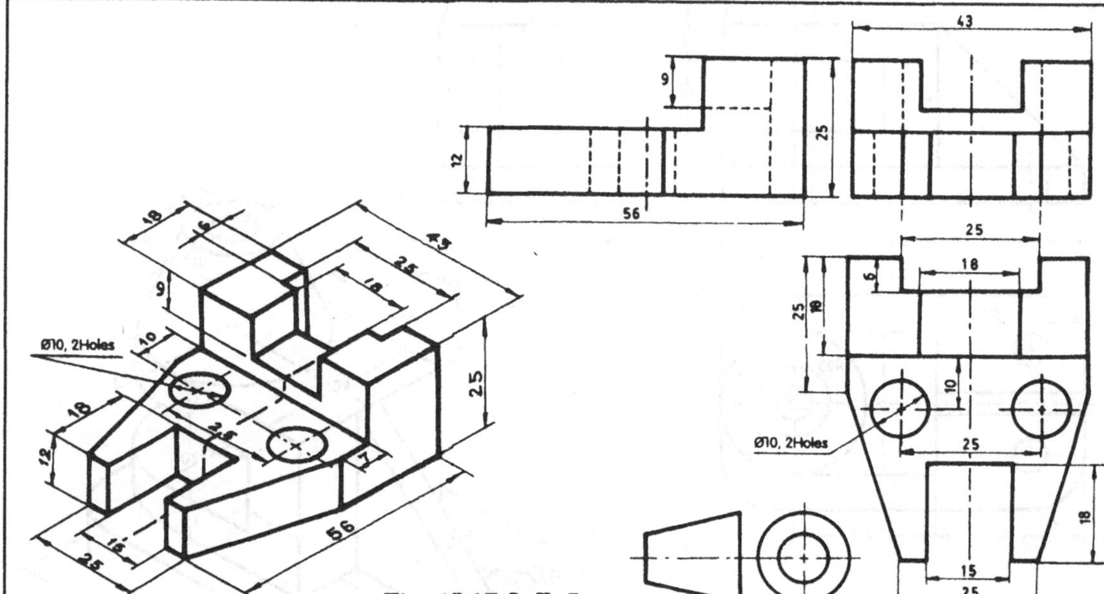

Fig. 15.17 S. E. 5.

S.E.6. In the Fig. No. 15.18, Pictorial View of the object is given. Draw (i) F.V. (ii) T.V. (iii) L.H.S.V. in the 1st Angle System. Do the dimensioning in The Unidirectional System.

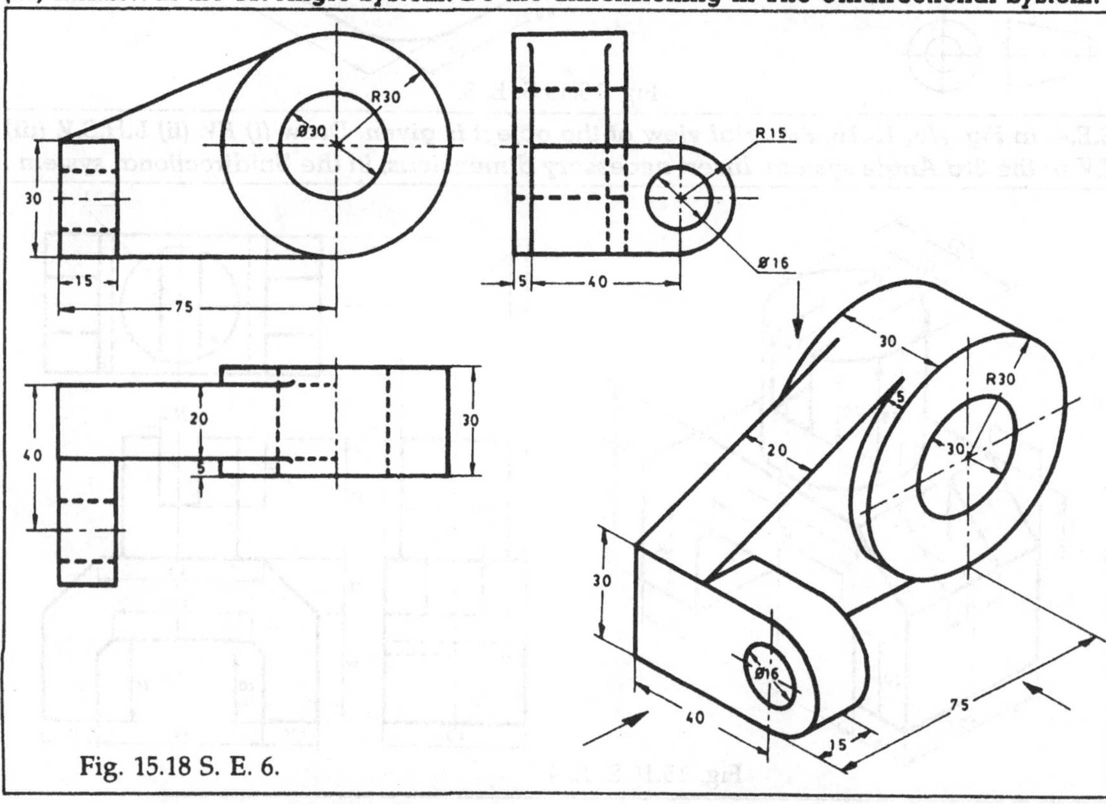

Fig. 15.18 S. E. 6.

S.E.7. In the Fig. NO. 15.19, Pictorial View of the object is given. Draw (i) F.V. (ii) L.H.S.V. in the 1st Angle System. Insert necessary dimensions in The Aligned System of Dimensioning.

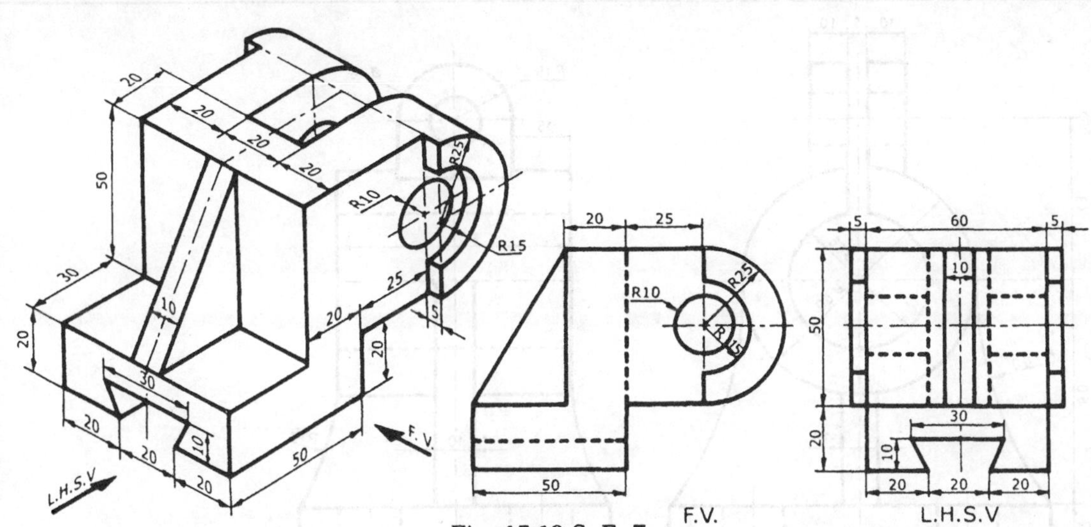

Fig. 15.19 S. E. 7.

S.E.8. In the Fig. No. 15.20, Pictorial view of the object is given. Draw (i) F.V. (ii) T.V. in The 3rd Angle System. Insert necessary dimensions in The Aligned System of Dimensioning.

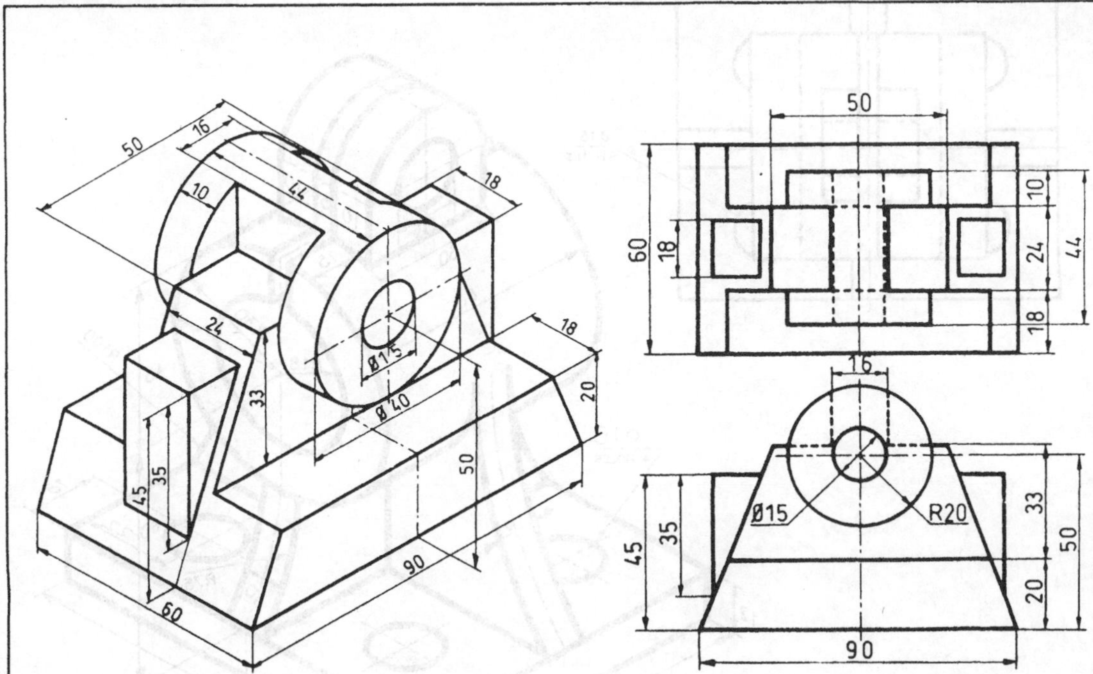

Fig. 15.20 S. E. 8.

S.E.9. In the Fig. NO. 15.21, The pictorial view of the object is given. Draw (i) F.V. (ii) T.V. (iii) L.H.S.V. in The 1st Angle System. Insert necessary dimensions in The Aligned System of Dimensioning.

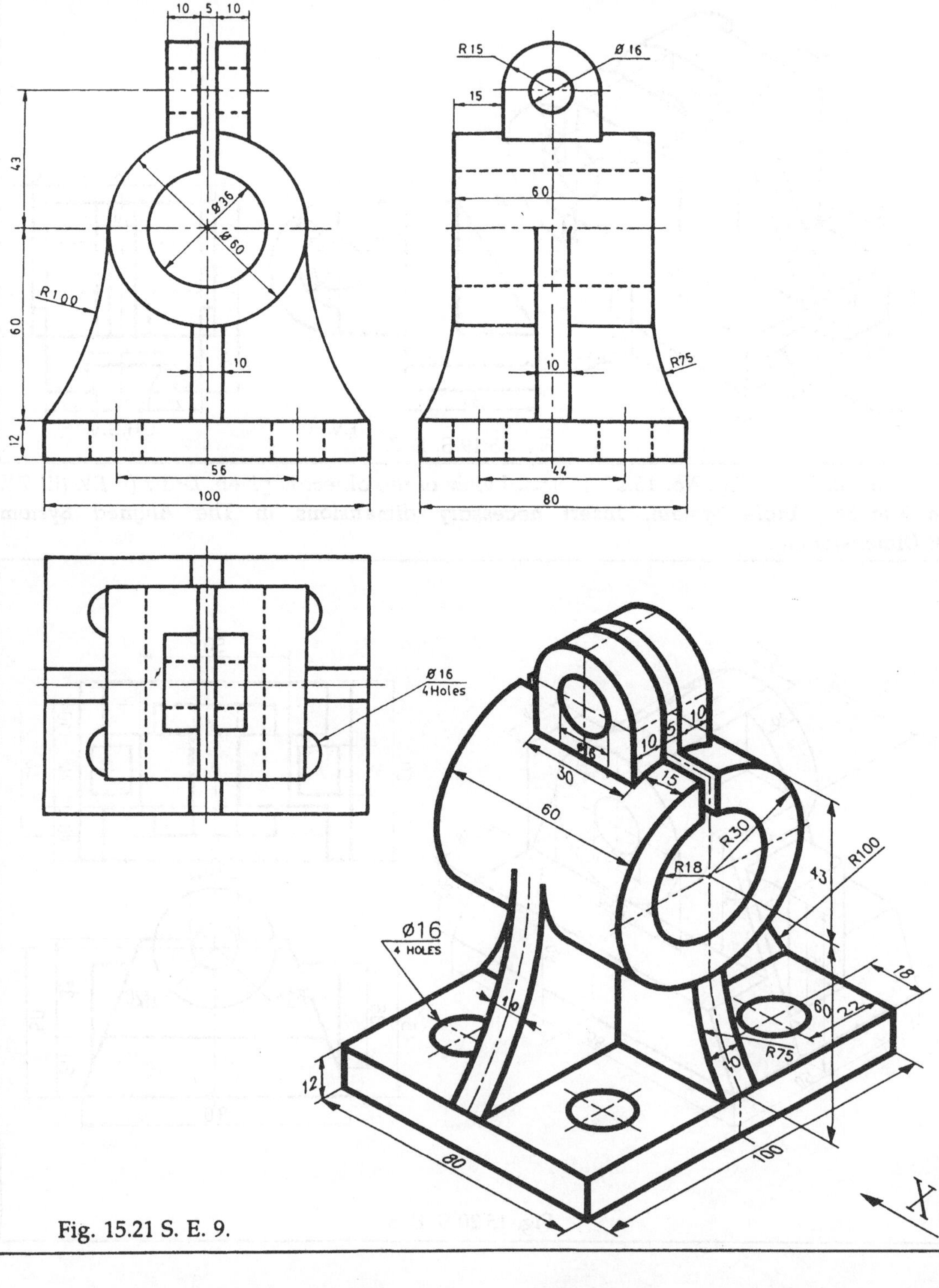

Fig. 15.21 S. E. 9.

Unsolved-Exercise

Note : Pictorial views of the objects are given in Fig. 15.22 to 15.35. Draw orthographic views as stated in each exercise. Insert necessary dimensions on the projections. Elevation or front view should be drawn as seen from the direction of arrow X. All dimensions are in mm. Assume dimensions, if not given. (Solutions of problems are given in last pages of this book. In solution dimensions are not given.)

Ex. 1. Fig 15.22. Draw (i) Front view (ii) Left hand side view (iii) Top plan. Use the 3rd angle system. **Fig. No. 15.22**

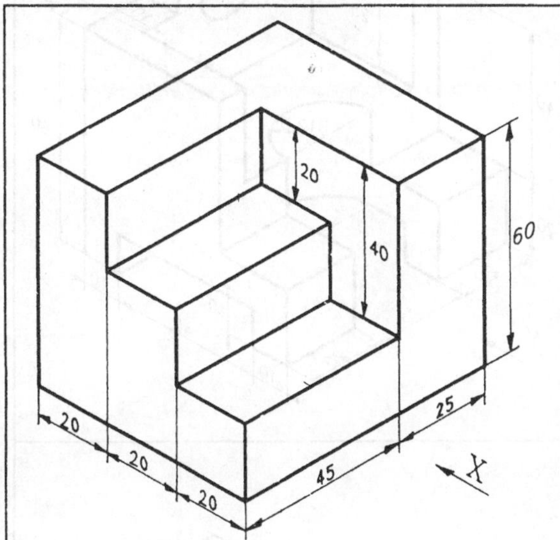

Ex. 3. Fig 15.24 Draw (i) Front view (ii) R.H. side view (iii) Top view. **Fig. No. 15.24**

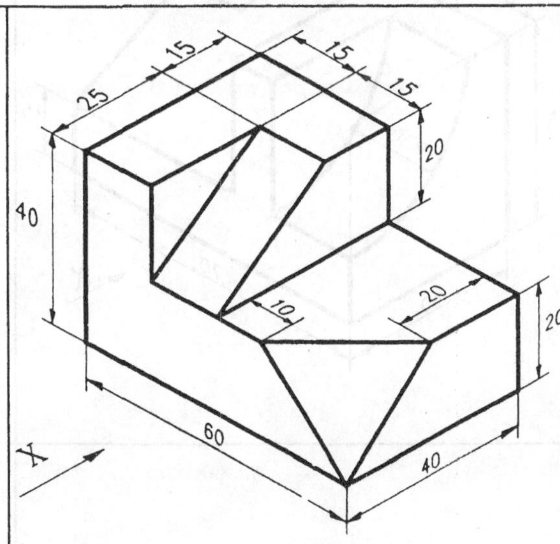

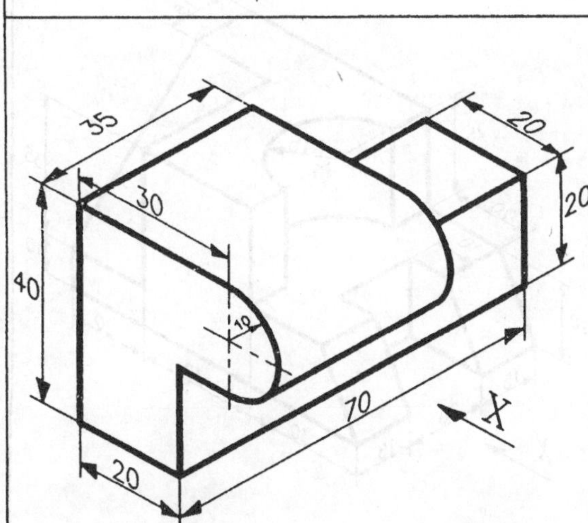

Fig. No. 15.23

Ex. 2. Fig 15.23. Draw (i) Front elevation (ii) Right hand side view (iii) Top plan. Use the 1st angle projection system.

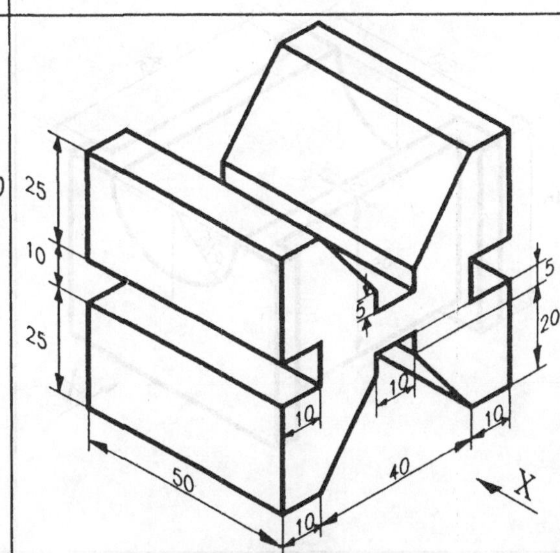

Fig. No. 15.25

Ex. 4. Fig. 15.25 Draw (i) Elevation (ii) L.H. side view (iii) Top plan.

Ex. 5. Fig 15.26 Draw (i) Front view (ii) R.H. side view (iii) Top plan (iv) L.H. side view. Use the 1st angle projection system. **Fig. No. 15.26**

Ex. 7. Fig 15.28. Draw (i) Front elevation (ii) R.H. side view (iii) Top plan. (iv) Bottom plan. **Fig. No. 15.28**

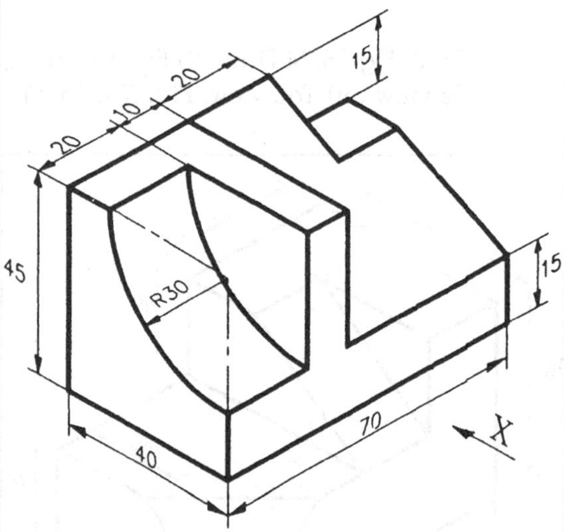

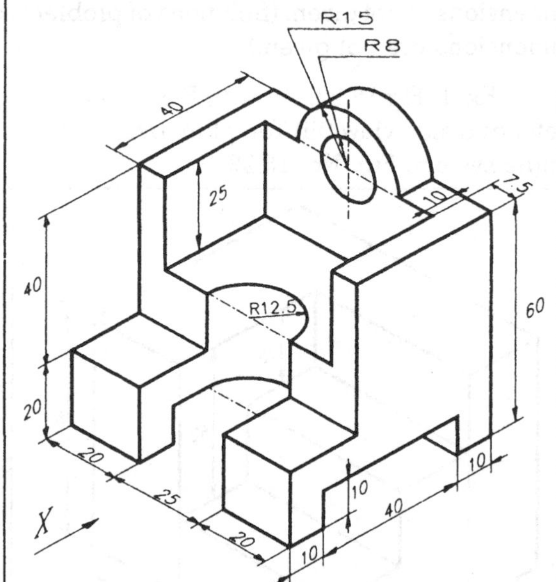

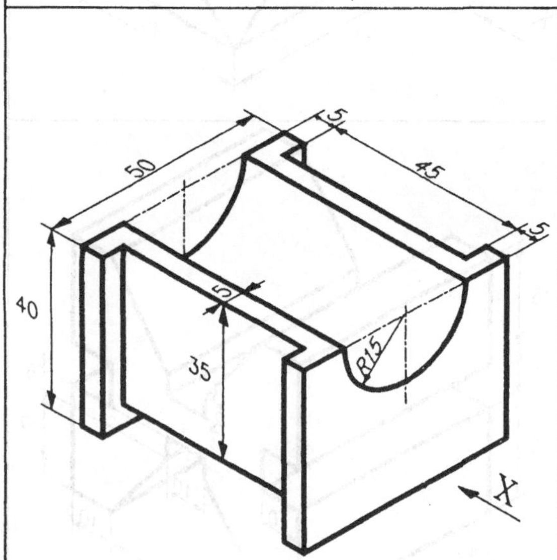

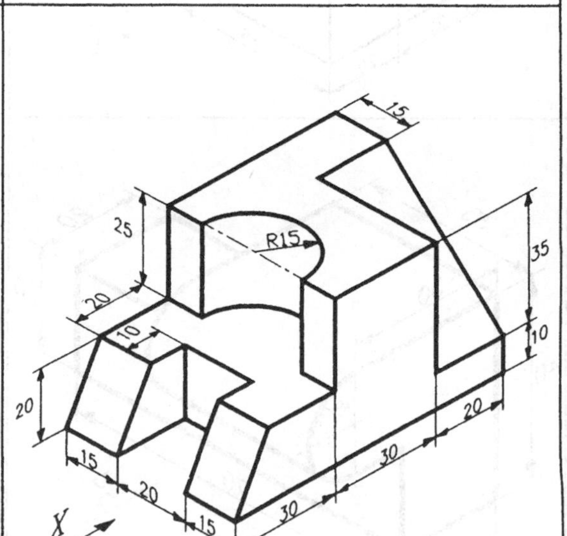

Fig. No. 15.27

Ex. 6. Fig 15.27. Draw (i) Front view (ii) Top plan (iii) L.H. side view. Use the 3rd angle projection system.

Fig. No. 15.29

Ex. 8. Fig 15.29. Draw (i) Front view (ii) R.H.S.V. (iii) Top view.

Ex. 9. Draw (i) Front elevation (ii) Top plan (iii) R.H.S.V. Use the 1st angle projection system. **Fig. No. 15.30**

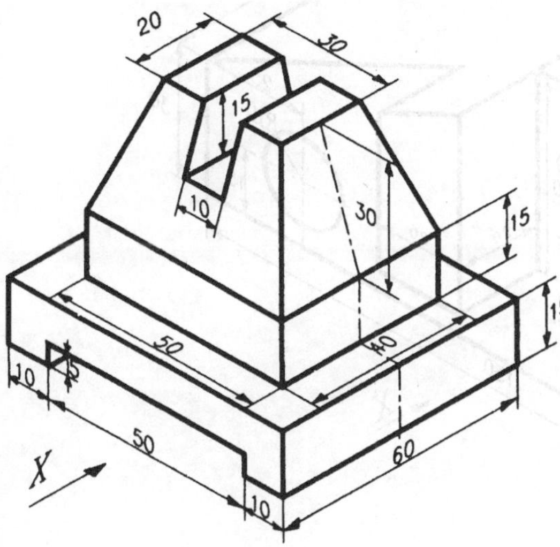

Ex. 10. Fig. 15.31. Draw (i) Front elevation (ii) R.H. side view (iii) Top plan. Use the 3rd angle projection system. **Fig. No. 15.31**

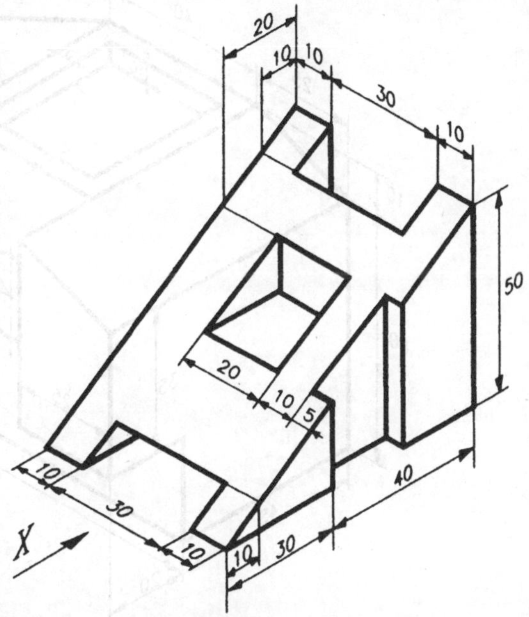

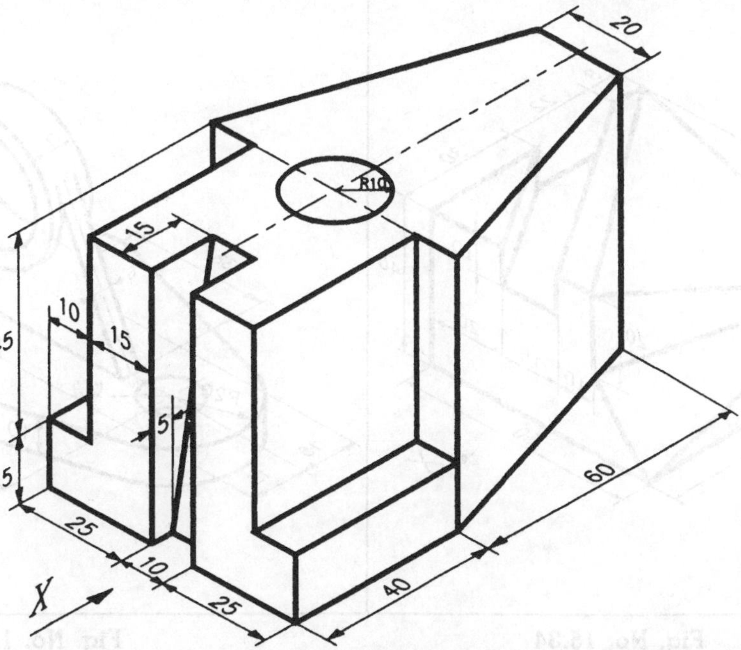

Fig. No. 15.32

Ex. 11. Fig. 15.32 Draw (i) Elevation (ii) R. H. S. V. (iii) Top Plan

Ex. 12. Fig. 15.33 Draw (i) Front view (ii) Top Plan (iii) R. H. S. V. (iv) L H. S. V. Use the 1st angle projection method.

Fig. No. 15.33

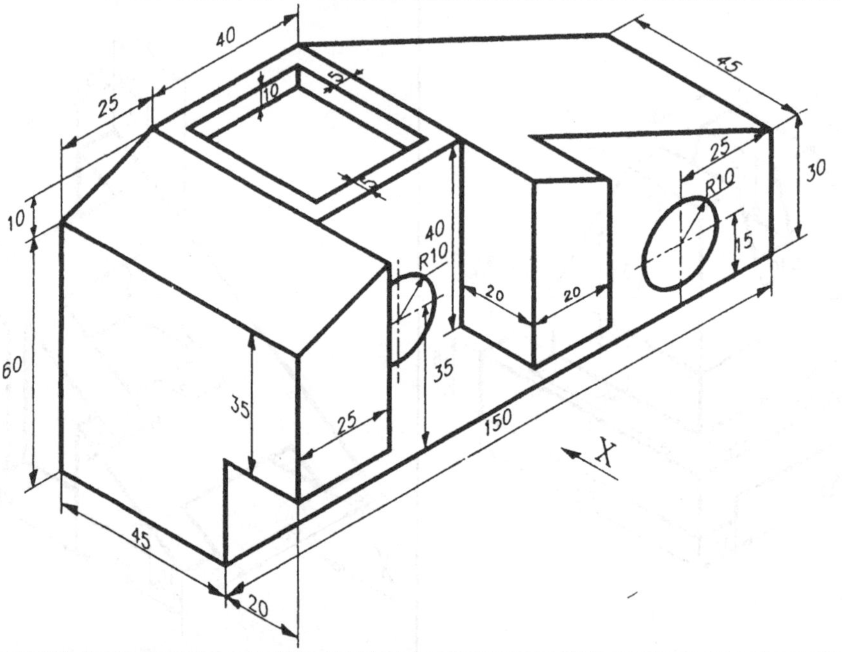

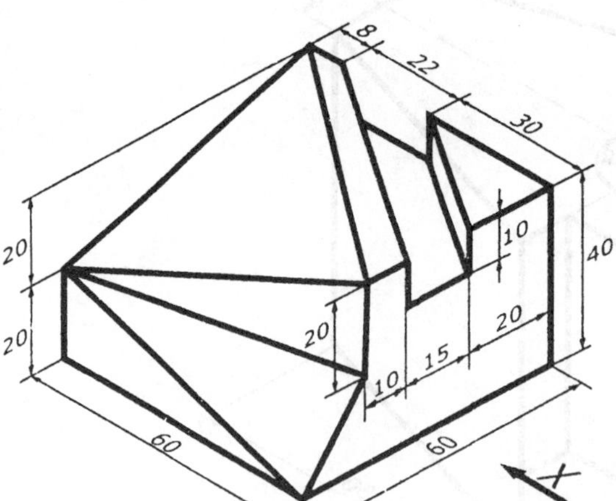

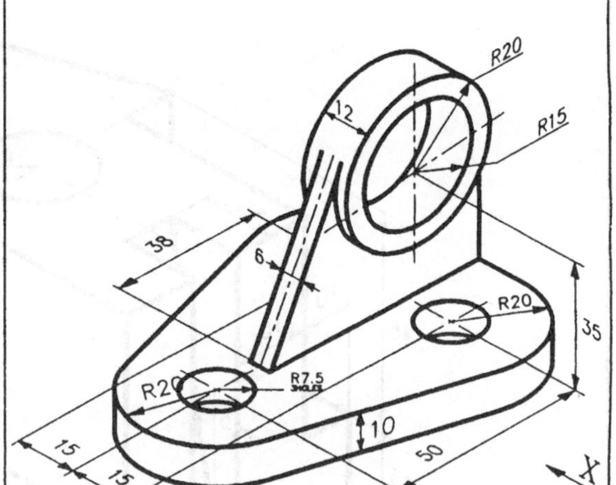

Fig. No. 15.34

Ex. 13. Fig. 15.34 Draw (i) Front elevation (ii) Side view from left (iii) S.V. from right (iv) Top plan.

Fig. No. 15.35

Ex. 14. Fig. 15.35 Draw (i) Front view (ii) Top view (iii) Left hand side view. Use the 3rd angle projection system.

CHAPTER 16
SECTIONAL ORTHOGRAPHIC PROJECTIONS

1. Introduction

In orthographic projections, details of the object which are not visible from outside are shown by means of dotted or dashed lines. When the object is complex from inside, there will be many dotted lines in the projections. It will be difficult to interpret or read the projections of such object when there are too many dotted lines inside projections.

Orthographic projections of such a complex object is drawn after imagining it as an object cut by an imaginary cutting plane/planes parallel to principal planes passing through the details of the object. An imaginary plane by which the object is cut is known as cutting plane or section plane. See Fig. 16.1. Portion of the object between cutting plane and an observer is imagined to be removed and then the object is viewed by the observer and projection is drawn. Projection of cut object is known as sectional view or sectional projection. See sectional T.V. in Fig. 16.1 (ii) a. Generally dotted or dashed lines are omitted in the sectional projection. In the sectional projection along with the shape of the section, visible edges and the contours of the object behind the section plane are shown.

When one view is drawn sectional, other views are drawn as if the object is not cut and the entire object exists as a whole. It means other views are not affected by taking a section in one or many projections. Only cutting plane line is shown to locate the section and arrows are drawn to show the direction of sight. See cutting plane line A-A in Fig. 16.1 other than sectional projection. The cutting plane line known as line of section or trace of the cutting plane is drawn with one long dash and one short dash alternately and evenly spaced. Ends of the cutting plane line is drawn thicker and arrows to indicate direction of sight is shown on these ends. See C.P. line A-A in Fig. 16.1. Change in the direction of cutting plane is also shown by a thicker line. See cutting plane line A-B, made thicker near centre, in Fig. 16.3 and Fig. 16.4

The elements of the object cut by cutting plane is made evident by drawing hatching lines in them. Hatching or section lines are drawn thinner than boundary lines and in normal circumstances they are drawn at 45°. Hatching lines are evenly spaced at about 2 to 3 mm apart.

The cutting plane need not be continuous but may be offset or turned through an angle thus giving rise to various types of sectional projections.

2. Full Section

When the section plane passes through the entire length of the object on the plane of symmetry, as shown in Figs. 16.1(i), 16.2(i), 16.3(i) and then if the view is taken of half or nearly half object then that view is known as full sectional view. If it happens to be elevation, it is known as full sectional elevation or sectional elevation. See Fig. 16.3(iv)(a). The word 'full' may be omitted. If it happens to be plan, it is known as full sectional plan or sectional plan, see Fig. 16.1(ii)(a). Similarly, there can be full sectional side view, see Fig. 16.2(iv).

3. Half Section

When the object is symmetrical about its centre line, from outside as well as from inside, half sectional view is preferred to that of full sectional view. It will show outside as well as inside details of the object in one view.

When two perpendicular cutting planes pass through the object on the planes of symmetry, as shown in Figs. 16.3(iii), 16.4(i), (iii) to remove one-quarter of the object and then if the view of three quarters of the object is taken then that view is known as half sectional view. See Fig. 16.3(iv) (c) and 16.4 (iv)(b). If it is elevation, it is called half sectional elevation. Similarly, we can have half sectional plan and half sectional end view or side view.

In half sectional view, the line, separating the sectional view and full view coincides with the centre line and as recommended by I.S. it is shown by dark continuous line. See Fig. 16.4(iv)(b).

4. Off-set Section

It is likely that one cutting plane need not pass through all axes of symmetry of different portions of the object. In such circumstances instead of cutting the object by one cutting plane the direction of cutting plane may be changed (i.e. off setted) in order to cut the object through maximum possible axes of symmetry and thereby show maximum details of the object in one offset sectional view. Here it is absolutely necessary to show the cutting plane line in other adjoining view or views. For illustration see Fig. 16.5 and Fig. 16.6.

5. Illustrative Examples

Now to understand and grasp the above three types of sectional views let us study a few illustrative examples.

Example 1 :

In Fig. 16.1.(i), object is shown in pictorial projection. Draw (i) Elevation (F.V.) and (ii) Full Sectional Plan (T.V.). Use 1st Angle System.

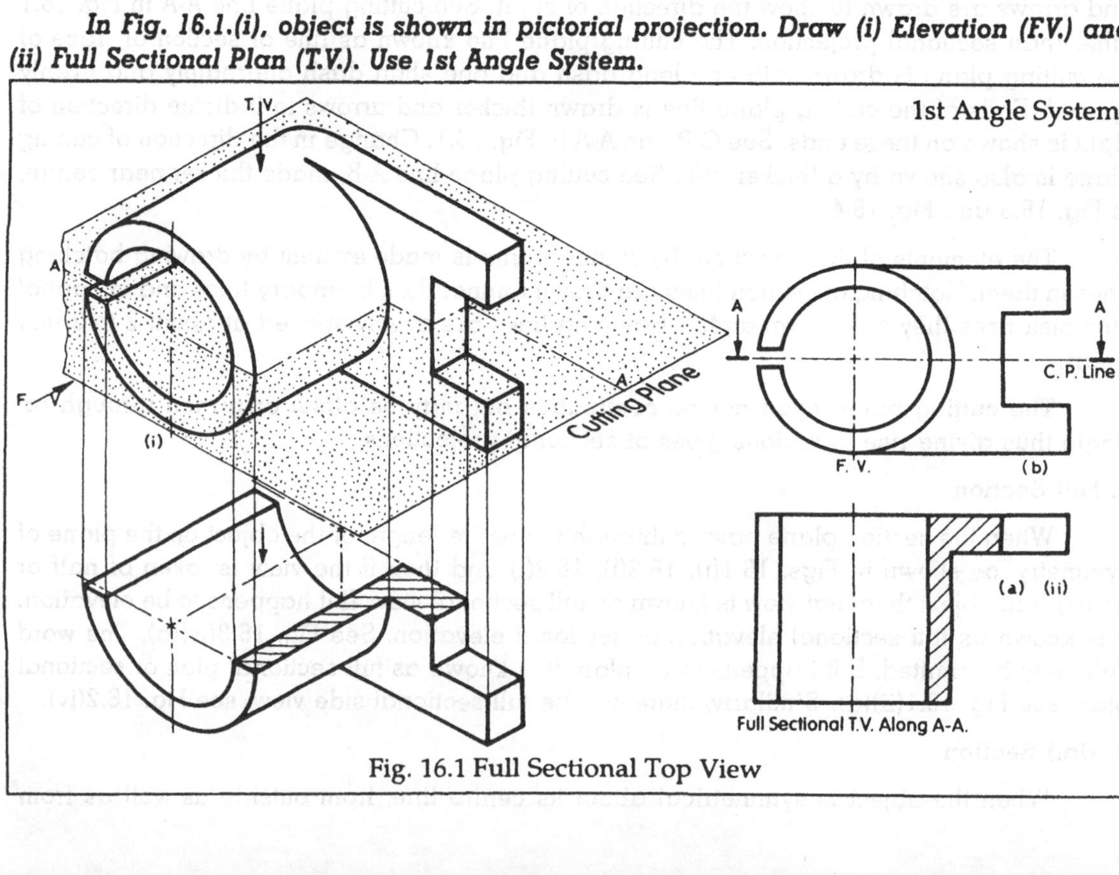

Fig. 16.1 Full Sectional Top View

In the figure, horizontal cutting plane is shown cutting the object into two equal halves. Upper half part is assumed to be removed and top view is taken from the top. Front view of complete object is taken from the front. Both front view and sectional top view are shown in Fig. 16.1(ii). Section lines or hatching lines are drawn where the material of the object is cut by cutting plane.

Example 2 :

In Fig. 16.2(i), object is shown in pictorial projection. Draw (i) Elevation (F.V.) (ii) Sectional Plan (T.V.) along B-B and (iii) Sectional Side View along A-A. Use 1st Angle System.

Draw first the elevation, as shown in Fig. 16.2(iv). Now imagine the object after cutting it along B-B to get full sectional plan. The object after cutting will look like Fig. 16.2(ii). Draw sectional plan. Similarly, imagine the object after cutting it along A-A to get full sectional side view. The object after cutting along A-A will look like Fig. 16.2 (iii). Take end view from left and draw it. All views are shown in Fig. 16.2(iv).

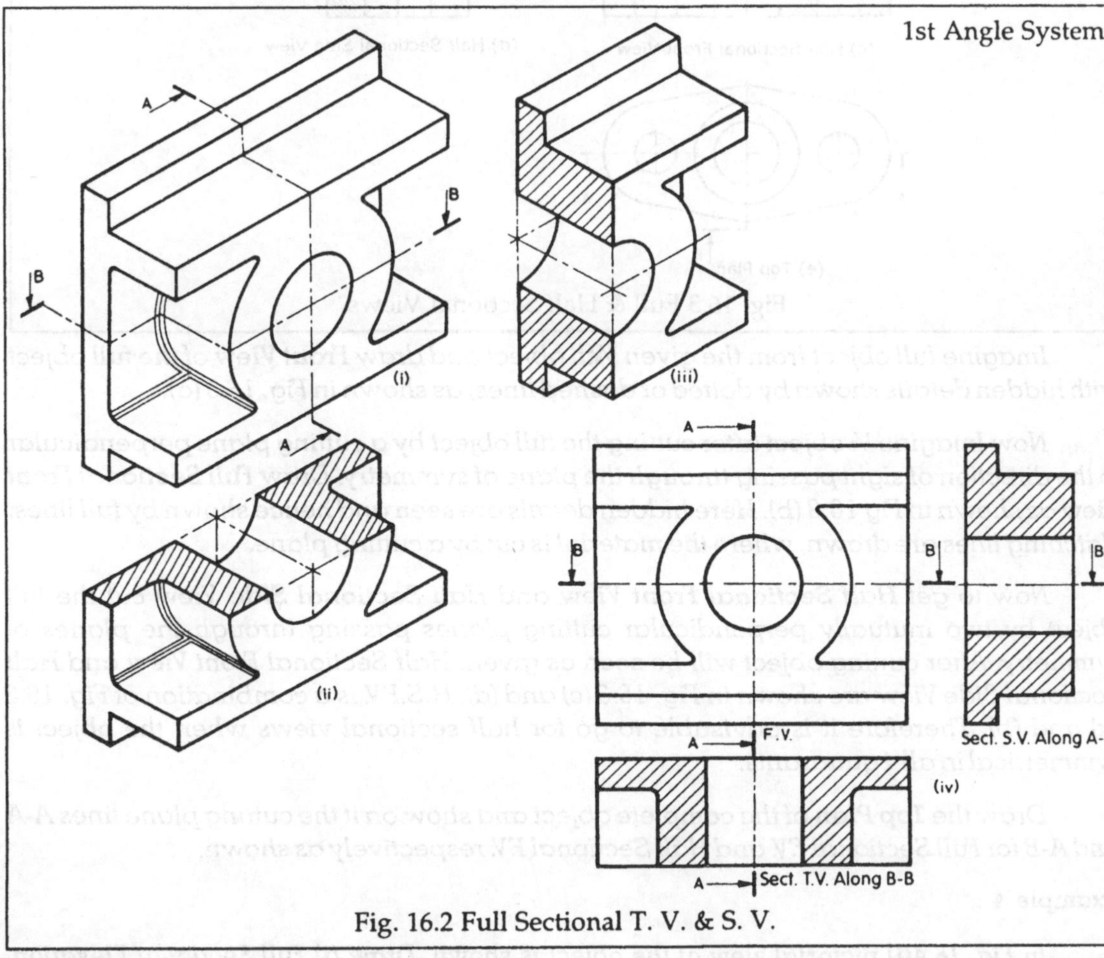

Fig. 16.2 Full Sectional T. V. & S. V.

Example 3 :

In Fig. 16.3 three dimensional isometric view of the cut ¾th object is given. Draw (a) Front View (b) Full Sectional Front View (c) Half Sectional Front View (d) Half Sectional Side View and (e) Top Plan. Use 1st Angle Projection System.

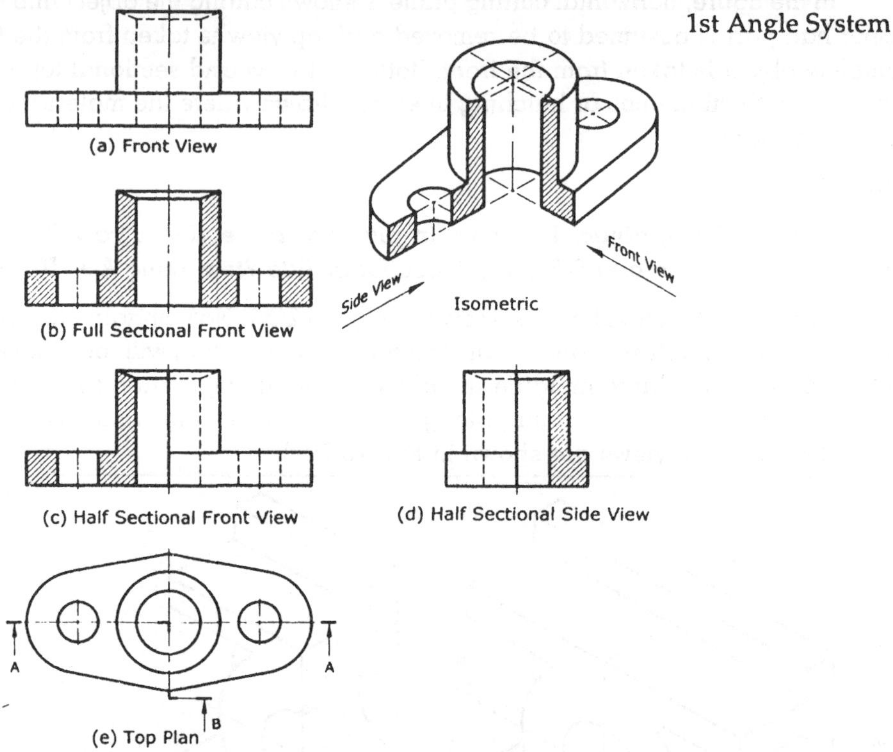

Fig. 16.3 Full & Half Sectional Views

Imagine full object from the given ¾th object and draw ***Front View*** *of the full object with hidden details shown by dotted or dashed lines, as shown in Fig. 16.3 (a).*

Now Imagine ½ object after cutting the full object by a cutting plane perpendicular to the direction of sight passing through the plane of symmetry. Draw ***Full Sectional Front View*** *as shown in Fig 16.3 (b). Here hidden details are seen and hence shown by full lines. Hatching lines are drawn, where the material is cut by a cutting plane.*

Now to get ***Half Sectional Front View*** *and* ***Half Sectional Side View*** *cut the full object by two mutually perpendicular cutting planes passing through the planes of symmetry. After cutting object will be seen as given. Half Sectional Front View and Half Sectional Side View are shown in Fig. 16.3 (c) and (d). H.S.F.V. is a combination of Fig. 16.3 (a) and (b). Therefore it is advisable to go for half sectional views when the object is symmetrical in all 4 quadrants.*

Draw the ***Top Plan*** *of the complete object and show on it the cutting plane lines A-A and A-B for Full Sectional F.V. and Half Sectional F.V. respectively as shown.*

Example 4 :

In Fig. 16.4(i) pictorial view of the object is shown. Draw (i) Full Sectional Elevation, (ii) Half Sectional Elevation and (iii) Top Plan. Use 1st Angle System.

To get the full sectional elevation from direction X cut the object by cutting plane A-A passing through the axis of symmetry and perpendicular to the direction X. After cutting, the half object will look like as shown in Fig. 16.4(ii). Take full sectional elevation from direction X and draw, as shown in Fig. 16.4(iv)(a).

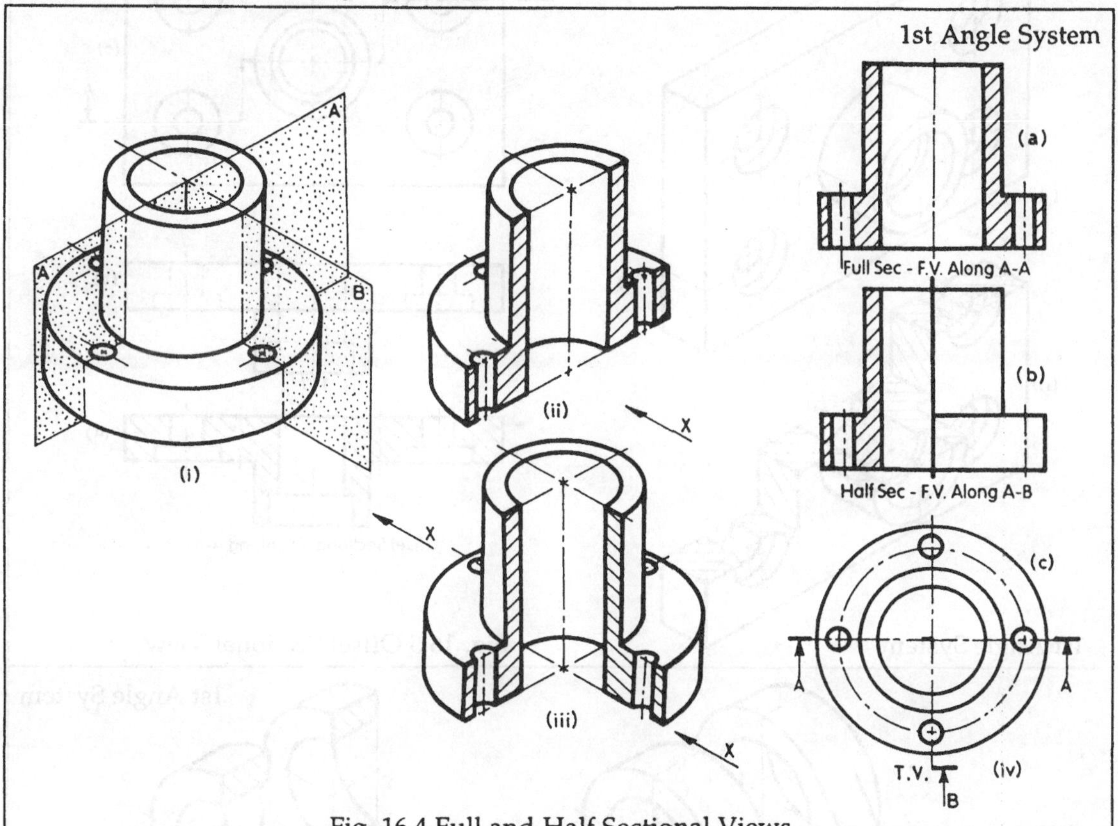

Fig. 16.4 Full and Half Sectional Views

To get the half sectional elevation from direction X cut the object by two mutually perpendicular planes A and B passing through axis of symmetry, as shown in Fig. 16.4(i). After cutting, the three quarter object will look like as shown in Fig. 16.4(iii). Take half sectional elevation from direction X and draw, as shown in Fig. 16.4(iv)(b).

Draw plan and show on it section plane line A-A for full sectional elevation and A-B for half sectional elevation. Draw hatching lines in areas where material is cut by section plane.

Example 5 :

Isometric projection of the object is given in Fig. 16.5(i). Draw (i) Elevation/F.V., (ii) Top Plan (T.V.) and (iii) Offset Sectional Plan (T.V.) along A-A. Use 1st Angle Projection System.

Here if section is not taken by offset plane then holes will not be cut and details about holes will not appear in sectional plan. So instead of taking one plane section, offset section is taken. After cutting the object along A-A and removing the upper portion, which is between the observer and the cutting plane, the object will look like as shown in Fig. 16.5(ii).

Draw first F.V. and T.V. of complete object, as shown in Fig. 16.5(iii) (c) and Fig. 16.5(iii)(b) respectively. Now draw offset sectional plan looking from top of the cut object shown in Fig. 16.5(ii). Offset sectional T.V. is shown in Fig. 16.5(iii) (a).

Do not forget to show cutting plane line A-A in elevation. See Fig. 16.5(iii)(c).

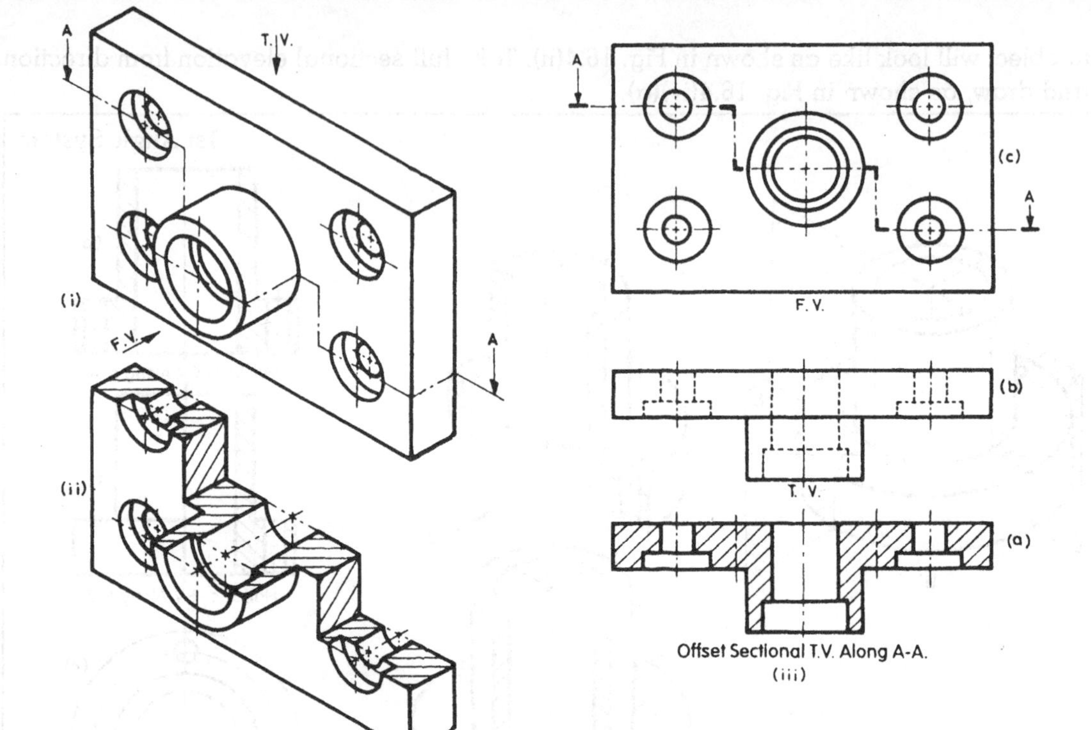

1st Angle System

Fig. 16.5 Offset Sectional View

1st Angle System

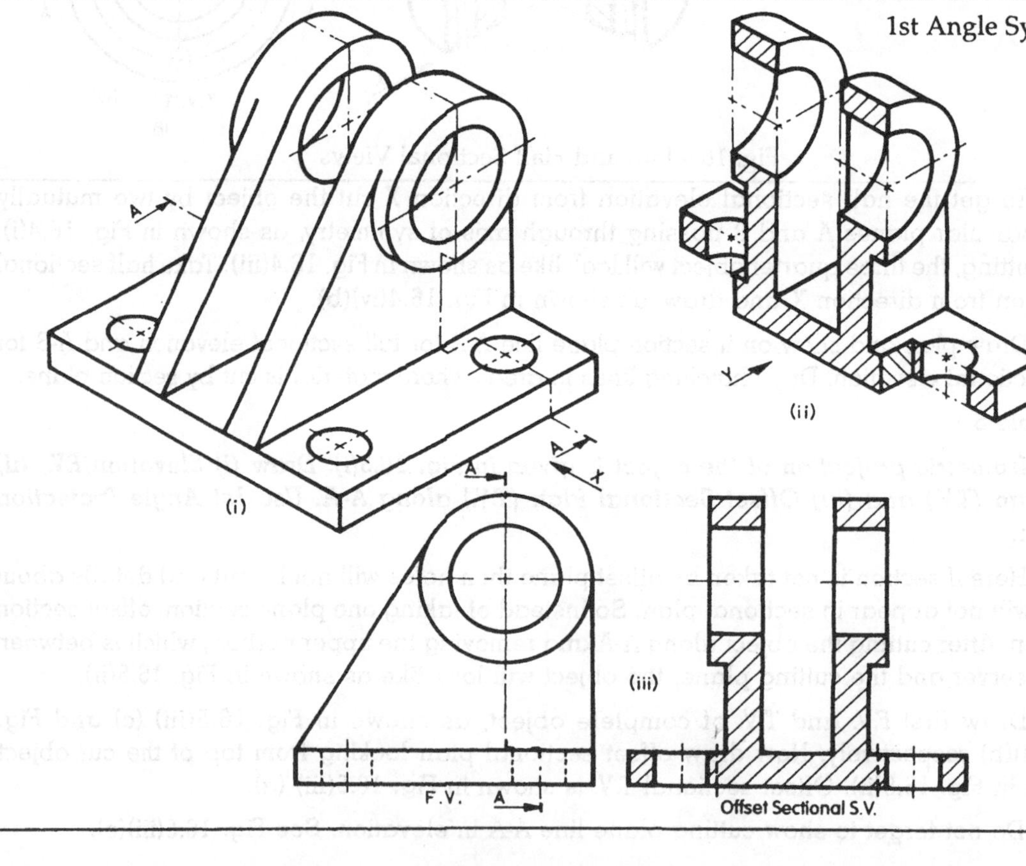

Fig. 16.6 Offset Sectional Side View

Example 6 :

Pictorial view of an object is given in Fig. 16.6(i). Draw (i) F.V. from direction X (ii) Offset Sectional S.V. from direction Y. Take section along A-A. Use 1st angle system.

After cutting the object along offset section A-A, it will look like Fig. 16.6(ii). Draw required views and do not forget to show section plane or cutting plane line A-A in F.V.

After solving the above six problems, students are requested to solve problems of solved exercises and unsolved exercises to grasp the subject and imbibe self confidence. See Previous Page No. 34.

6. Partial or Local or Broken Out Section

If it is required to show only isolated inside details or one of the important features of a component or of an assembly, this type of section is used instead of full or half section. The imaginary cutting plane passes through particular important features, exposing its interior to view and leaving an irregular break line surrounding important feature as boundary. See Fig. 16.7(a)(b) (c) and (d)

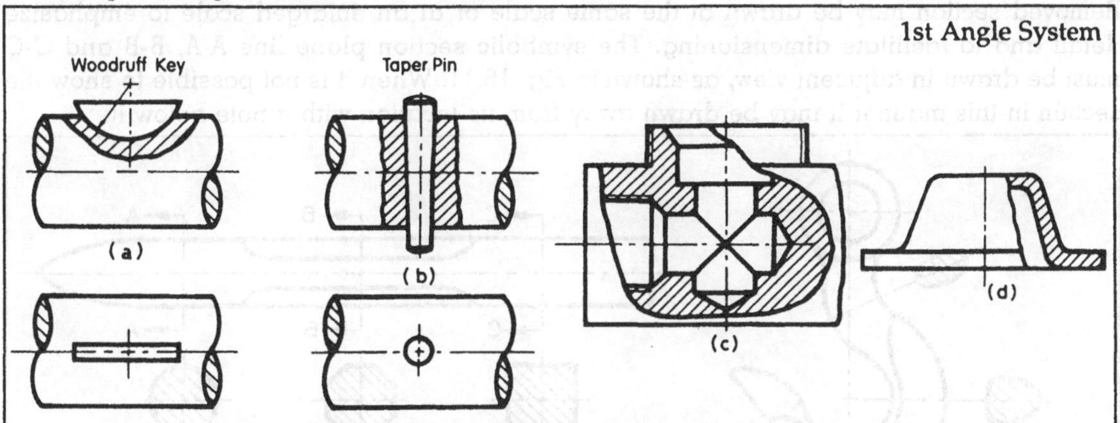

Fig. 16.7 Local - Partial - Broken out section

In Fig. 16.7(a) position of key and in Fig. 16.7(b) position of pin is shown by taking this type of section. Further in Fig. 16.7(c) inside details of the valve are also shown by this type of section. In Fig. 16.7(d) thickness of the vessel is shown.

7. Revolved Section

Revolved section is used to show the cross-section of objects of uniform section like arms, spokes, structural sections, bars, etc. Imaginary cutting plane is passed through the object perpendicular to the profile or longitudinal axis and then revolved on the centre line of the section by 90° to bring it in the plane of the paper. By this type of view we can avoid other view. See Fig. 16.8 for round section, I section and elliptical section.

Sometimes the view may be broken on both the sides of a revolved section, as shown in Figs. 16.8 and 16.9 to make the section more clear. I-Section of Fig. 16.8 and all sections of Fig. 16.9 are shown as broken revolved section.

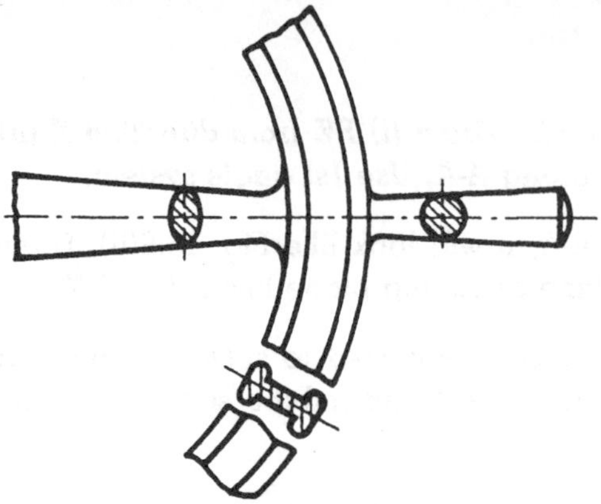

Fig. 16.8 Revolved Section

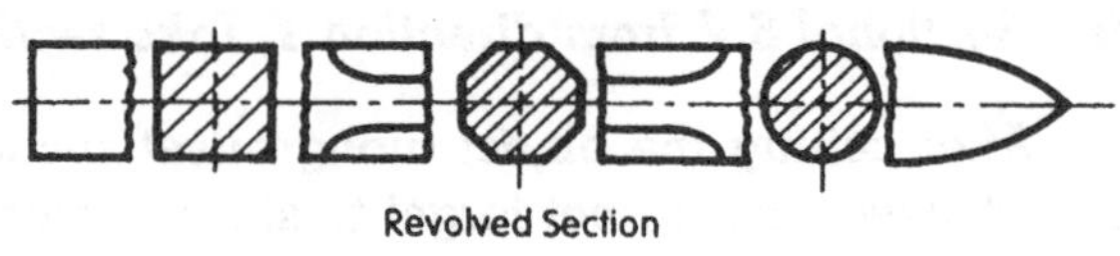

Fig. 16.9 Broken Revolved Section

8. Details or Removed Section

In this type of section, section is taken out of the view to make it more clear. It also shows the cross-section of the object like revolved section. It is mostly drawn keeping the axis of section on the extension of the cutting plane line, as shown in Fig. 16.10 or 16.11. Removed section may be drawn at the same scale or at an enlarged scale to emphasize detail and to facilitate dimensioning. The symbolic section plane line A-A, B-B and C-C must be drawn in adjacent view, as shown in Fig. 16.11. When it is not possible to show the section in this manner it may be drawn away from its location with a note below it.

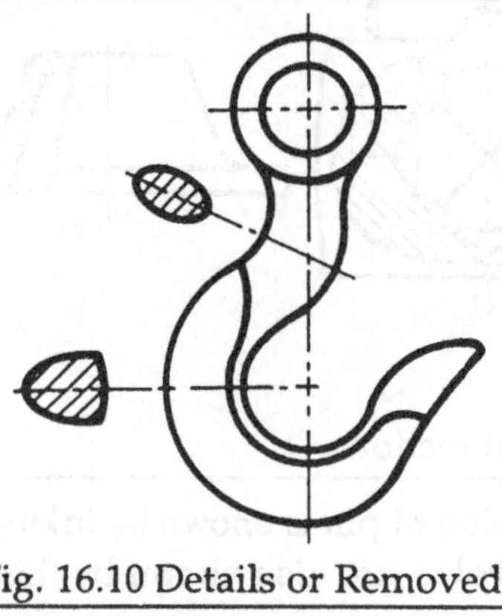

Fig. 16.10 Details or Removed Section

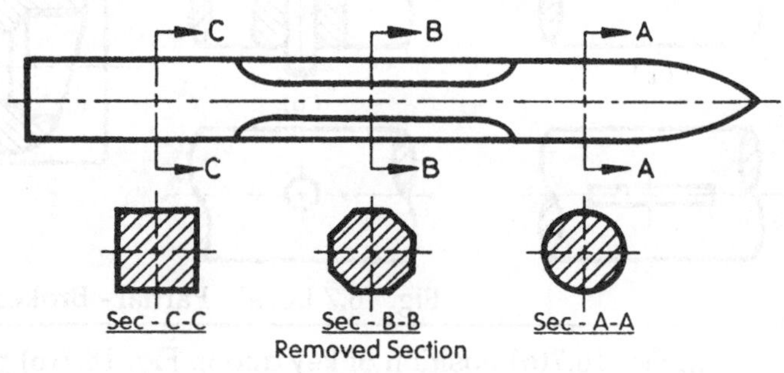

Fig. 16.11 Removed Section

9. Auxiliary Section

A sectional view taken on auxiliary plane is called auxiliary section. See Fig. 16.12. An auxiliary section may be of any type studied so far i. e. full, half, removed, revolved, etc.

10. Hatching

Following rules are adopted for hatching :

(1) Area of a metal or a material, cut by a section plane is shown by hatching lines in projection. It is executed by thin lines normally at an angle of 45° to the axis or to the main outline or boundary of the sectional area. See Fig. 16.13, part A and C.

(2) If the axis or the boundary of section is at 45° then the horizontal or vertical hatching lines are drawn. See Fig. 16.13, part B and D.

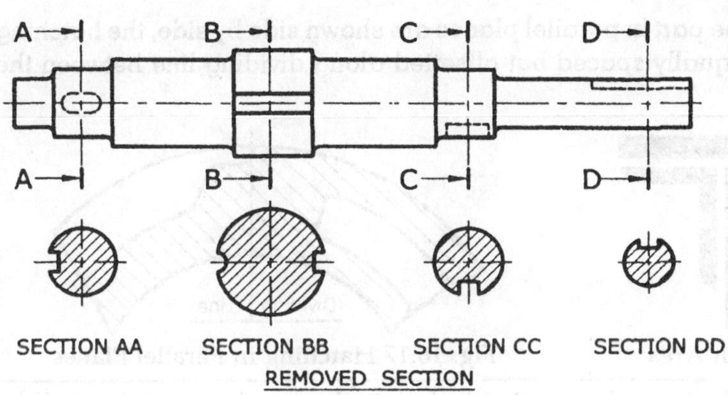

Fig. 16.11 Removed Section

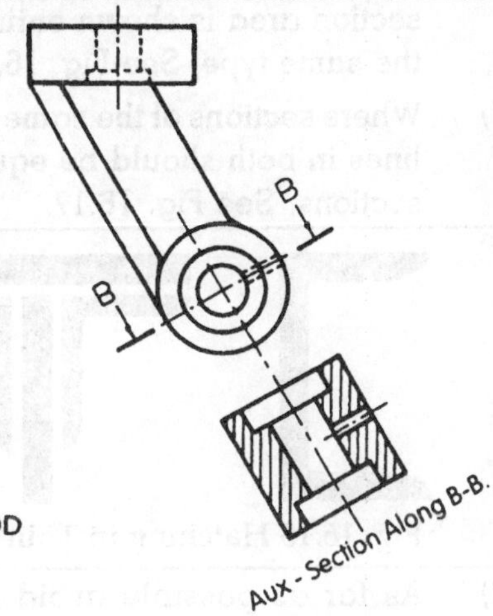

Fig. 16.12 Auxiliary Section

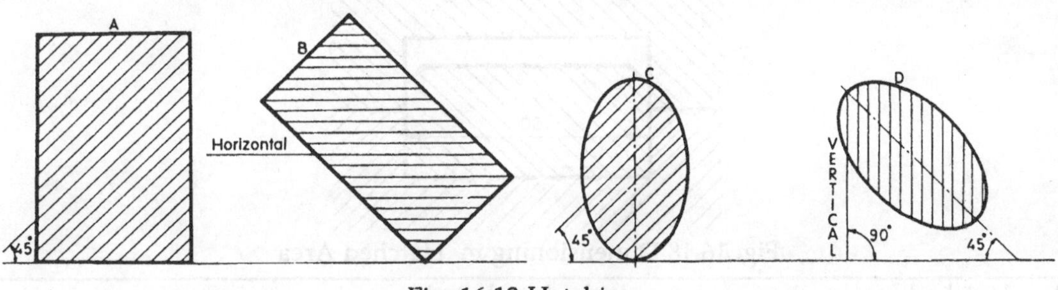

Fig. 16.13 Hatching

(3) Separated areas of a section of a single component are hatched in the same manner. Hatching lines in different projections for the same component, are drawn in the same manner. See Fig. 16.2(iv).

(4) Hatching lines on a second part, adjacent to the first are drawn at an angle of 45° but in the opposite direction. See Fig. 16.14, part A and B. Hatching lines on a third part adjacent to first two are drawn at an angle of 30° or 60°. See Fig. 16.14, part C adjacent to A and B. Sometimes spacing or pitch of the hatching lines is varied to separate it from the adjacent parts. See Fig. 16.14, part D, adjacent to parts A, B and C.

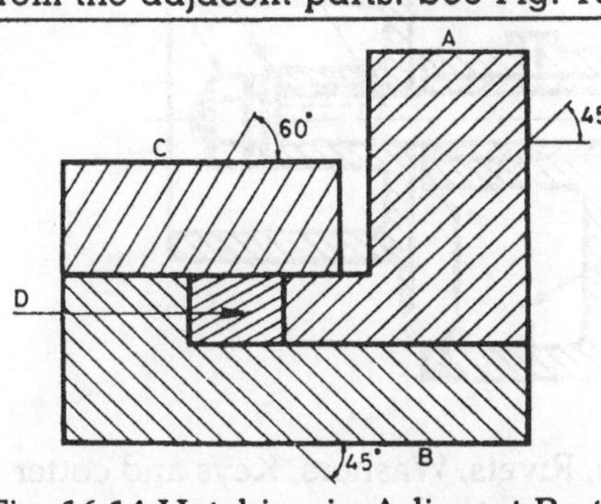

Fig. 16.14 Hatching in Adjacent Parts

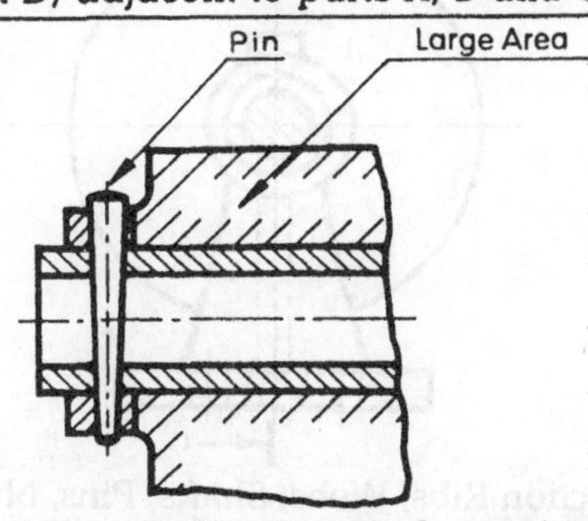

Fig. 16.15 Hatching in Large Area

(5) It hatching is required on a large section area it is avoided. It is limited to a zone following the contour of the section area required to be hatched. See Fig. 16.15.

(6) Similarly, if hatching is required on a very thin section area it is avoided. Complete section area is shown entirely black. Thin space is left between adjacent sections of the same type. See Fig. 16.16.

(7) Where sections of the same part in parallel planes are shown side by side, the hatching lines in both should be equally spaced but offsetted along dividing line between the sections. See Fig. 16.17.

Fig. 16.16 Hatching in Thin Area

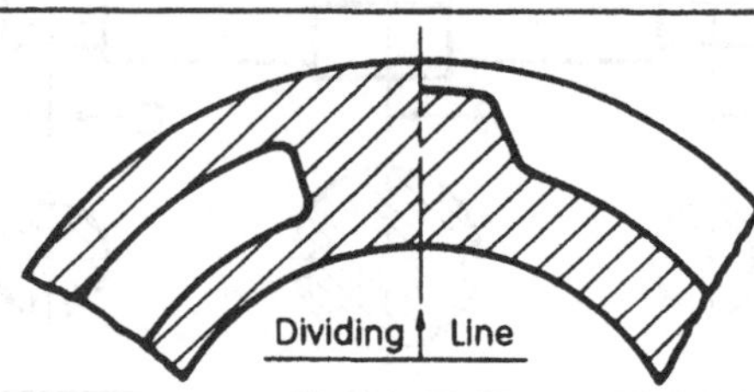

Fig. 16.17 Hatching in Parallel Planes

(8) As far as possible avoid dimensioning inside hatched area, but, if not possible, hatching lines may be interrupted for the dimensioning. See Fig. 16.18.

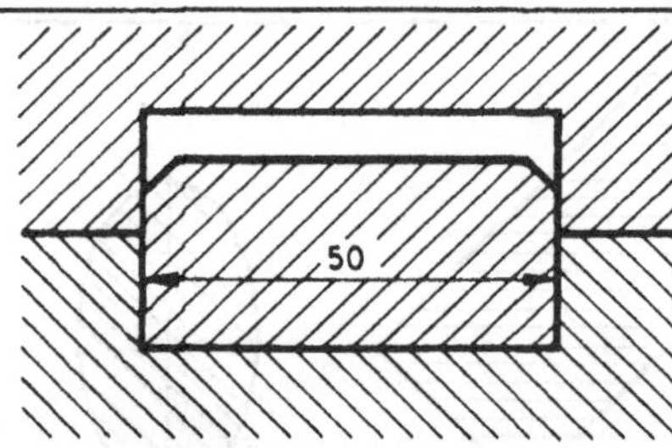

Fig. 16.18 Dimensioning in Hatched Area

(9) Hatched area can never have dashed or dotted boundary.

Having studied the major important sectional views like full, half, offset, partial, revolved, removed, etc. now we shall study various conventional practices required to be followed as per Indian Standards.

11. Sectioning Conventions for - Ribs Webs, Shafts, Pins, Nuts, Bolts, Rivets, Washers, Keys, Cotters, Rods, Pulley Arms, etc.

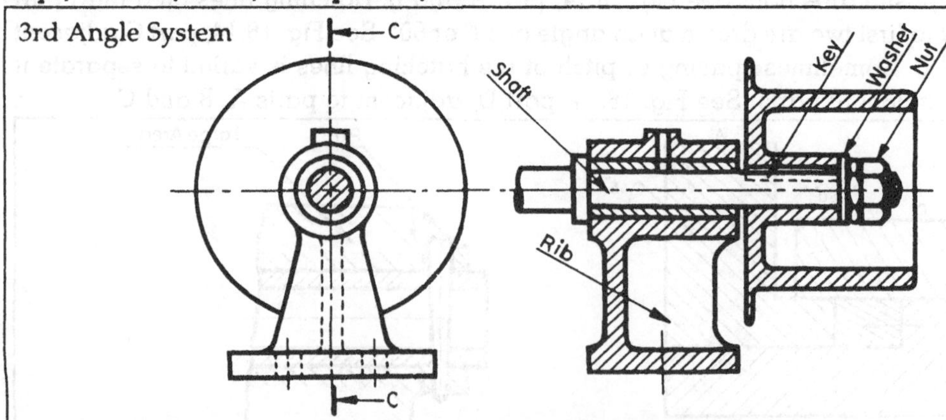

Don't Section Ribs, Webs, Shafts, Pins, Nuts, Bolts, Rivets, Washers, Keys and cotter when cutting plane is along axis.

Fig. 16.19 Sectioning Conventions - Rib, Shaft, Key, Washer, Nut, Etc.

Whenever, the cutting plane is passing through above machine elements and if C.P. contains the axis of above machine elements then those machine elements are not sectioned. See Figs. 16.19, 16.20, 16.21, 16.7(a) & (b) and 16.22. However, when the cutting plane is cutting, elements perpendicular to the axis of the above machine elements, section is taken and in projection hatching lines are drawn, see section lines in rib in plan of Fig. 16.21 and section lines in shaft in S.V. of Fig. 16.19.

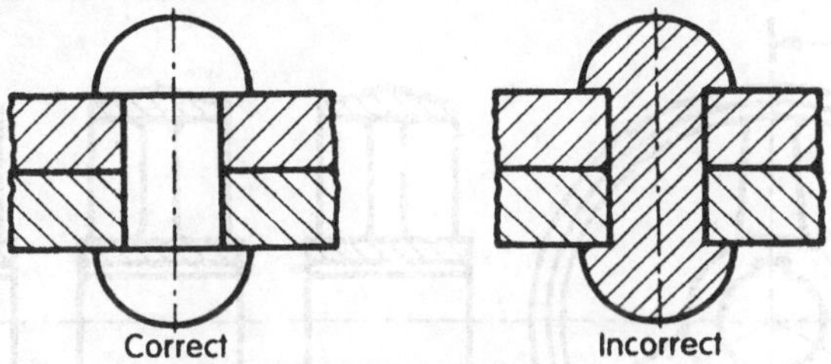

Fig. 16.20 Rivet in Section

These machine elements are simple, they do not have inner details and they are more easily recognised by their simple projection and hence they are not sectioned.

1st Angle System

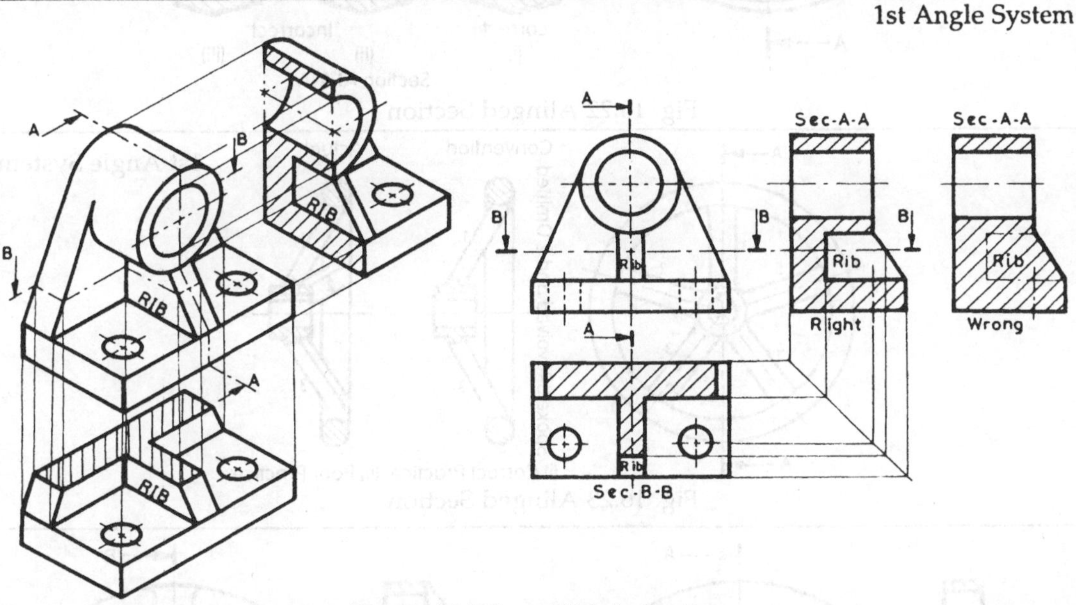

Fig. 16.21 Rib or Web in Section

Rib or web is a flat plate like part of the object used only to strengthen the object. Rib or web is not sectioned when cutting plane passes longitudinally through the centre of the rib or web. If section is taken in rib and if hatching lines are drawn then it will create wrong impression about the thickness or solidity of the component. See Fig. 16.21 side view with section lines in rib. However, if the cutting plane cuts the rib or web at right angle to the longitudinal surface of rib then section is taken and hatching lines are drawn in projection, see rib in section along B-B in plan in Fig. 16.21.

Spokes or arms of wheels, pulleys or flywheels are never sectioned when cutting plane passes through the axis of the arms, see Fig. 16.22(i). Here cutting plane A-A is passing through the axis of the arms and hence not sectioned in correct side view. Incorrect end views are also shown in Fig. 16.22(ii) and (iii).

12. Aligned Sections

Any part having odd number of spokes, ribs or holes will give an unsymmetrical and misleading section if the rules of projection are strictly followed. See Fig. 16.22(ii) and Fig. 16.23(ii). In such cases, to avoid confusion, the inclined or unsymmetrically placed elements should be rotated in the plane of projection in one view to bring it into the cutting plane. See Fig. 16.22(i) and Fig. 16.23(i).

1st Angle System

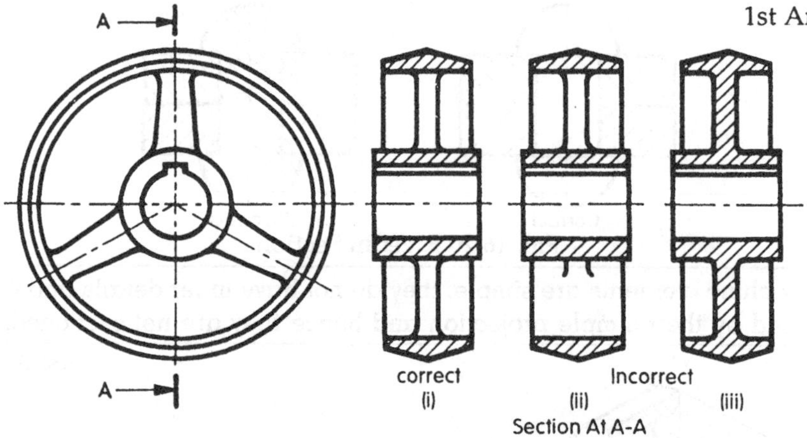

Fig. 16.22 Alinged Section

1st Angle System

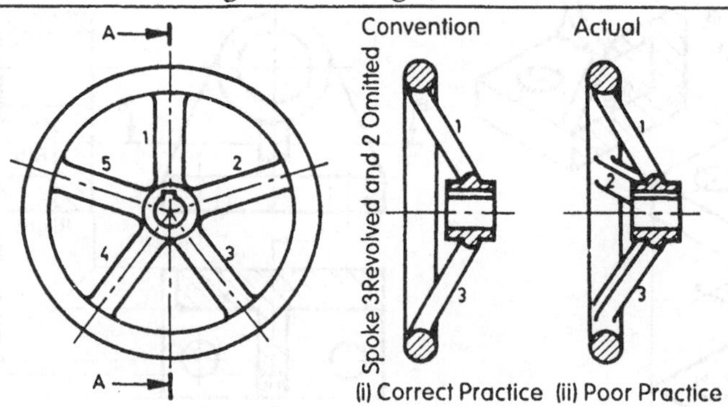

Fig. 16.23 Alinged Section

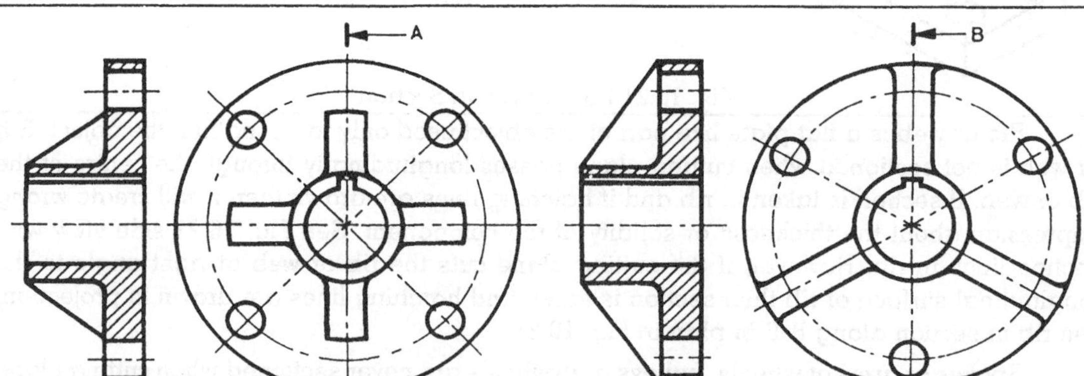

1st Angle System

Fig. 16.24 Align Section - Symmetrical Appearance

In Fig. 16.24 two flanges are shown one with even number of ribs and holes and another with odd number of ribs and holes. In such cases whether the cutting plane passes through the hole or not they are rotated in the plane of projection to bring it into the cutting plane and shown in sectional view. Similarly, ribs are also rotated and shown in sectional view. This will create symmetrical appearance in sectional view and true positions of the ribs and the holes will be seen in adjacent view. See Fig. 16.24.

13. Conventional Breaks

Long members of uniform cross section are not drawn of full lengthwise since it occupies more space unnecessarily. Conventional break is given along the length, as shown in Fig. 16.25. Eventhough the drawing is drawn of lesser length, actual dimension is given on the drawing of the member. Earlier, we have seen such a break is given in revolved section.

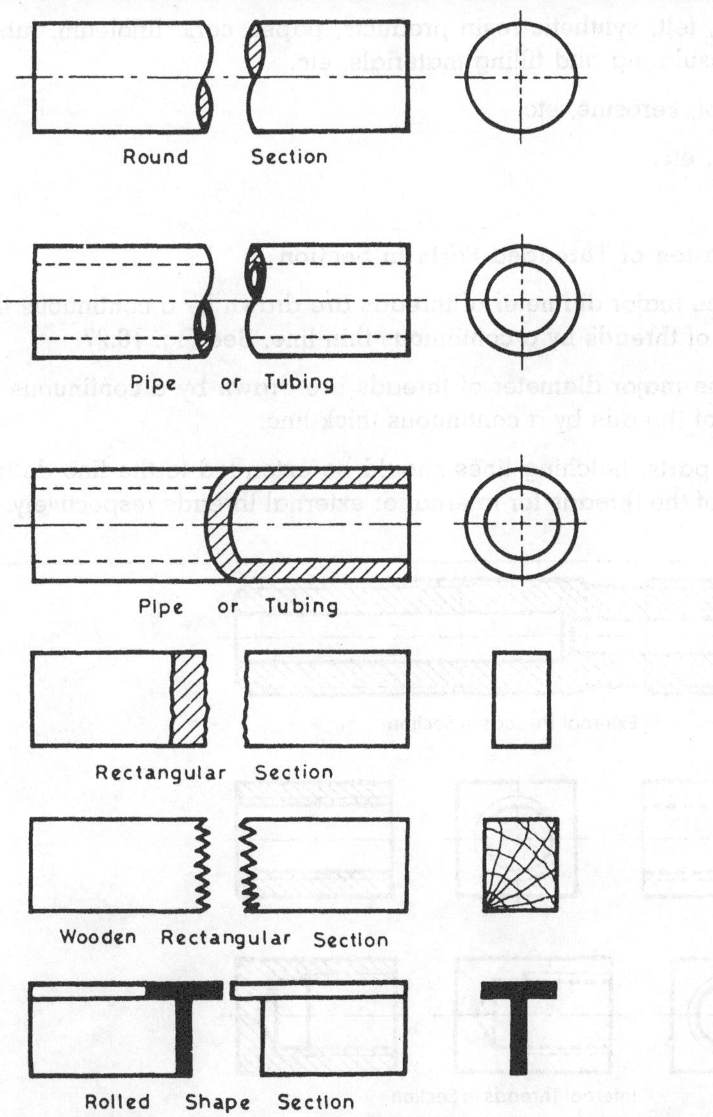

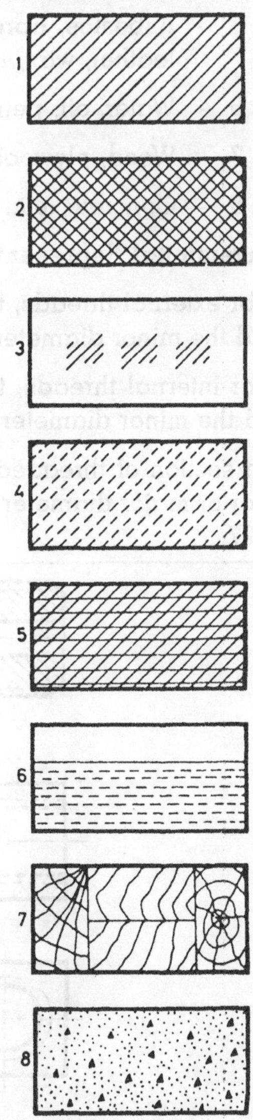

Fig. 16.25 Conventional Breaks Fig. 16.26 Convention for Material Symbols

In Fig. 16.25 conventional methods, adopted to show broken ends of (i) round bars or shafts, (ii) pipes or tubes (iii) square or rectangular bars and (iv) wood scantlings, (v) Rolled shape section are given.

14. Convention for Material Symbols

Different metals and materials are shown symbolically in different manner in section. There are 8 different symbols adopted, as per I. S. as shown in Fig. 16.26 :

Symbol 1 - Used for steel, C. I, copper and its alloys, aluminium and its alloys, etc.

Symbol 2 - Used for lead, zinc, tin, white metal, etc.

Symbol 3 - Used for glass.

Symbol 4 - Glass, porcelain, stone ware, marble, slate, etc.

Symbol 5 - Asbestos, fibre, felt, synthetic resin products, paper, cork, linoleum, rubber, leather, wax, insulating and filling materials, etc.

Symbol 6 - Water, oil, petrol, kerosine, etc.

Symbol 7 - Wood, plywood, etc.

Symbol 8 - Concrete, etc.

15. Conventional Representation of Threaded Parts in Section

For external threads, the major diameter of threads are drawn by a continuous thick line, and the minor diameter of threads by a continuous thin line. See Fig. 16.27.

For internal threads, the major diameter of threads are drawn by a continuous thin line and the minor diameter of threads by a continuous thick line.

In section of threaded parts, hatching lines should be extended to the line defining the minor or major diameter of the threads for internal or external threads respectively. See Fig. 16.27.

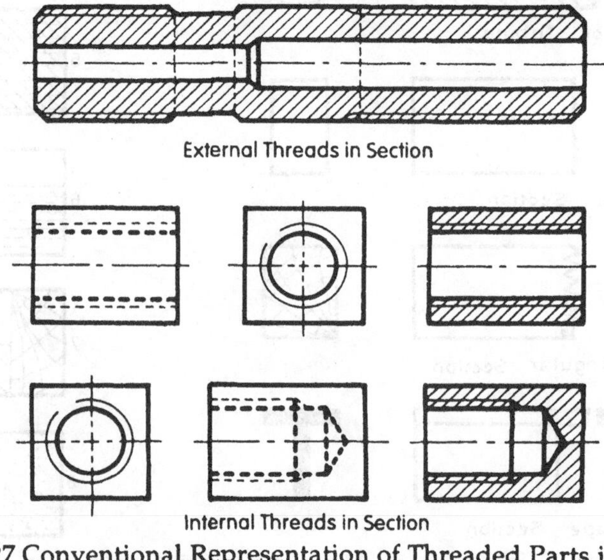

Fig. 16.27 Conventional Representation of Threaded Parts in Section

16. Cutting Planes / Section Planes.

The following three types of cutting planes are used for the purpose of section in drawings :

(i) One plane section, Fig. 16.28.

(ii) Section in two parallel planes, Fig. 16.29.

(iii) Section in three continuous planes, Fig. 16.30.

(i) If important features of the object are in one plane then one plane section (like A-A) is used to get sectional view. See Fig. 16.28, section plane line A-A.

(ii) If important features of the object are on two parallel planes then two parallel planes are used to get sectional view. See Fig. 16.29, section plane line B-B.

(iii) If important features of the object are on three continuous planes then three continuous plane section is used to get sectional view. See Fig. 16.30, section plane line C-C.

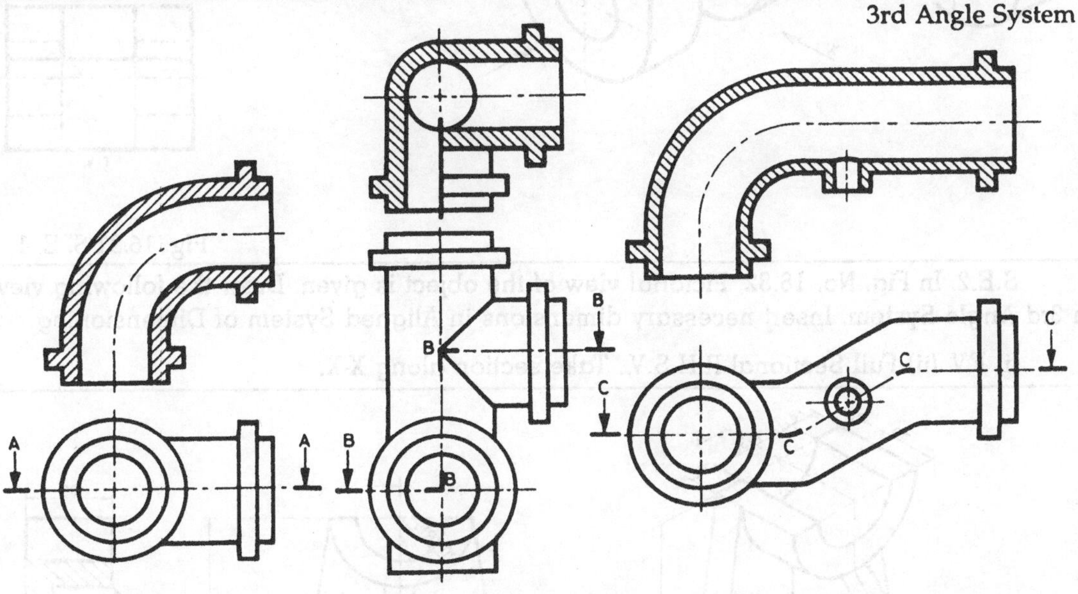

One Plane Section
Fig. 16.28

Two Parallel Planes Section
Fig. 16.29

Three Continuous Planes Section
Fig. 16.30

Solved Exercise

S.E.1. In Fig. No. 16.31, Pictorial View of the object is given. Draw (i) F.V. (ii) Full sectional Right Hand Side View along A-A. (iii) T.V. Use 1st Angle System.

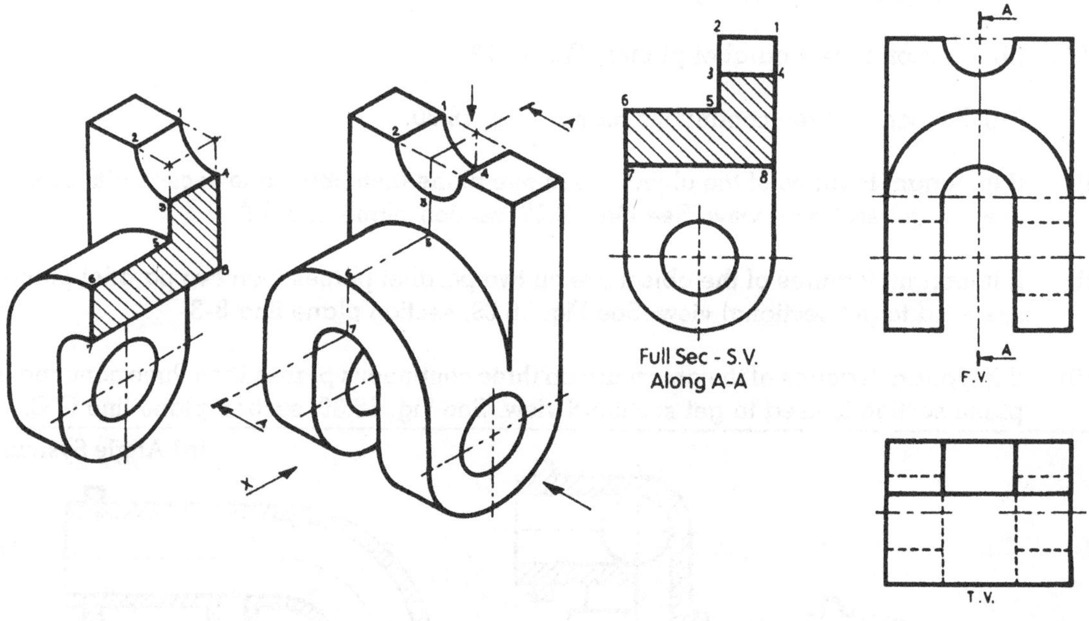

Fig. 16.31 S. E. 1

S.E.2. In Fig. No. 16.32, Pictorial view of the object is given. Draw the following views in 3rd Angle System. Insert necessary dimensions in Aligned System of Dimensioning.

(i) F.V. (ii) Full Sectional R.H.S.V.. Take section along X-X.

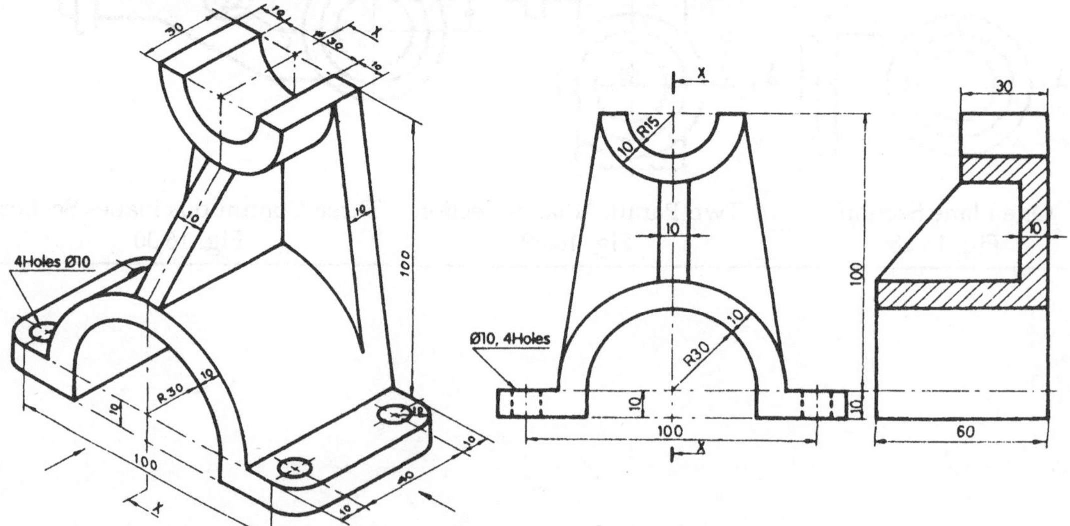

Fig. 16.32 S. E. 2

S.E.3. In Fig. No. 16.33, Pictorial View of the object is given. Draw (i) Full Sectional F.V. along B-B. (ii) T.V. (iii) L.H.S.V. Use 1st Angle System.

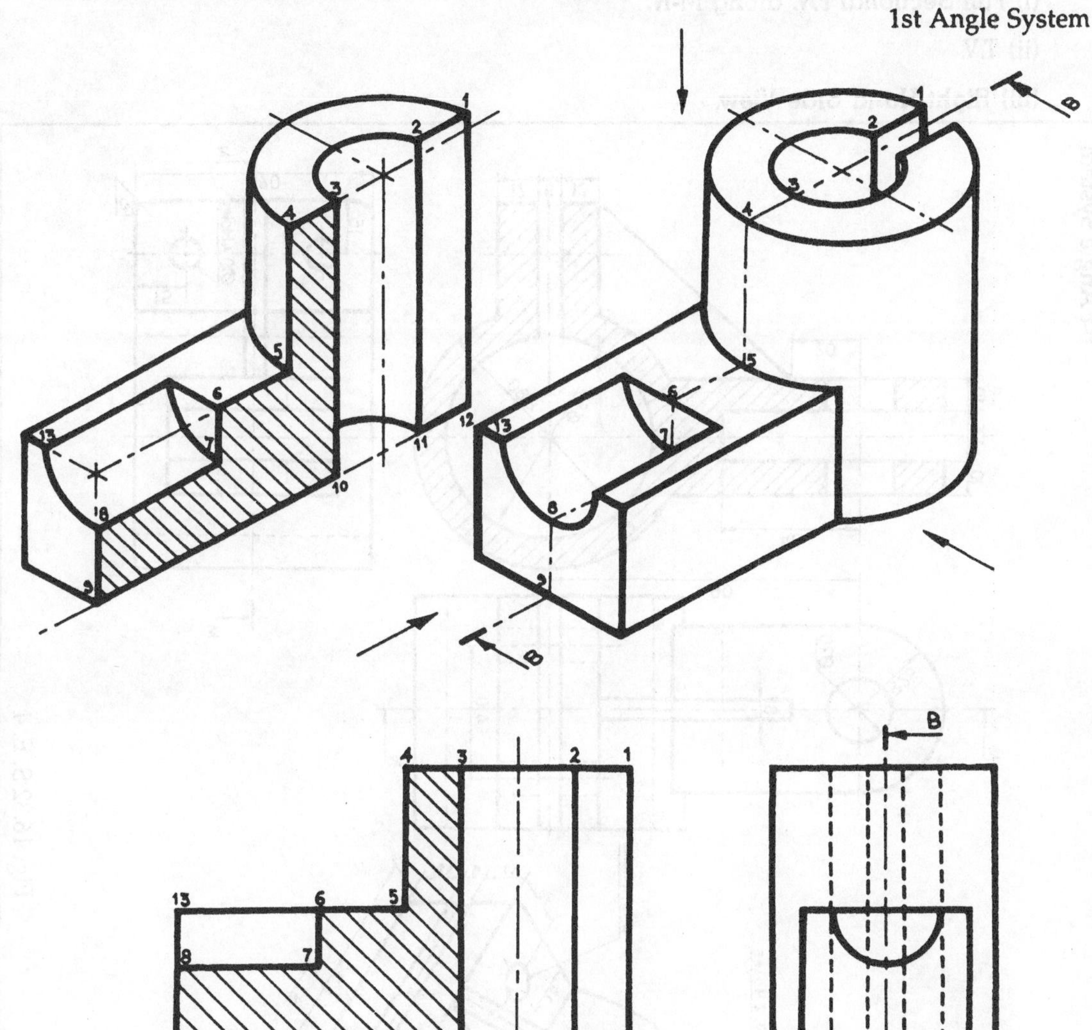

Fig. 16.33 S. E. 3

S.E.4. In Fig. No. 16.34, Pictorial view of the object is given. Draw following views in 1st Angle System and insert necessary dimensions in the projections.

(i) Full Sectional F.V. along M-N.

(ii) T.V.

(iii) Right Hand Side View.

1st Angle System

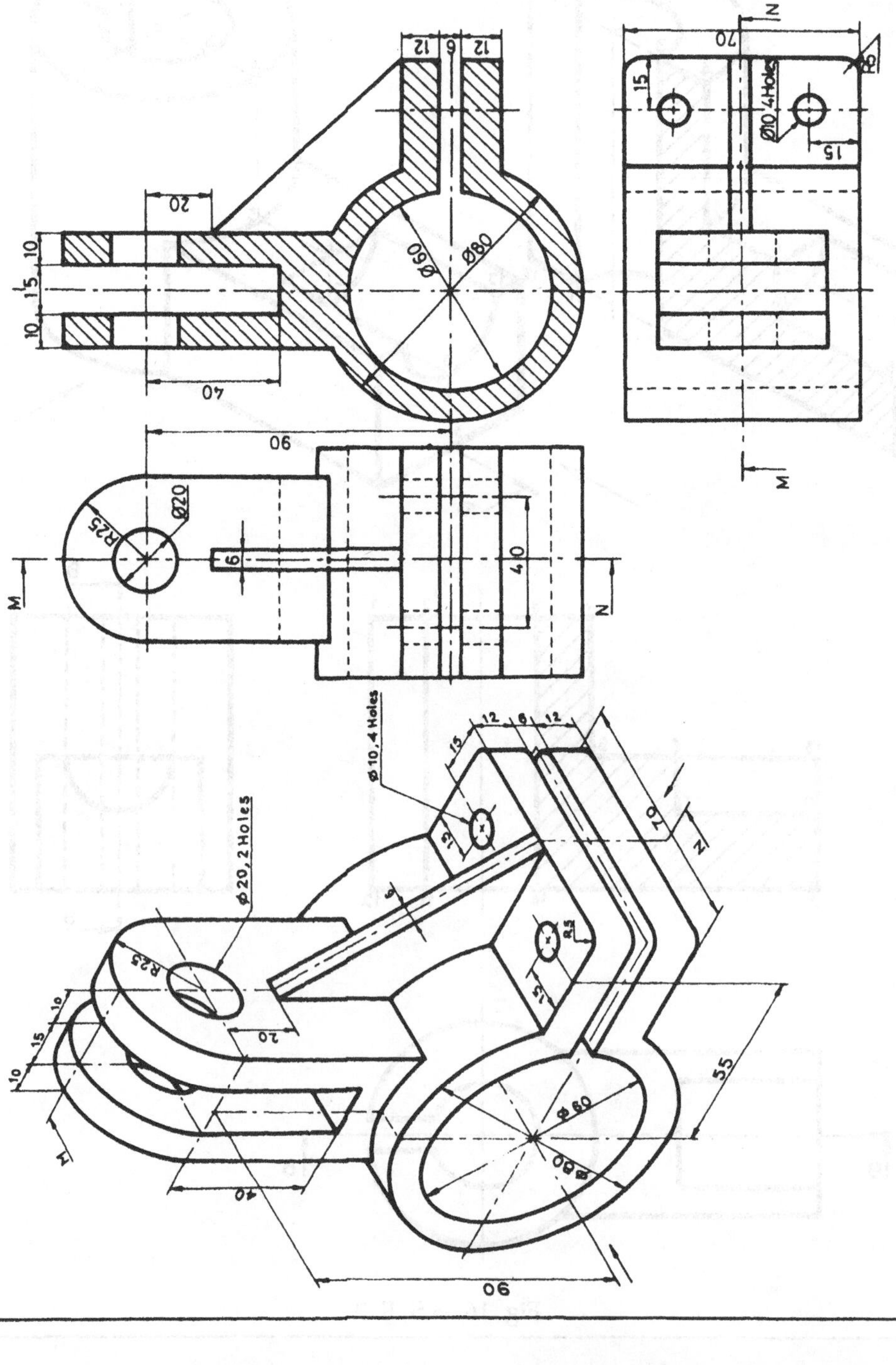

Fig. 16.32 S. E. 4

S.E.5. In Fig. No. 16.35 Pictorial view of the 3/4th object alongwith 1/4th cut portion of the object is given. Draw the following views in 1st Angle Projection System. Insert necessary dimensions (i) Half Sectional F.V. (ii) Half Sectional L.H.S.V. (iii) T.V. (Full)

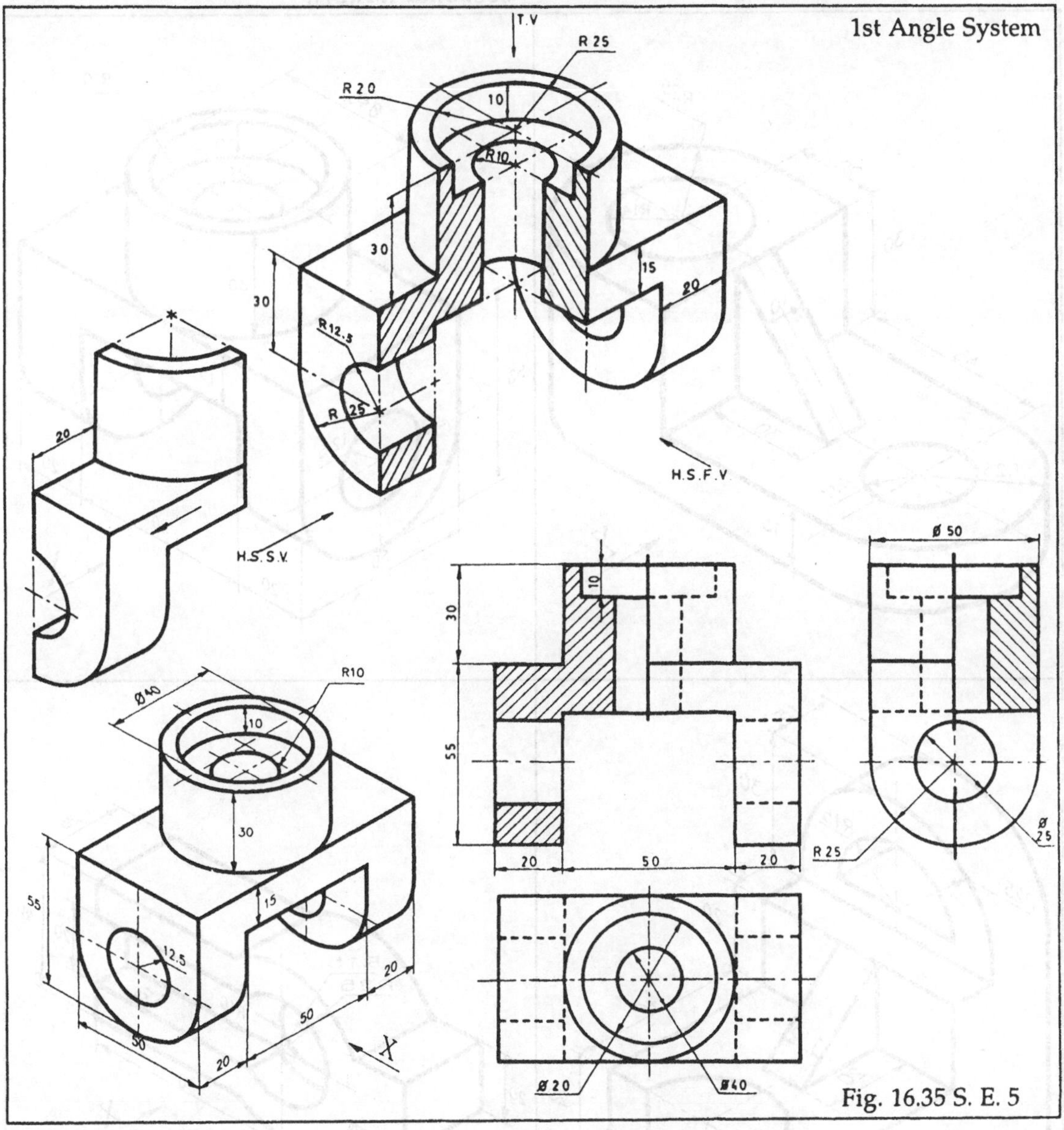

Fig. 16.35 S. E. 5

Unsolved - Exercise

Draw orthographic views, as stated in each exercise, from the pictorial views given in Figs. 16.36 to 16.62. Show cutting plane line/lines in respective view/views. Use conventional practices given in theory of this chapter as well as given in Indian Standards for engineering drawing. Insert dimensions in the views. Assume proportionate dimensions, if missing. Answers are given in last pages of this book.

Ex. 1. Fig. 16.36 Draw (i) Full sectional elevation (ii) Top plan (iii) L.H.S.V.

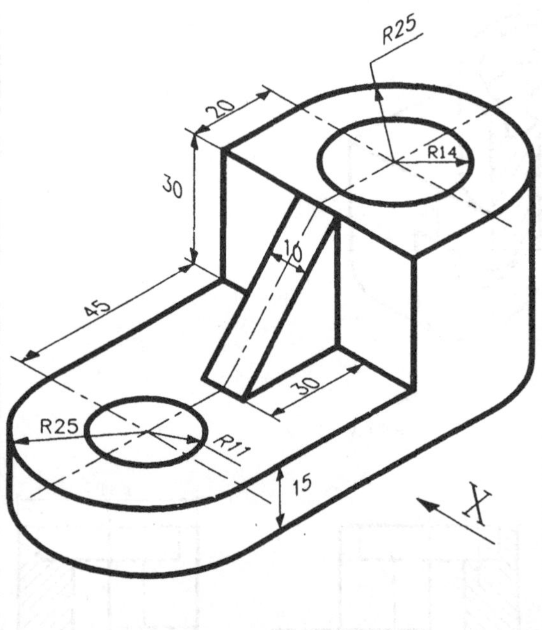

Ex. 3. Fig. 16.38. Draw (i) Half sectional elevation (ii) L.H.S.V. (iii) Top plan (iv) Half sectional R.H.S.V.

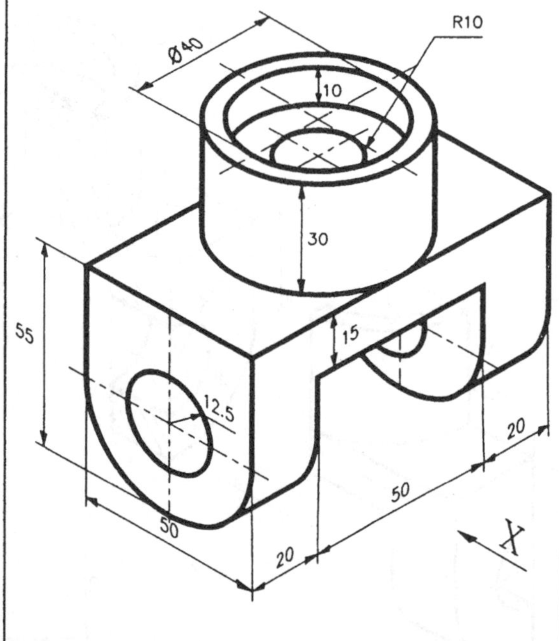

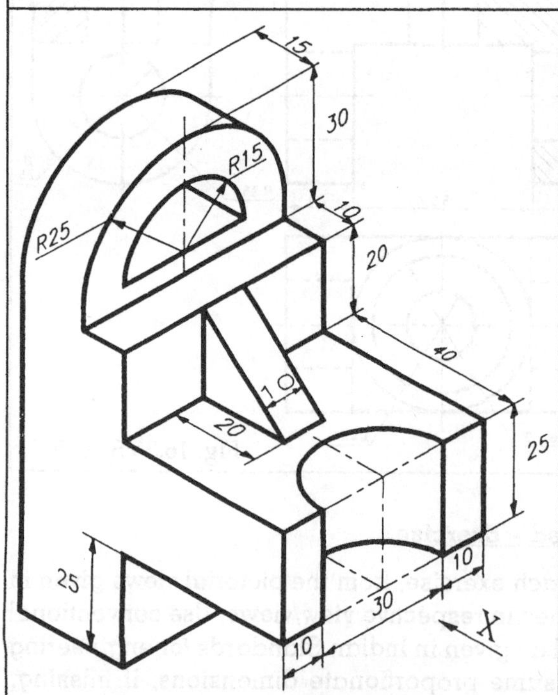

Ex. 2. Fig. 16.37. Draw (i) Front view (ii) Full sectional L.H. side view (iii) Top plan.

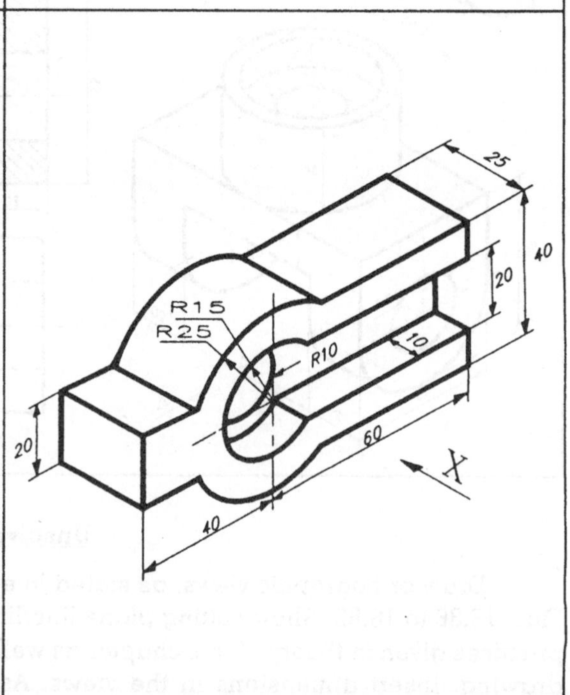

Ex. 4. Fig. 16.39. Draw (i) Front view (ii) Full sectional plan (iii) R.H. side view. (iv) L.H.S.V.

Ex. 5. Fig. 16.40. Draw (i) Front elevation (ii) Full sectional side view (iii) Top Plan.

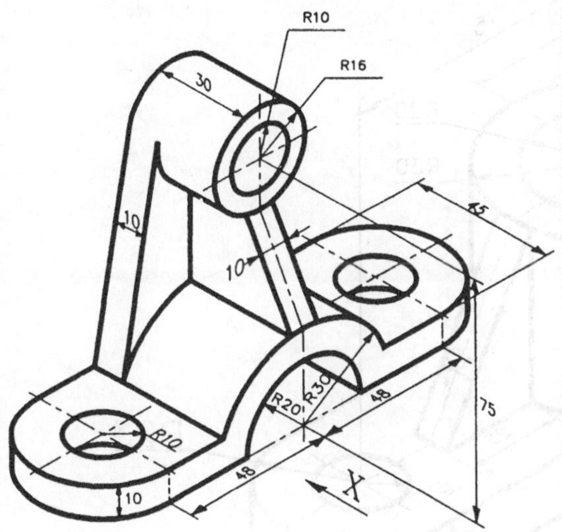

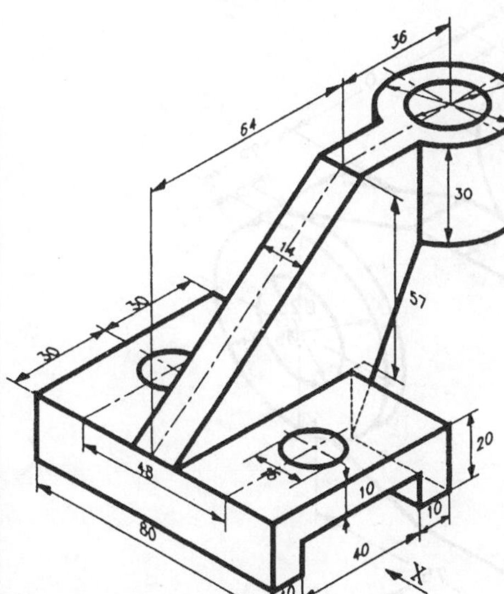

Ex. 6. Fig. 16.41. Draw (i) Full sectional elevation (ii) L.H. side view (iii) R.H. side view (iv) Top plan.

Ex. 7. Fig. 16.42 Draw (i) Elevation (ii) R.H. sectional side view (iii) Rear view (iv) Top plan.

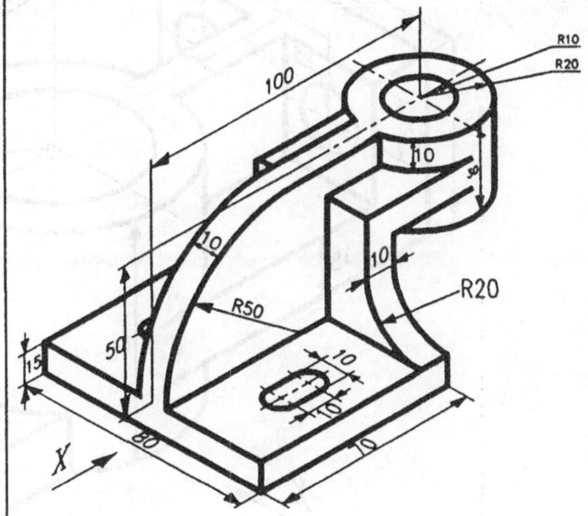

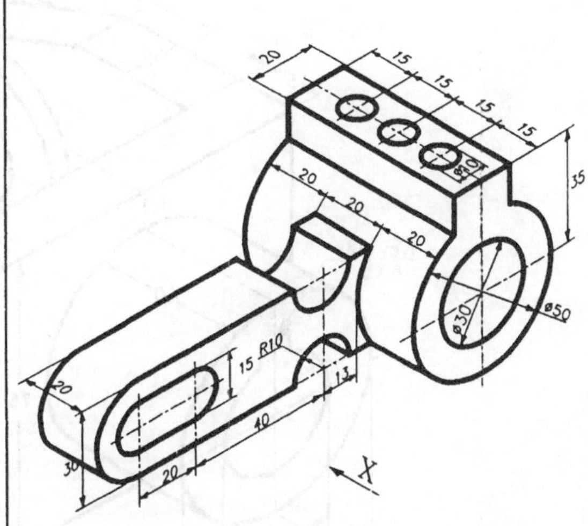

Ex. 8. Fig. 16.43. Draw (i) Full sectional elevation (ii) L.H. side view (iii) Top plan.

Ex. 9. Fig. 16.44. Draw (i) Sectional elevation take section along A-B (ii) R.H.S.V. (iii) Top plan. **[Gujarat University Nov., 1991]**

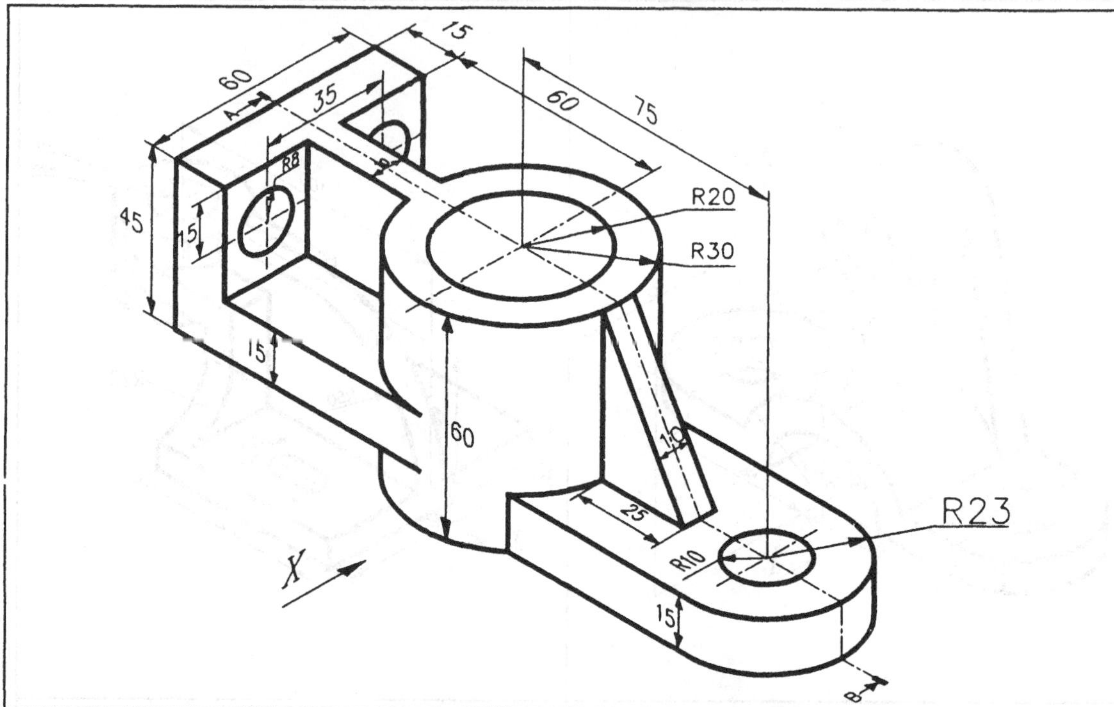

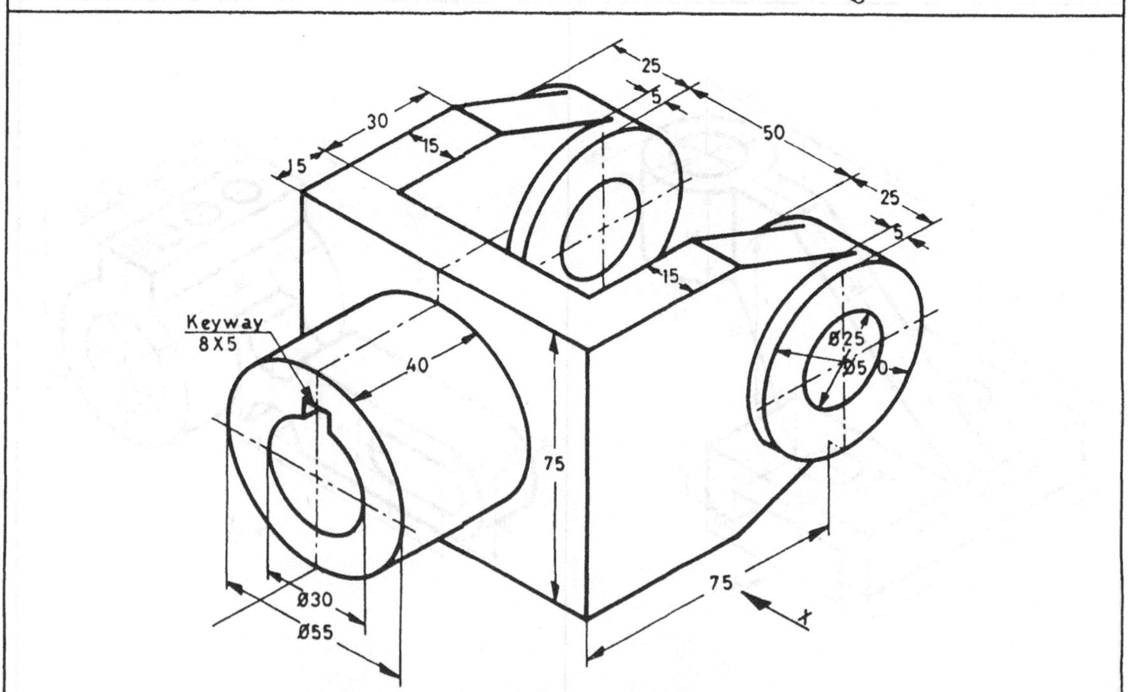

Ex. 10. Fig. 16.45. Draw (i) Full sectional front view (ii) Full sectional top view (iii) L.H.S.V. (iv) R.H.S.V.

Ex. 11. Fig. 16.46. Draw (i) Elevation (ii) Full sectional side view (iii) Top plan

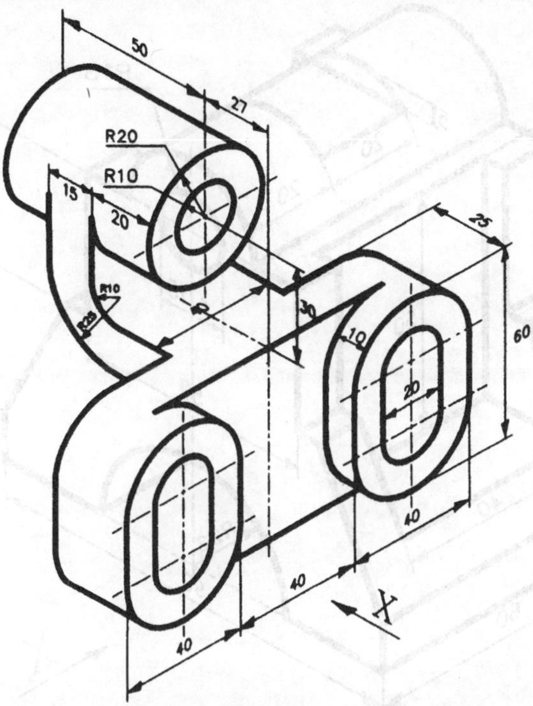

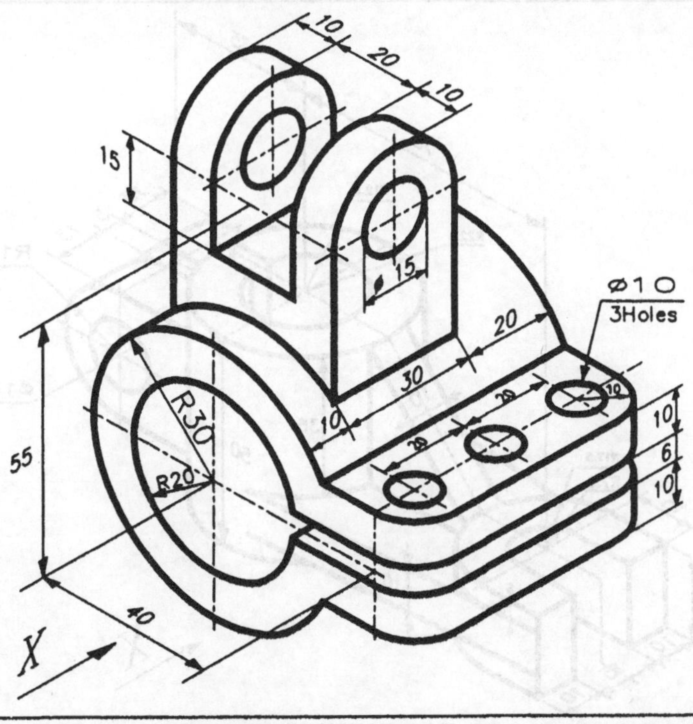

Ex. 12. Fig. 16.47. Draw (i) Offset sectional elevation, select suitable section.
(ii) R.H. side view (iii) Top plan

Ex. 13. Fig. 16.48. Draw (i) Elevation (ii) L.H. sectional side view (iii) Top plan

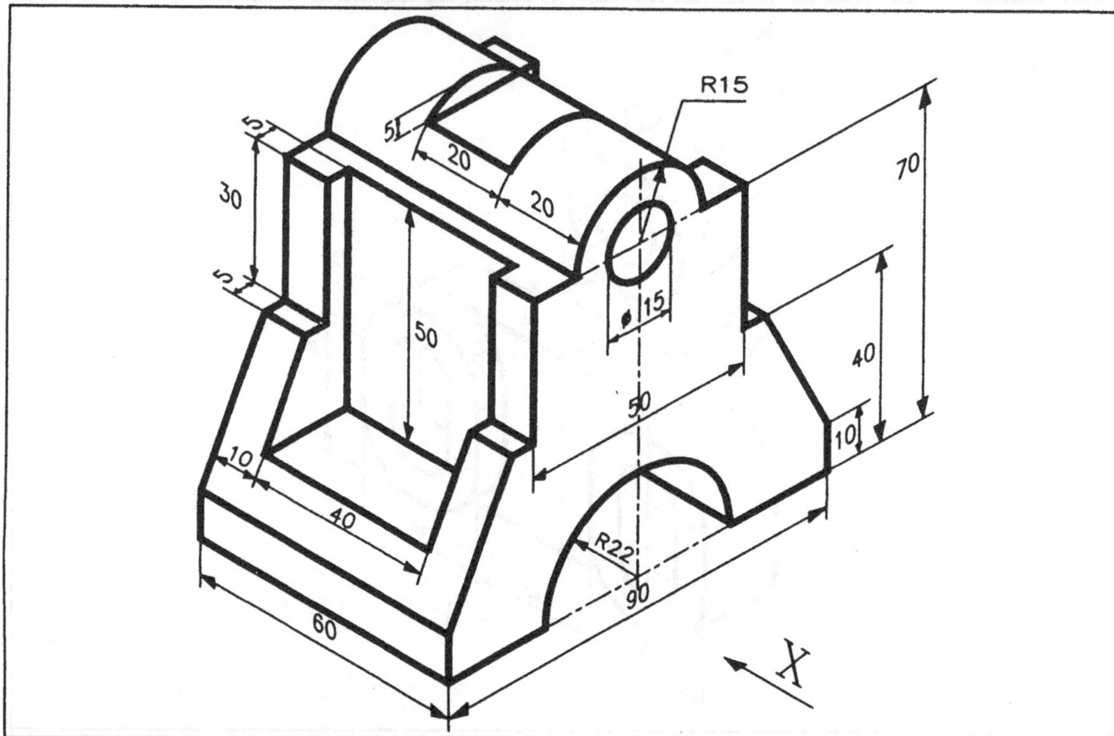

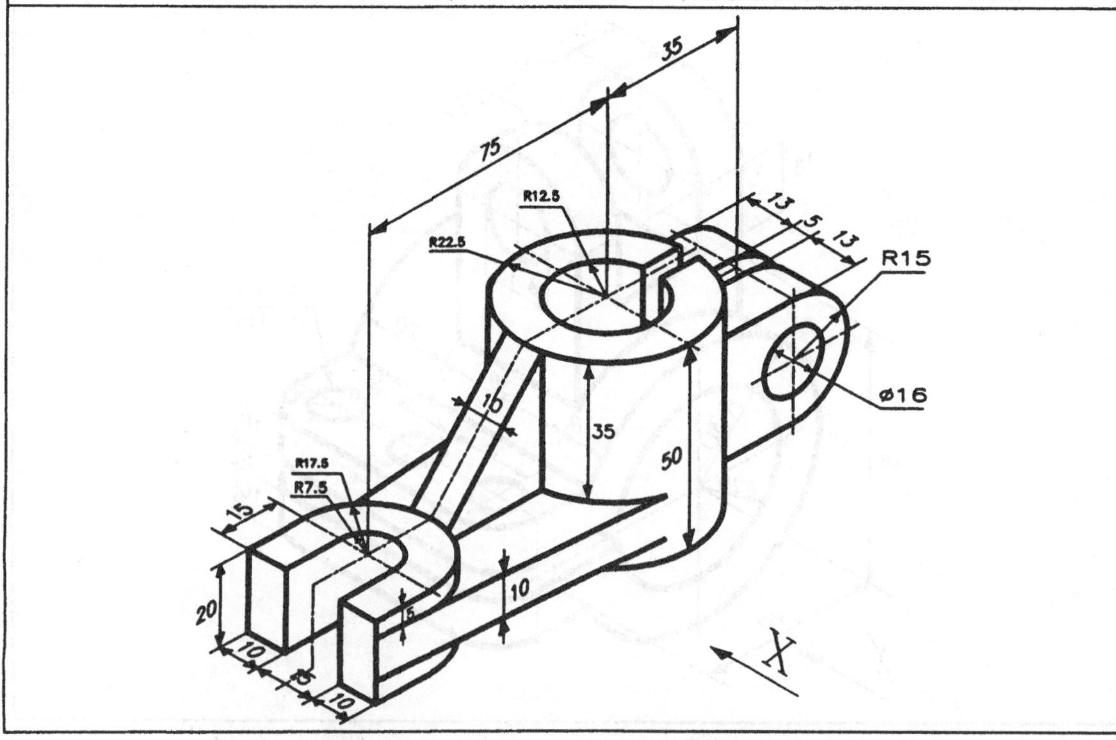

Ex. 14. Fig. 16.49. Draw (i) Full sectional elevation (ii) Top plan (iii) R.H. side view (iv) L.H. side view.

Ex. 15. Fig. 16.50. Draw (i) Half sectional elevation (ii) Half sectional side view from left (iii) R.H.S.V. (iv) Top plan (complete).

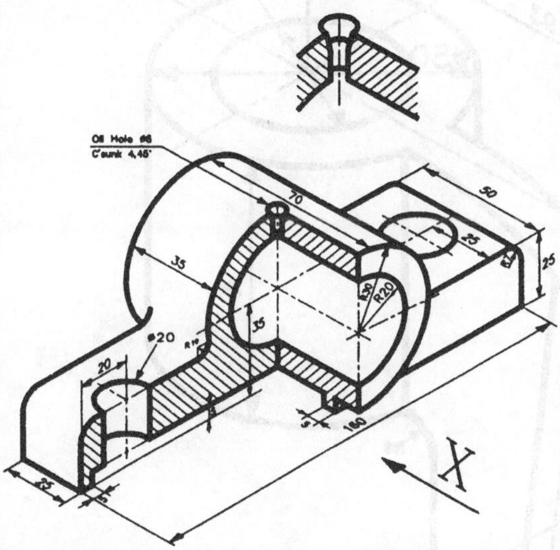

Ex. 16. Fig. 16.51. Draw (i) Full sectional elevation (ii) R.H.S.V. (iii) L.H.S.V. (iv) Top plan.

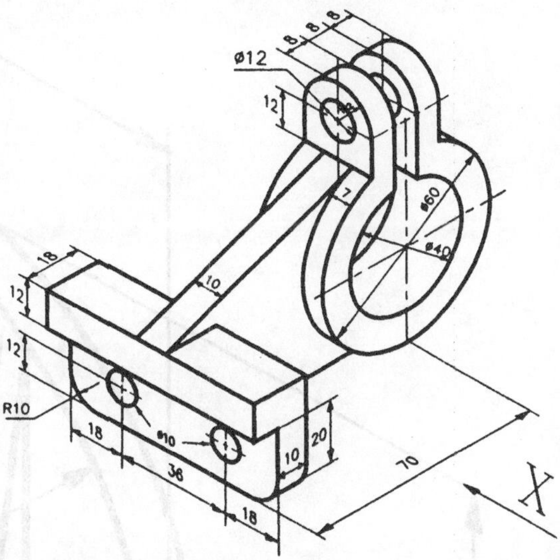

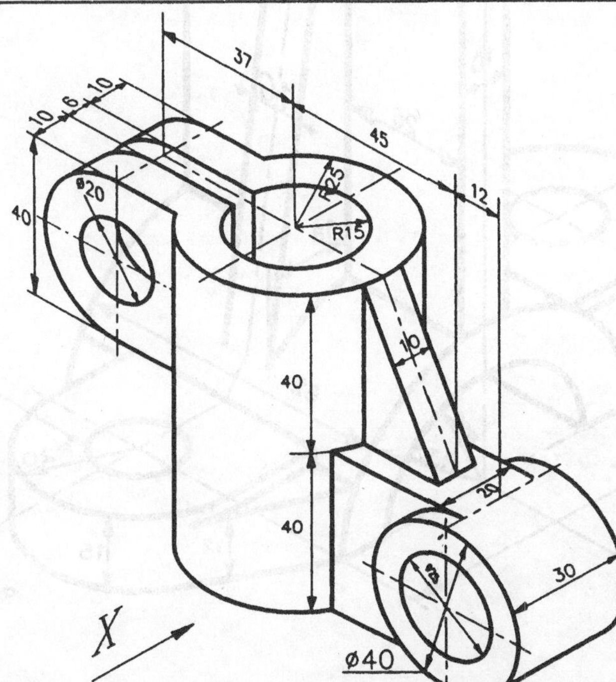

Ex. 17. Fig. 16.52. Draw (i) Full sectional front view (ii) R.H.S.V. (iii) L.H.S.V. (iv) Top plan

Ex. 18. Fig. 16.53. Draw (i) Elevation (ii) Sectional side view from left (Y) take section along B-B (iii) Top plan [Gujarat University June 1990]

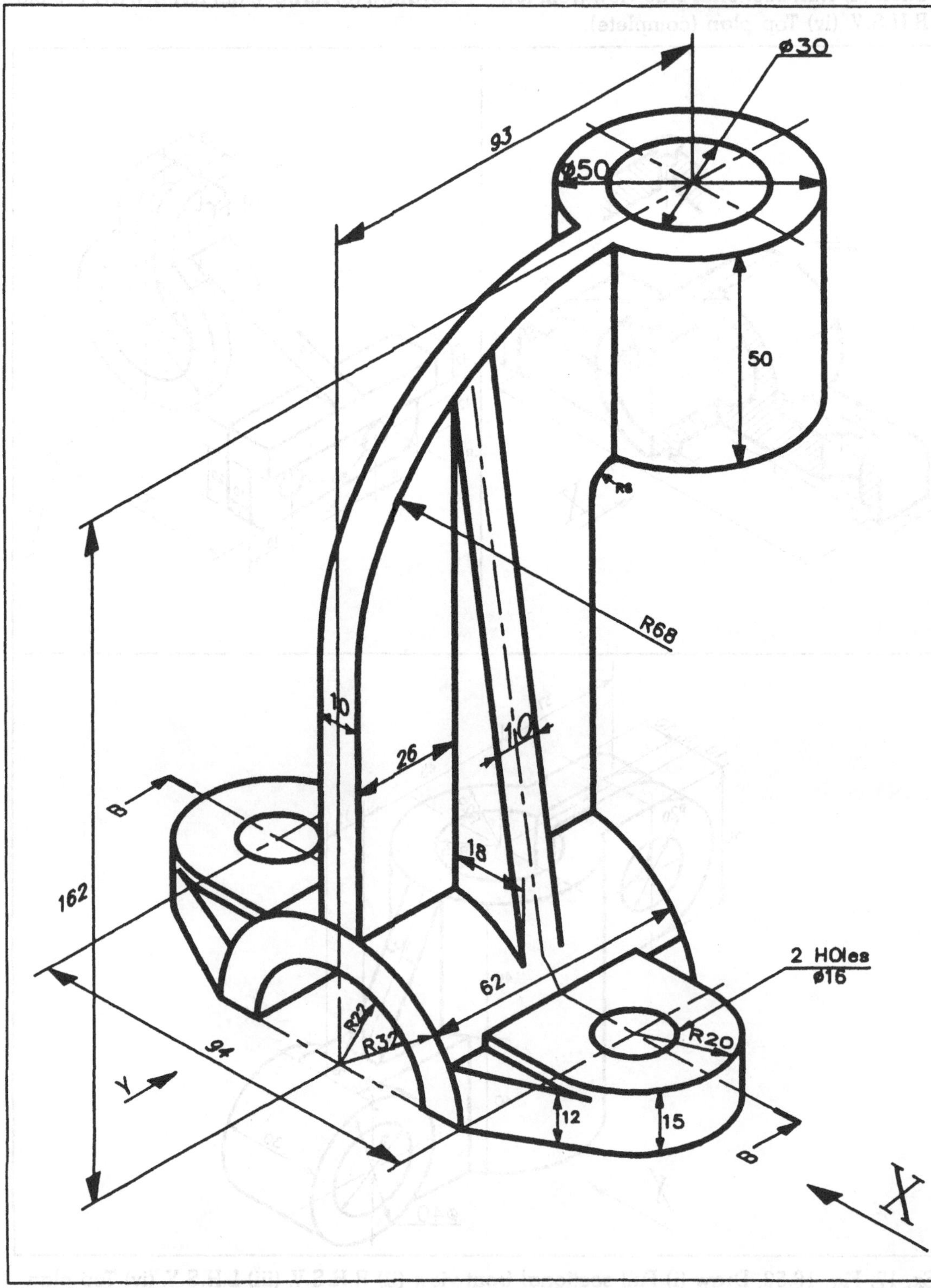

Ex. 19. Fig. 16.54. Draw (i) Front view (ii) Full sectional side view from right (iii) Top plan

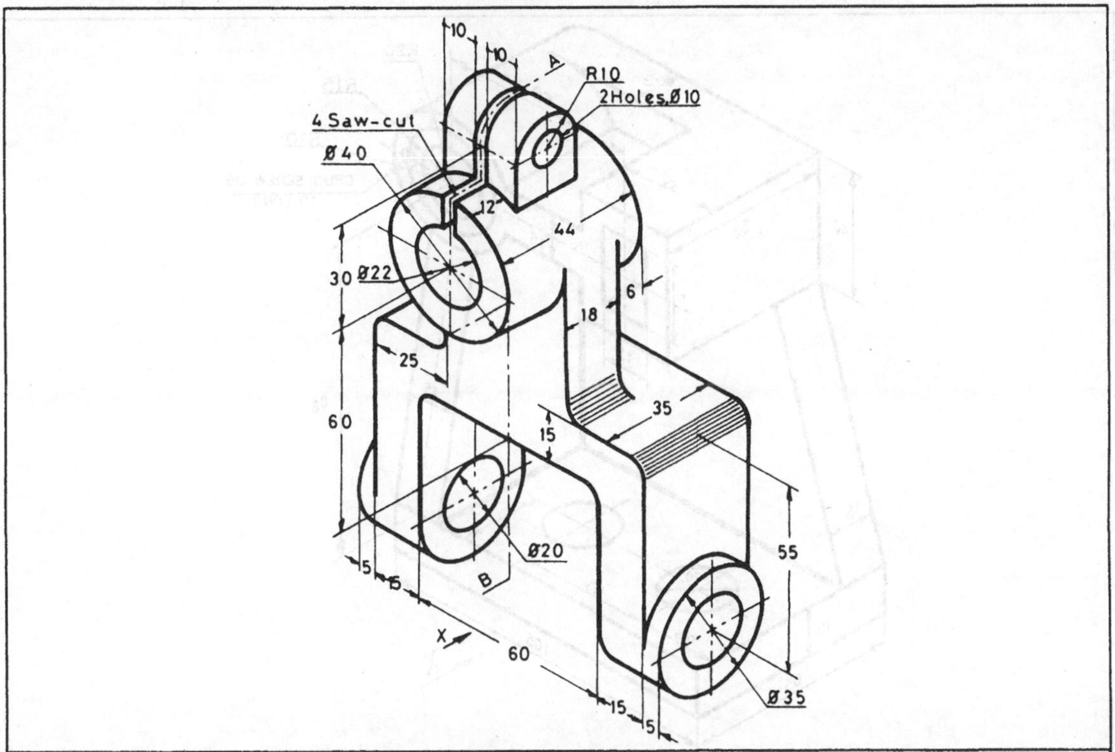

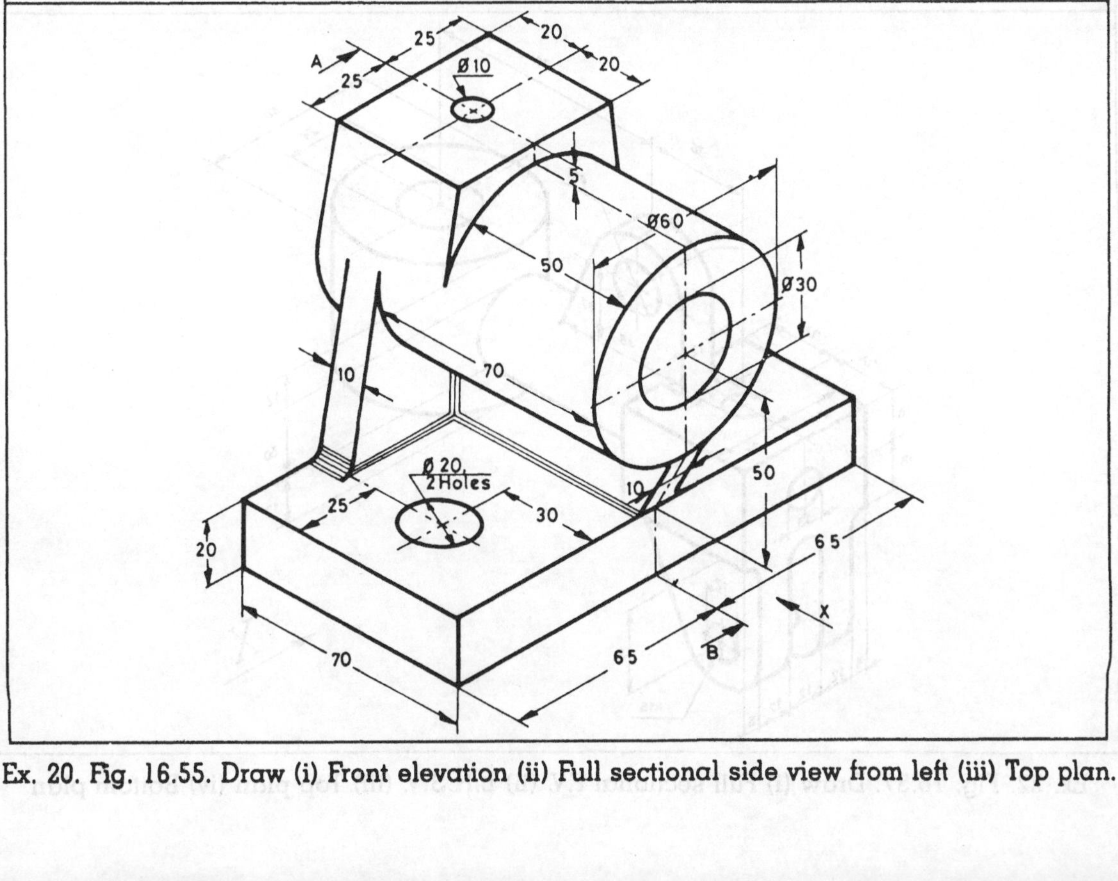

Ex. 20. Fig. 16.55. Draw (i) Front elevation (ii) Full sectional side view from left (iii) Top plan.

Ex. 21. Fig. 16.56. Draw (i) Front view (ii) Side view from left (iii) Top plan.

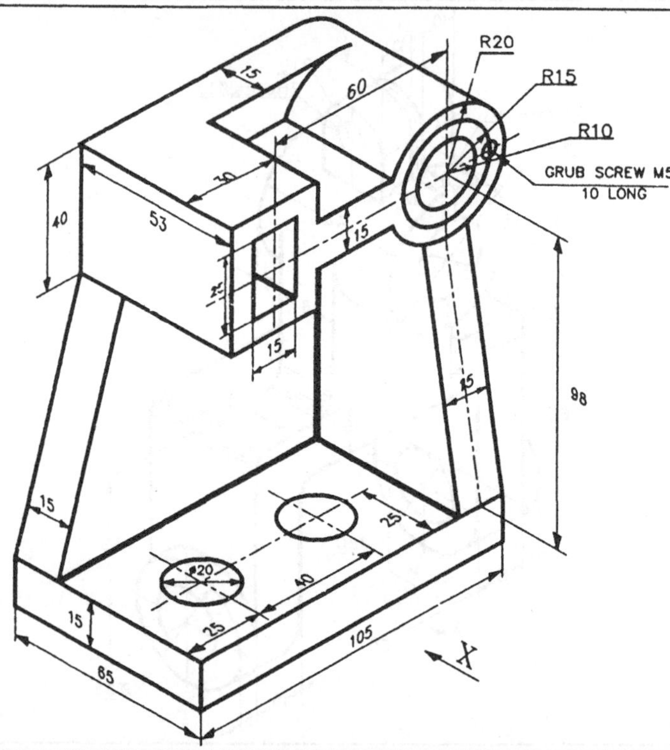

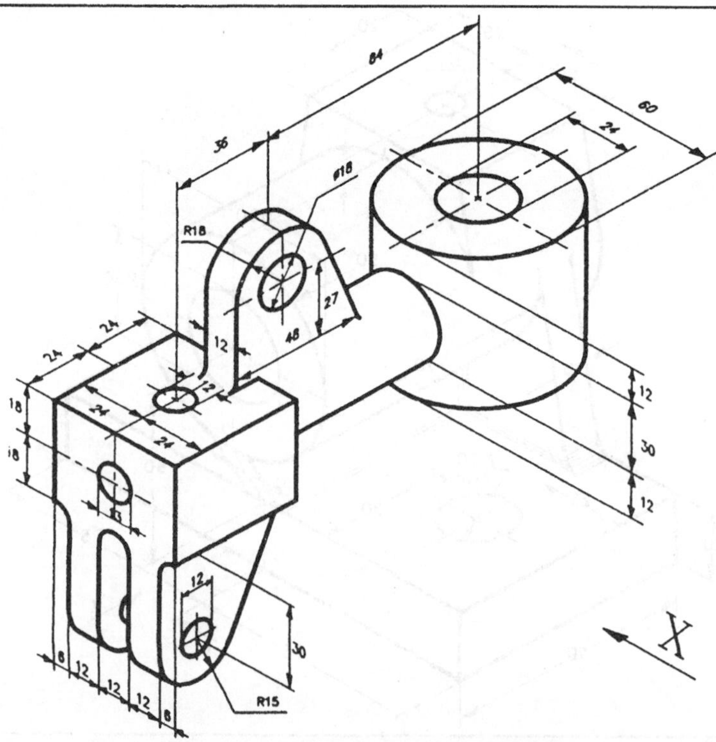

Ex. 22. Fig. 16.57. Draw (i) Full sectional F.V. (ii) L.H.S.V. (iii) Top plan (iv) Bottom plan

Ex. 23. Fig. 16.58. Draw (i) Front elevation (ii) Sectional side view from right (iii) Top plan (iv) L.H.S.V.

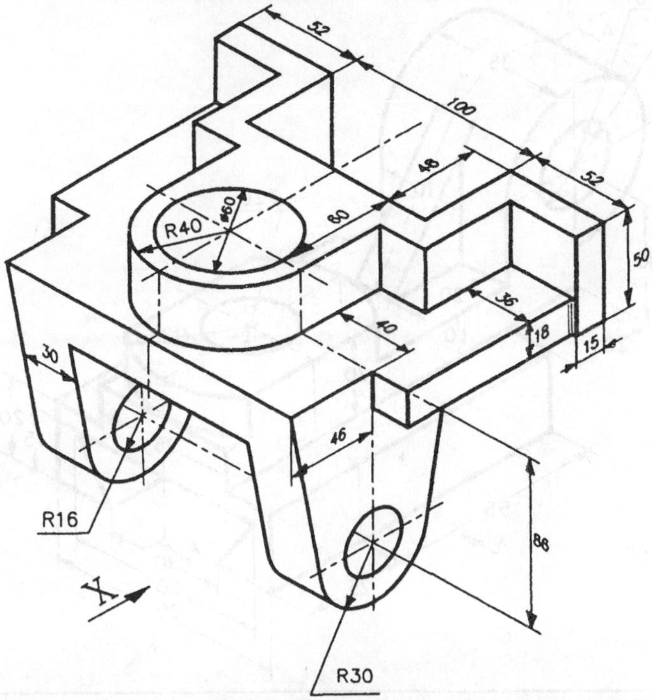

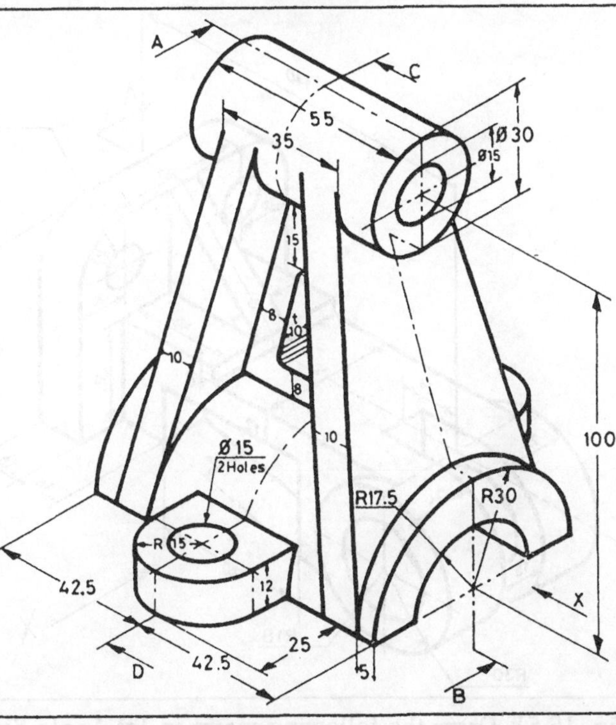

Ex. 24. Fig. 16.59. Draw (i) Sectional front view, take section along C-D (ii) Sectional side view from left, take section along A-B (iii) Top plan

Ex. 25. Fig. 16.60. Draw (i) Full sectional front view along A-A. (ii) Right hand side view (iii) Top plan. Show all dimensions. **[Gujarat University August 1993]**

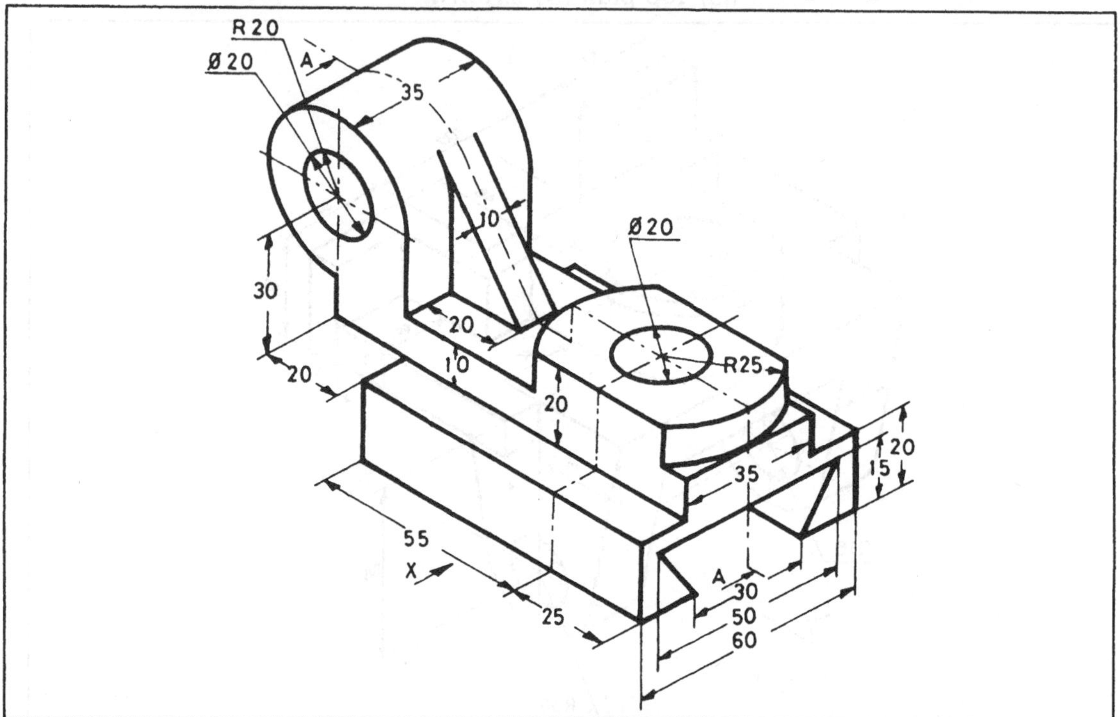

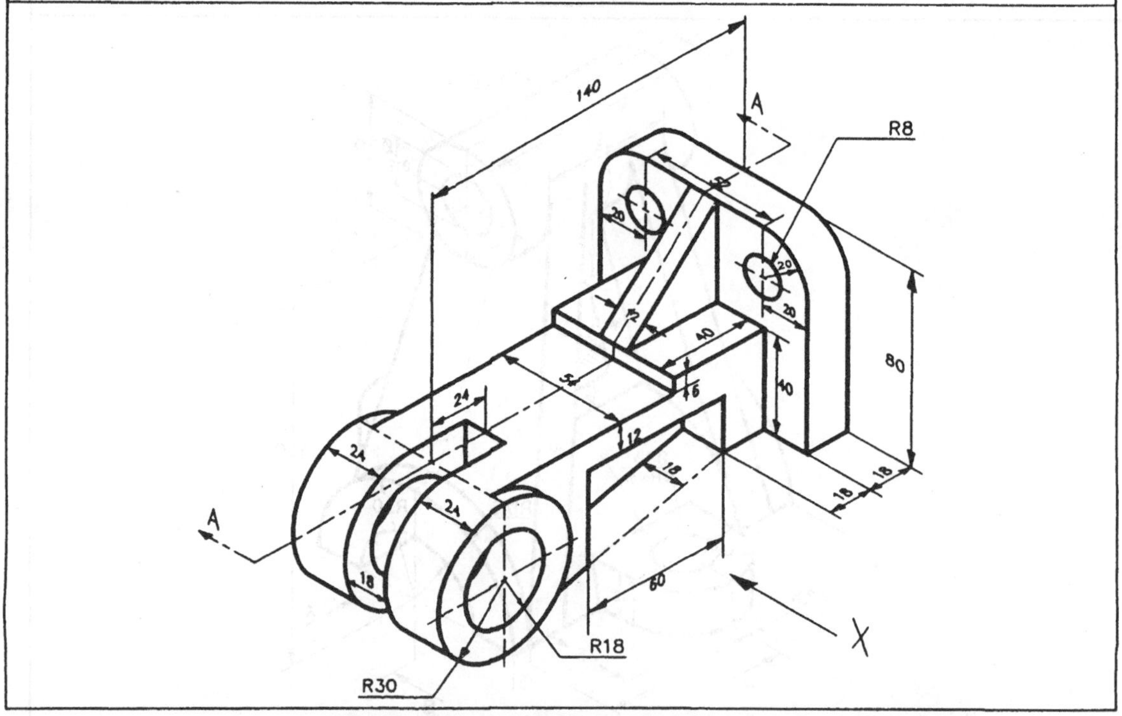

Ex. 26. Fig. 16.61. Draw the following views in 1st Angle System
(i) Sectional Elevation-Section A-A. (ii) Side elevation (iii) Plan. Insert all dimensions.

Draw 6 Principal Orthographic Projections for the two pictorial views given.

(1) In First Angle Projection System and (2) In Third Angle Projection System.

Draw also symbol for the system of projection.

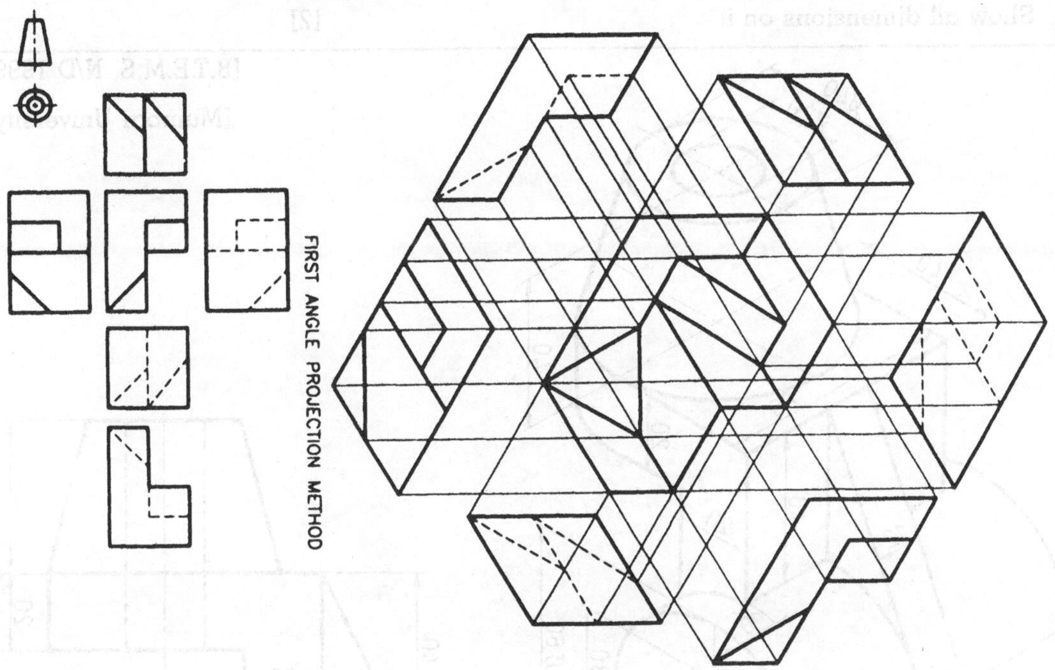

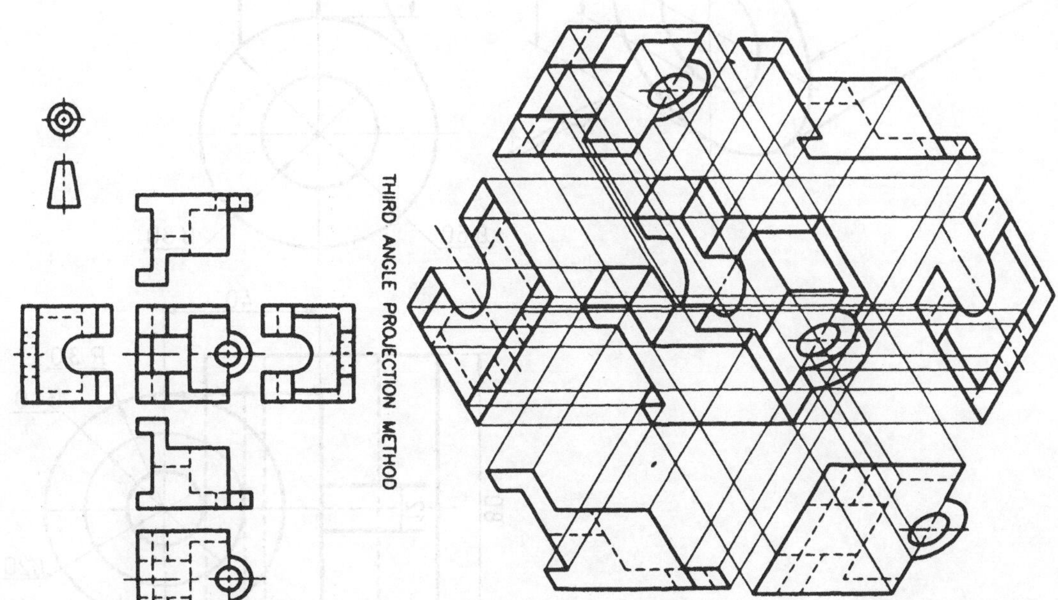

Fig. shows a pictorial view of A.C.I. BLOCK. Draw to scale full size, the following views by using first angle method of projection.

(a) F. V. looking in the direction of arrow X [8]

(b) Top View [6]

(c) Show all dimensions on it [2]

[B.T.E.M.S. N/D 1999]

[Mumbai University]

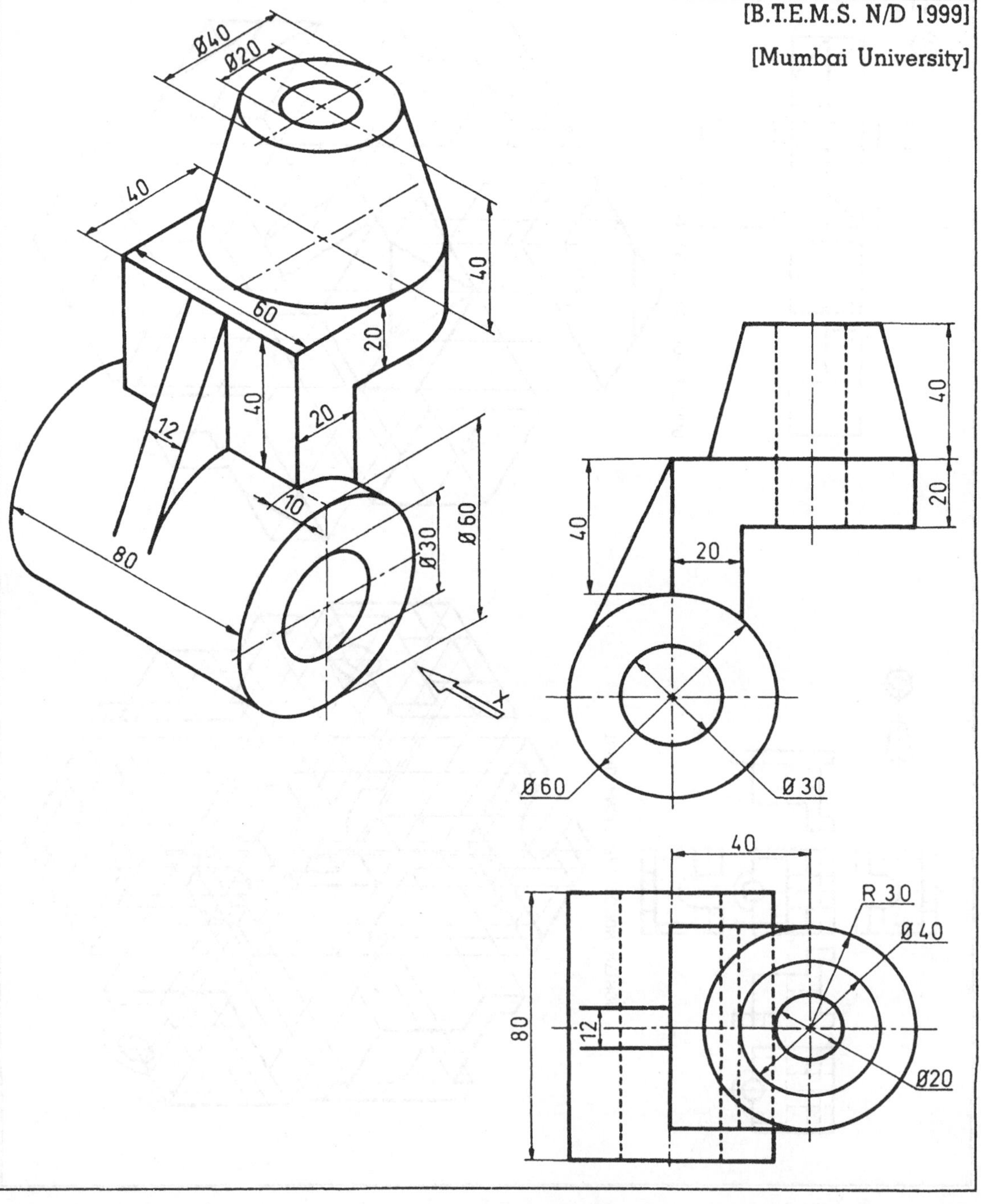

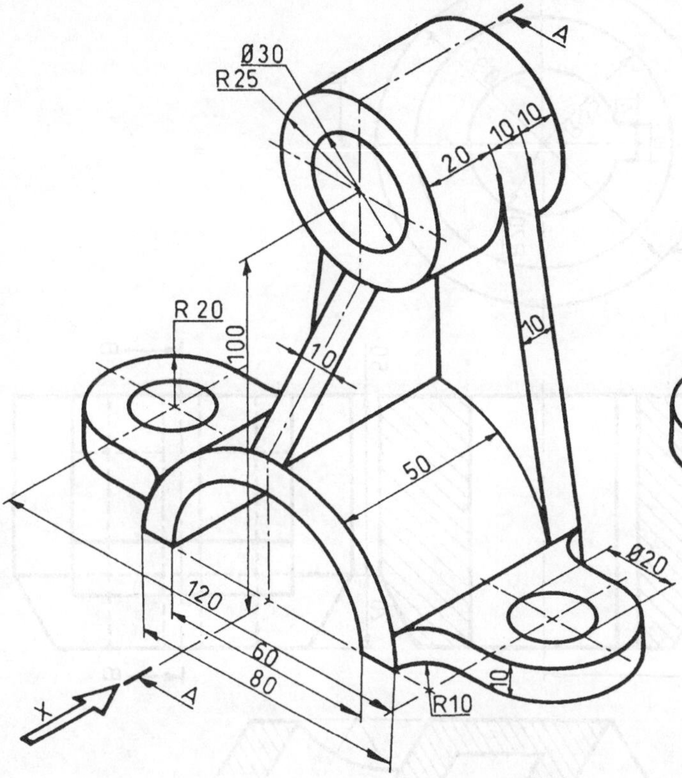

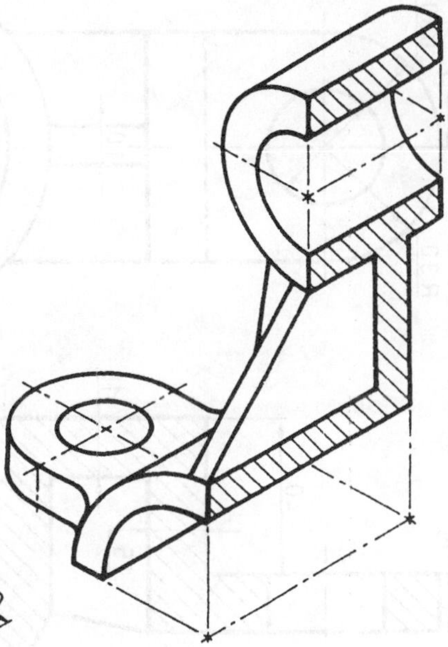

Fig. Shows a pictorial View of an object. Using first angle method, draw

(a) Front View [5]

(b) Top View [5]

(c) Sectional side view, section along A-A. [6]

Give all dimensions.

[B.T.E.M.S. A/M 2000]

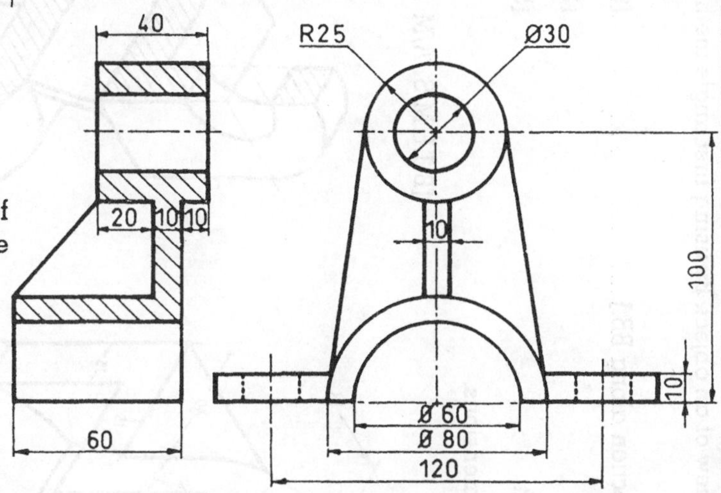

Fig. shows a pictorial view of an object. By using first angle method of projection draw

(a) Sectional F. V., [Section along BB.] [6]

(b) Top View [5]

(c) Side View from left [5]

Give important dimensions.

[B.T.E.M.S. A/M 1998]

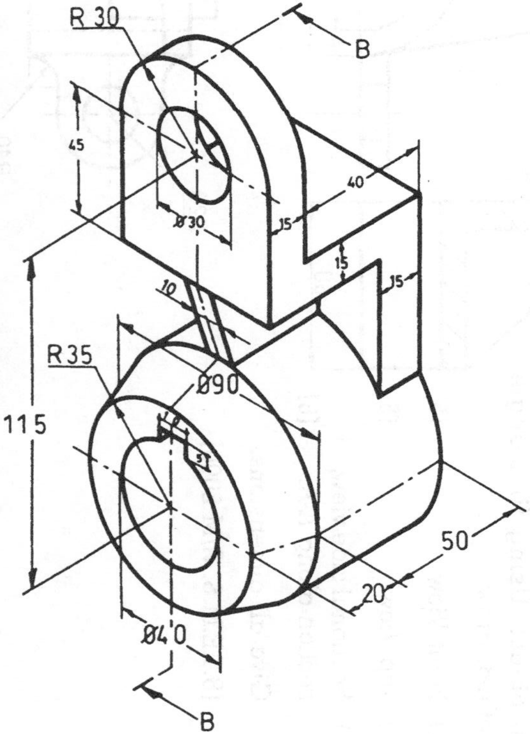

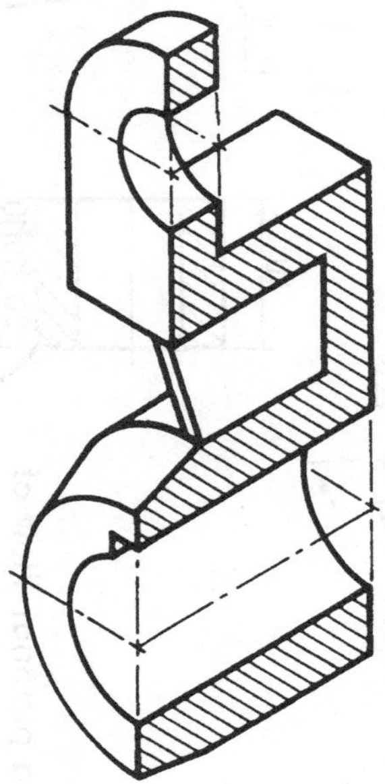

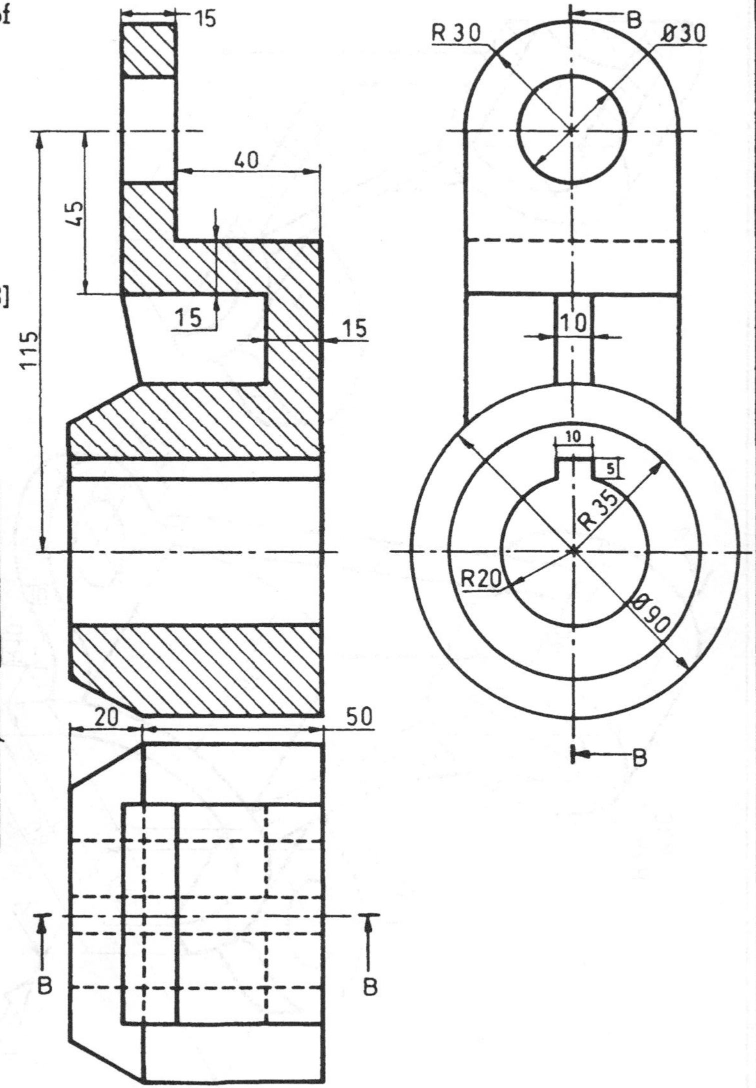

Fig. shows a pictorial view of A MACHINE PART. Draw to scale full size, the following views by using the first angle method of projection :

(a) Sectional F. V. in the direction of arrow X taking section along A-A.

(b) Sectional L.H.S.V. taking section along B-B.

[Mumbai University, December, 1993]

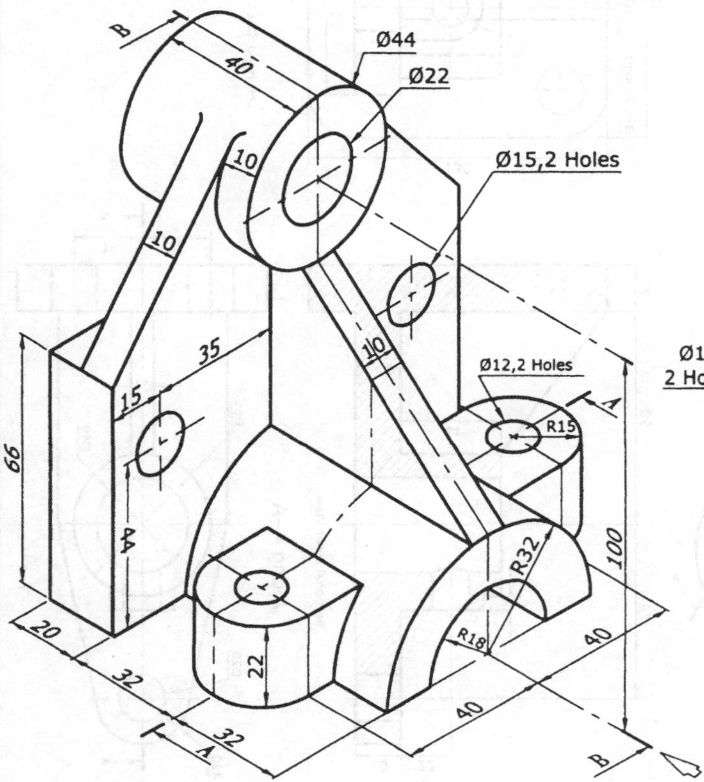

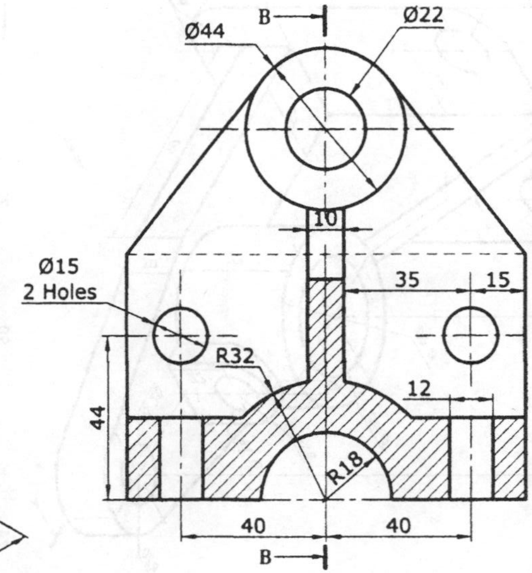

Sectional Front View

Along A–A.

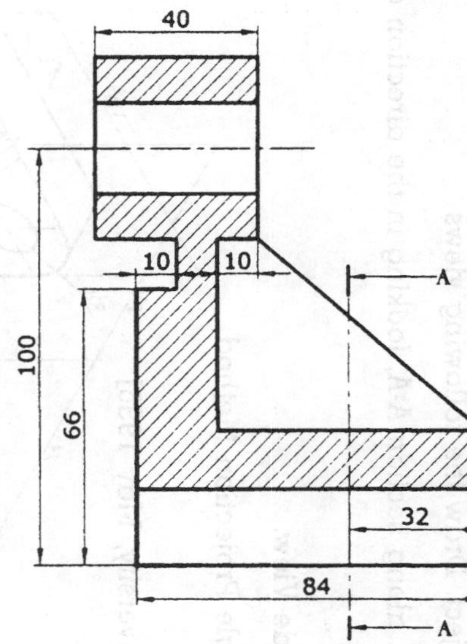

Sectional Left Hand Side View

Along B–B.

Fig. shows an object. Draw the following views

(a) Sectional F.V. along section A-A, looking in the direction of arrow X.

(b) Top View

(c) Left Hand Side View.

Use First Angle Projection Method only.

[Mumbai University, May 1996]

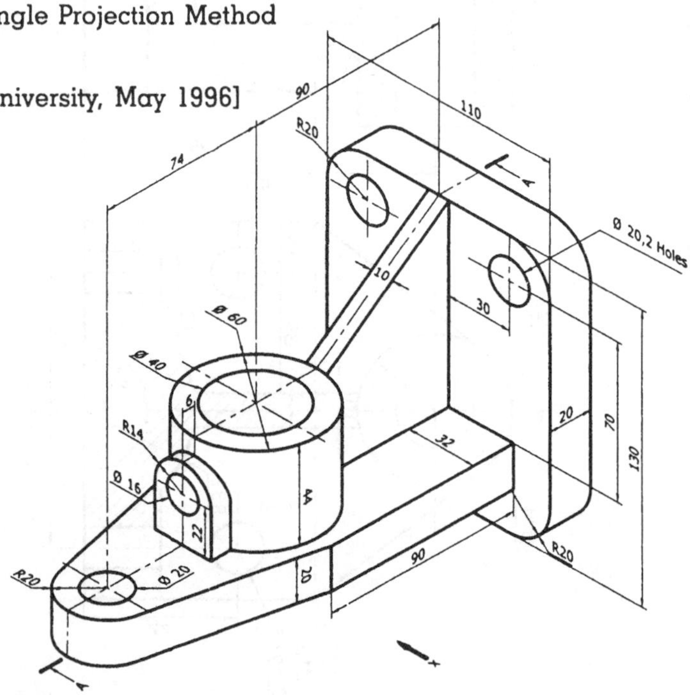

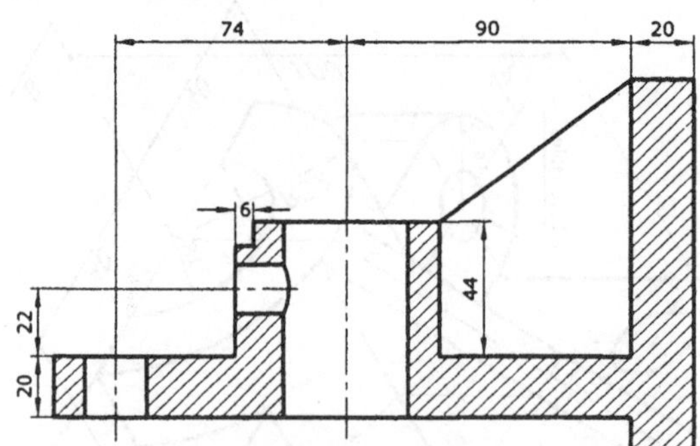

Sectional Front View

Along A–A

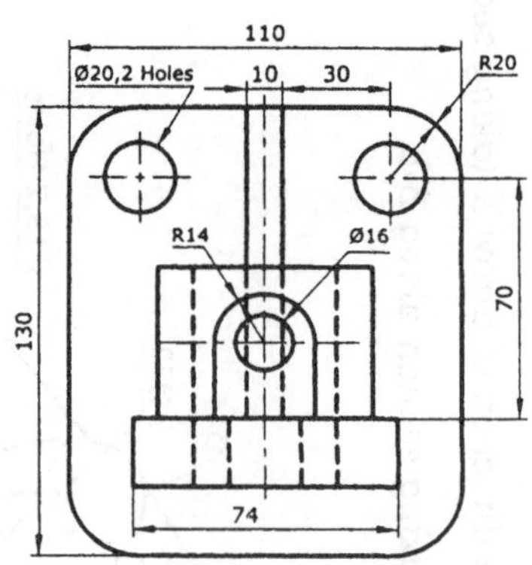

Left Hand Side View

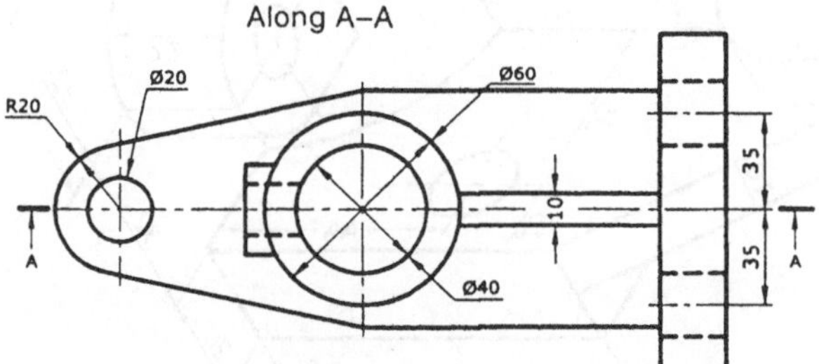

Top View

Fig. shows a pictorial view of A GUIDE-BRACKET. Draw to scale full size the following views.

(i) Sectional elevation on the section plane A-A in the direction of arrow-X

(ii) End view in the direction of arrow Y.

(iii) Plan.

[Mumbai University]

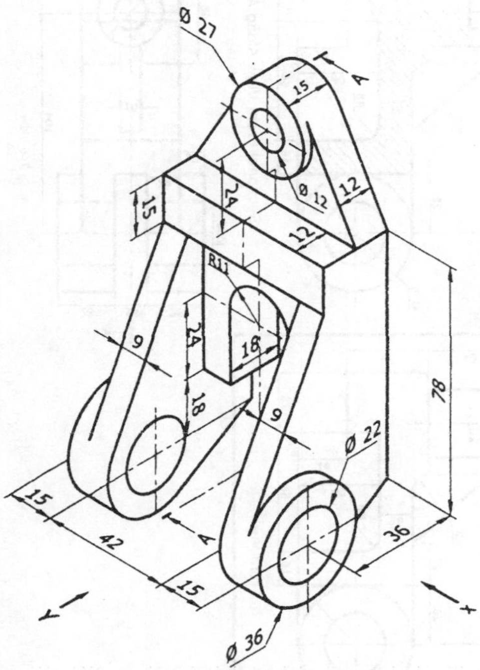

GUIDE BRACKET

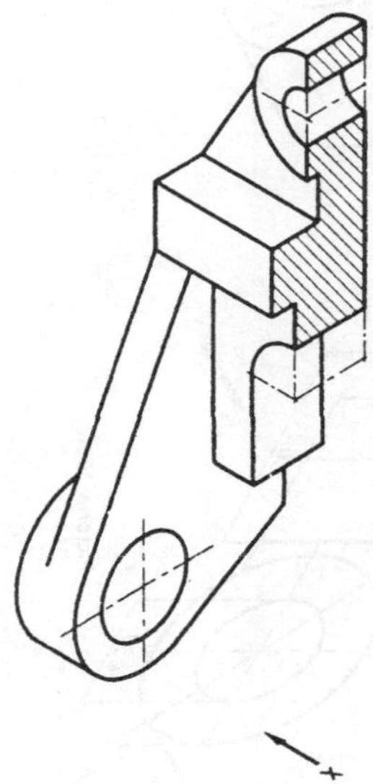

SECTIONAL VIEW ALONG A-A

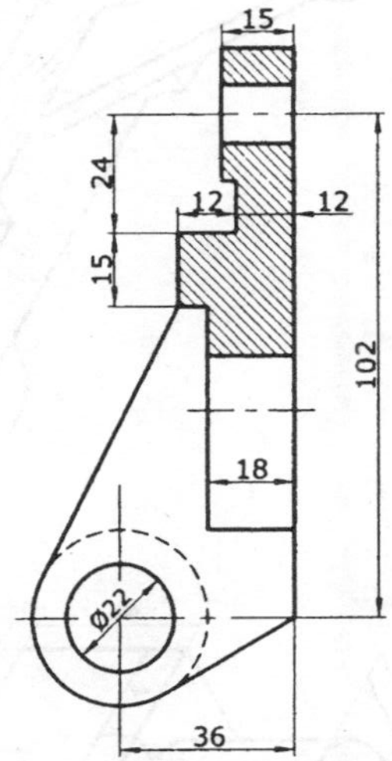

SECTIONAL ELEVATION ALONG A-A

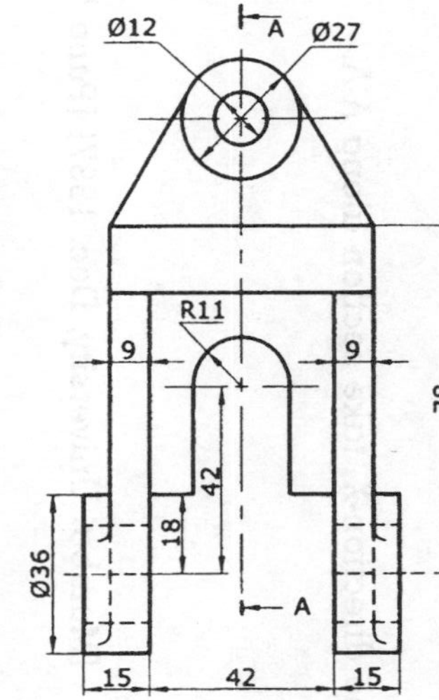

END-VIEW

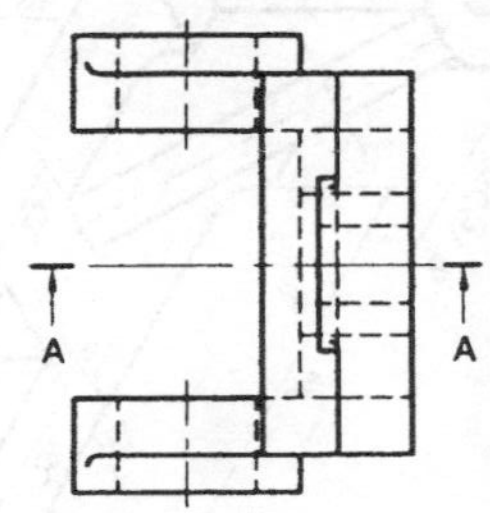

PLAN

GUIDE BRACKET

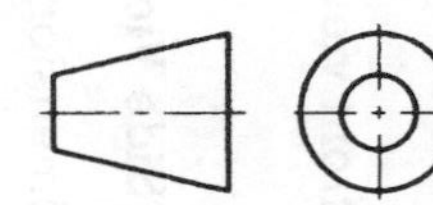

A pictorial view of A PRESS-BRACKET is shown in Fig. below. Draw by first angle method of projection.

(a) Sectional front view, looking in the direction-X, take section along A-A.

(b) Top View

(c) Left Hand Side View.

Give all dimensions.

[Mumbai University, Dec. 1997] [Pune University]

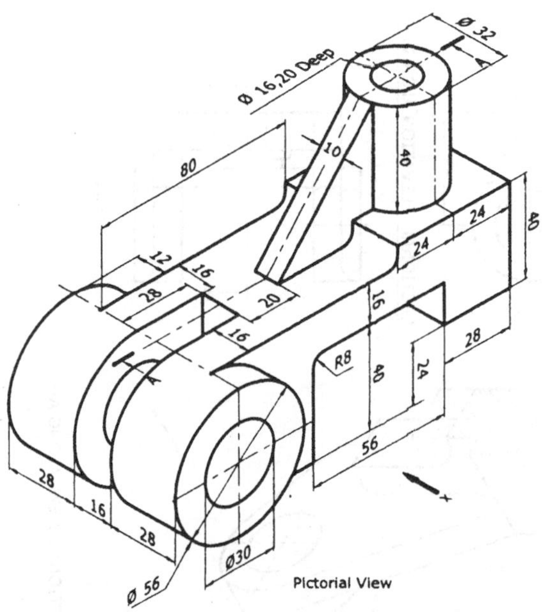

Pictorial View

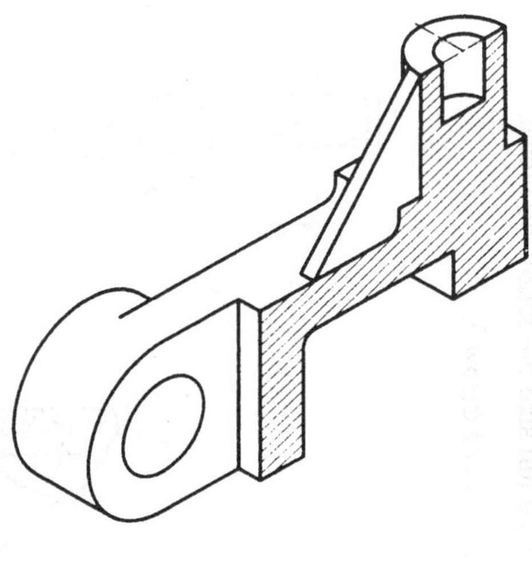

Pictorial View With Section

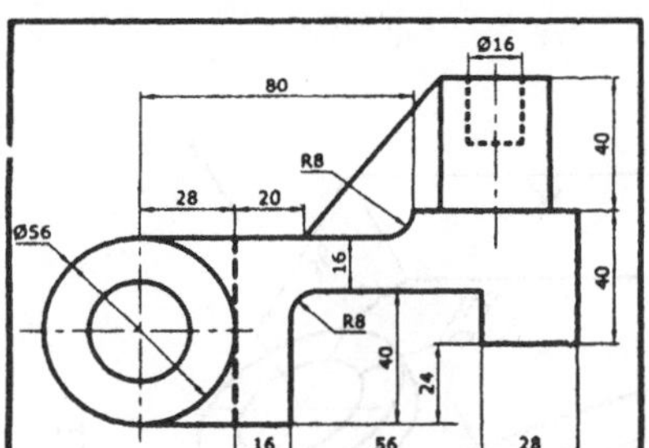

Front View

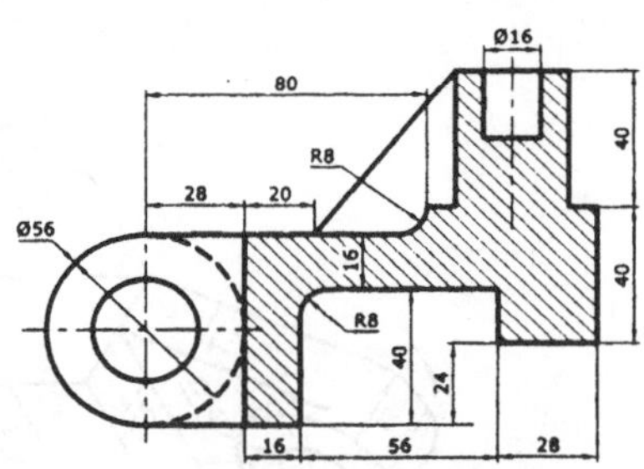

Sectional Front View Along A-A

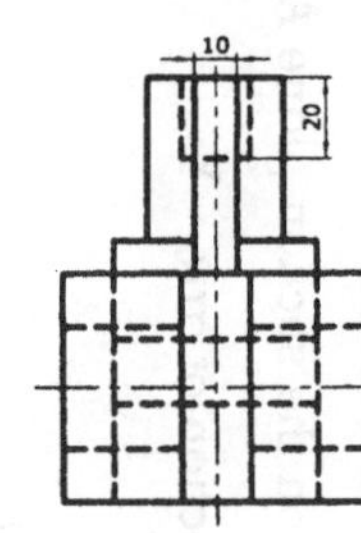

Left Hand Side View

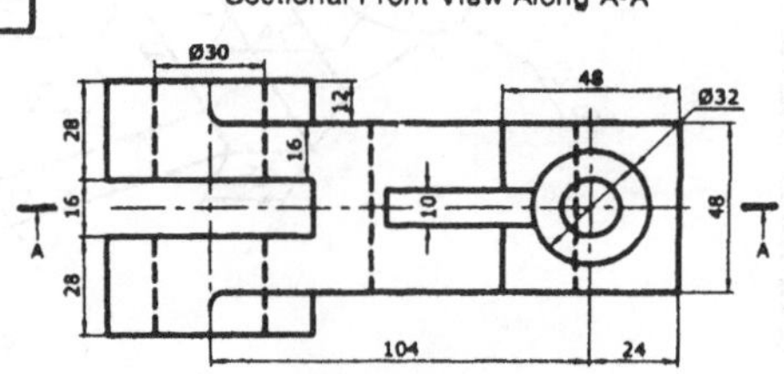

Top View

Fig. shows a pictorial view of A SPINDLE BEARING. Draw to scale full size, the following views by using First Angle Method of Projection.

(a) Sectional F.V along Section A-A.

(b) Left Hand Side View.

(c) Top View.

Dimension the views.

[Mumbai University, Nov. 1987] [B.T.E.M.S.]

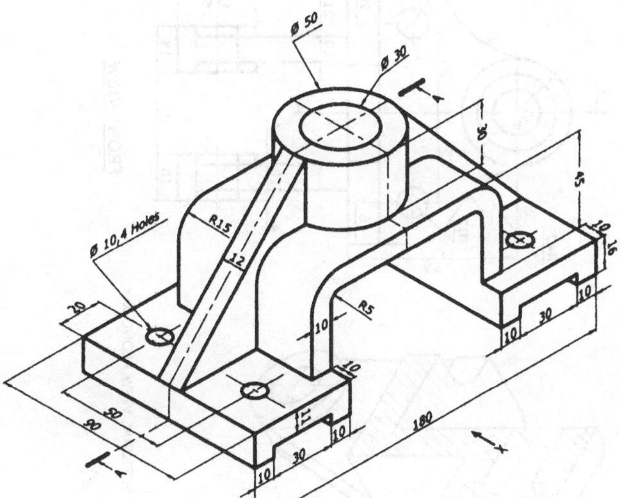

Pictorial View

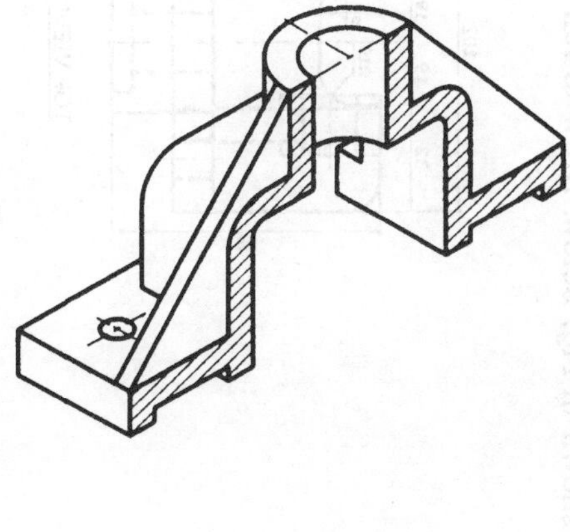

Pictorial View With Section

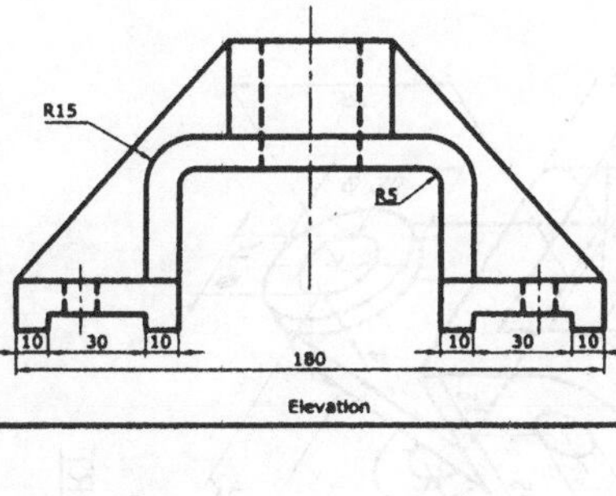

Elevation

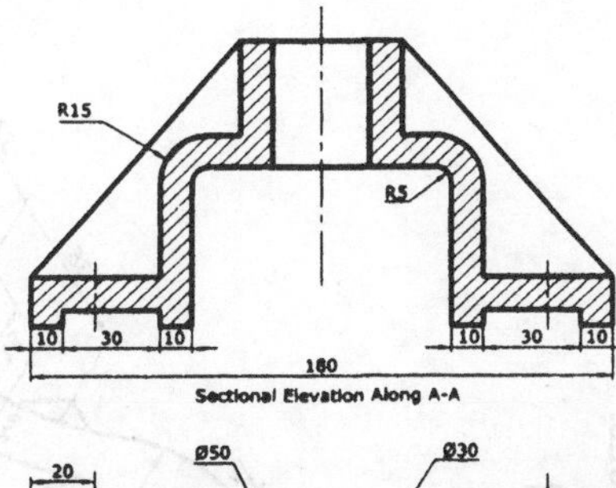

Sectional Elevation Along A-A

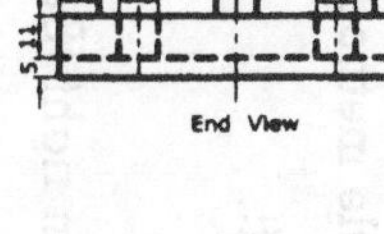

End View

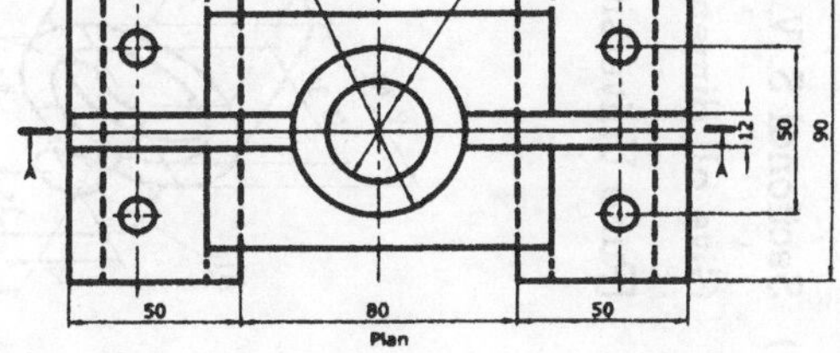

Plan

A pictorial view of AN ADJUSTABLE SHAFT SUPPORT is shown in Fig. below. Draw to full size by using third angle method of projection.

(a) Front-View from left.

(b) Top-View.

(c) Sectional S. V. from right [Section Along X-X]

Give all dimensions.

[Pune University]

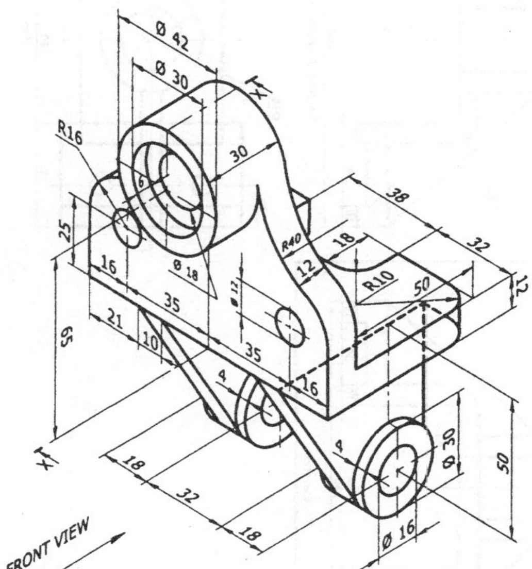

SHAFT SUPPORT

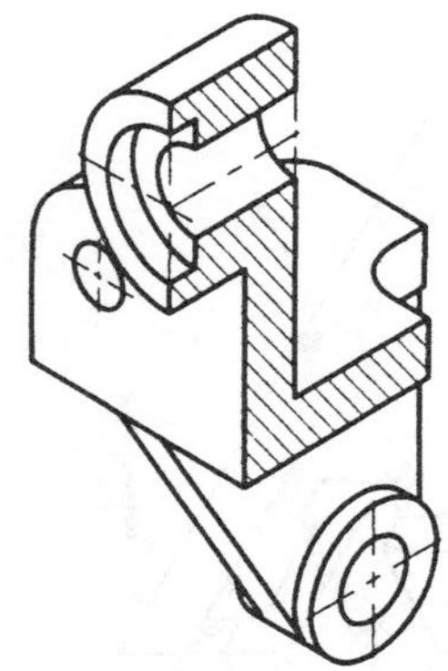

SECTIONAL VIEW ALONG X-X

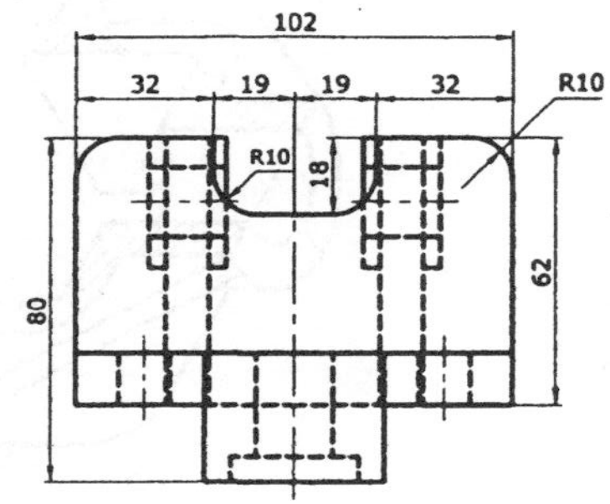

TOP VIEW

SHAFT SUPPORT

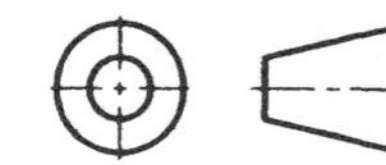

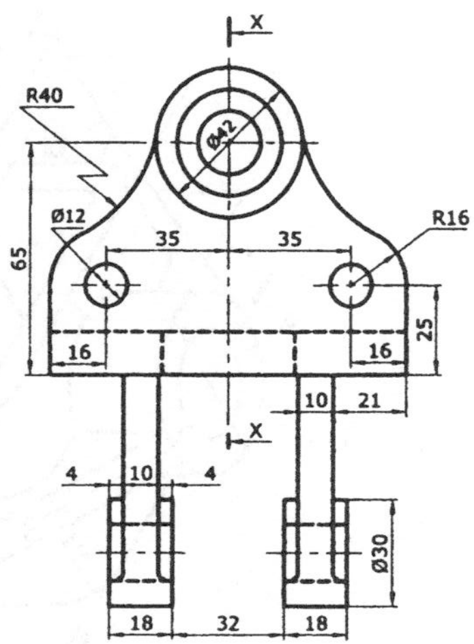

FRONT VIEW

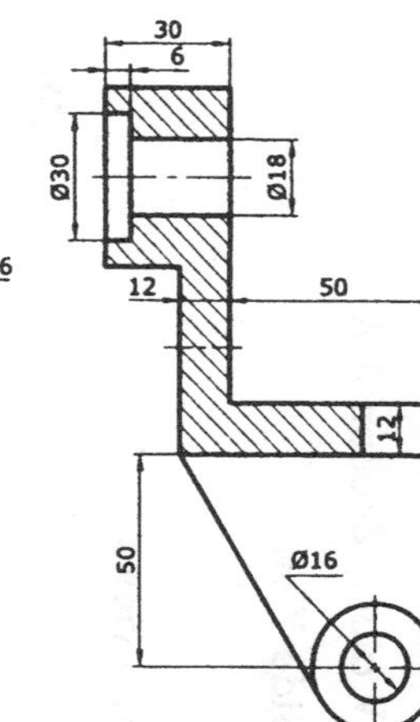

SECTIONAL SIDE VIEW ALONG X-X

Fig. shows a pictorial view of A BRACKET. Using third angle method of projection, draw the following views.

(a) Sectional front view, looking along direction X.

(b) Top View.

(c) Left Hand Side View.

Give all necessary dimensions.

[Pune University]

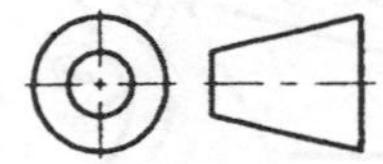

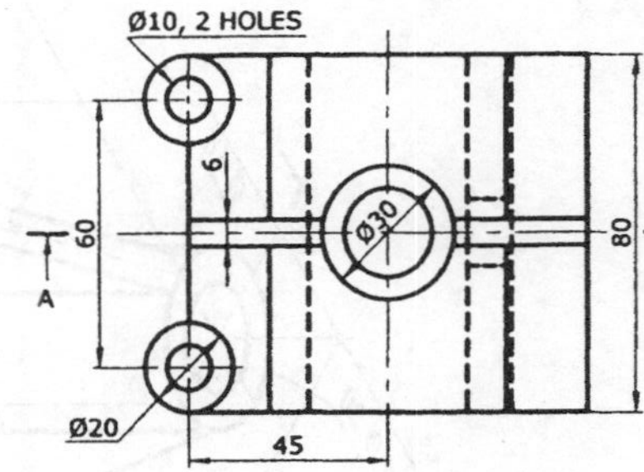

TOP - VIEW

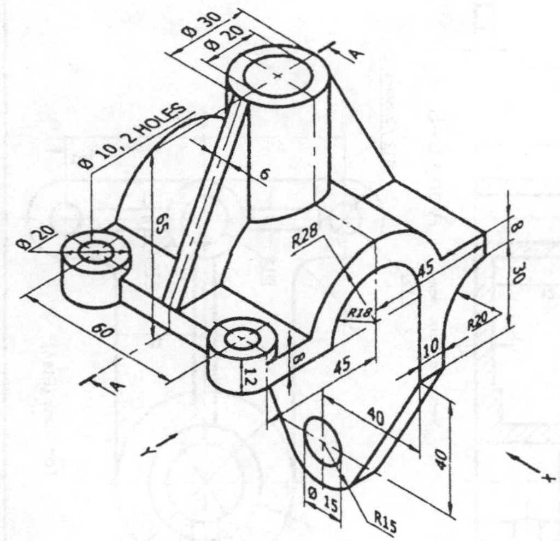

BRACKET

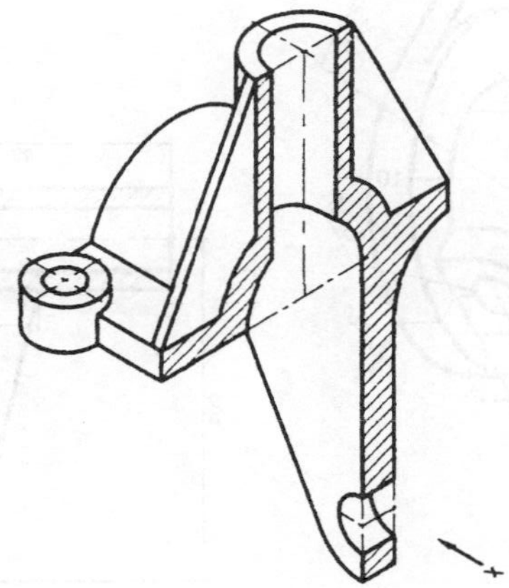

SECTIONAL VIEW ALONG A-A

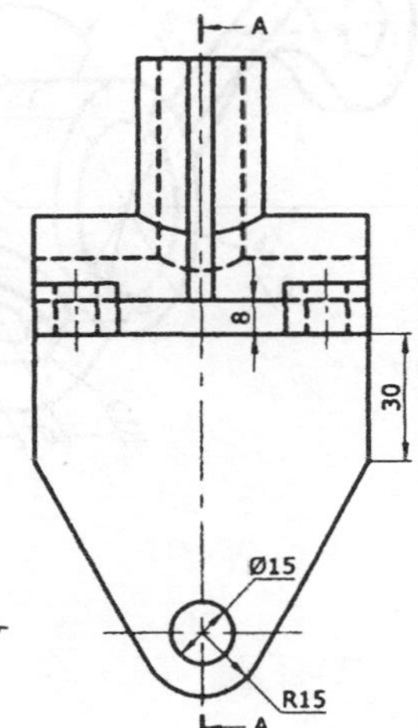

LEFT HAND SIDE-VIEW

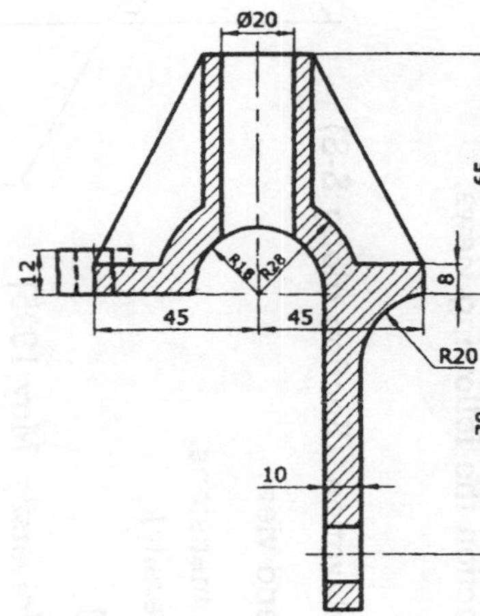

SECTIONAL FRONT-VIEW ALONG A-A

Fig. shows a pictorial view of A MACHINE COMPONENT. Draw to full scale, by first angle method of projection the following views.

(a) Plan

(b) Sectional elevation. (Section along S-S)

(c) Left hand end view.

Show all dimensions.
[Pune University]

[B.T.E.M.S.]

[Mumbai University, May 1995]

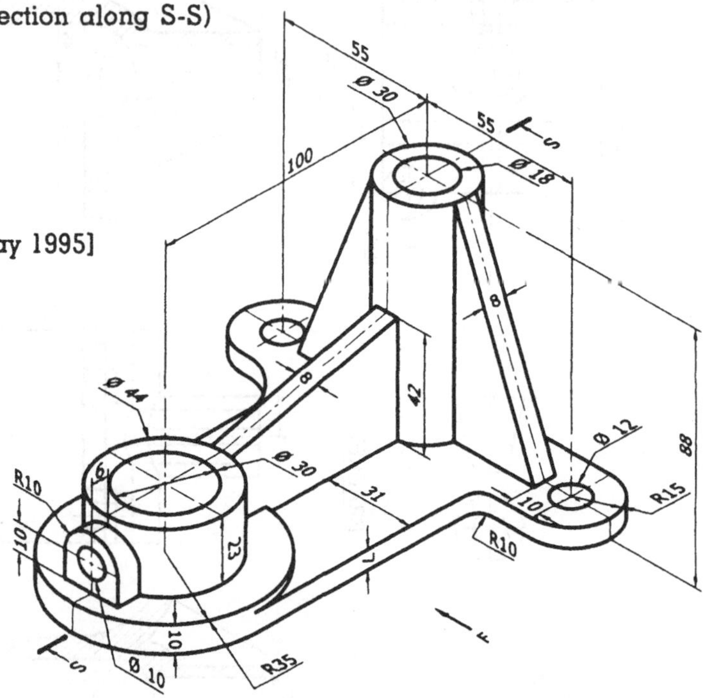

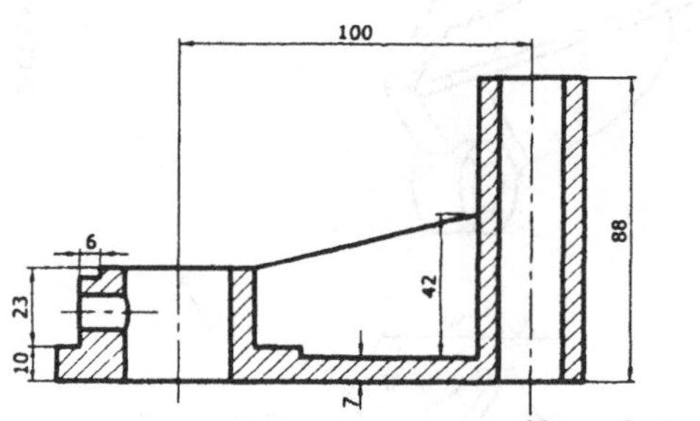

Sectional Front View Along S–S.

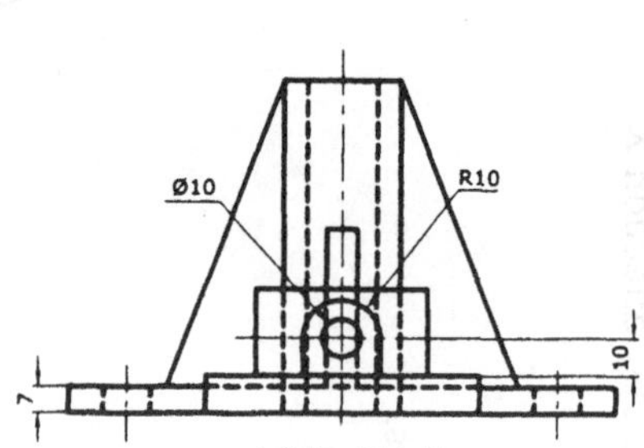

Left Hand Side View

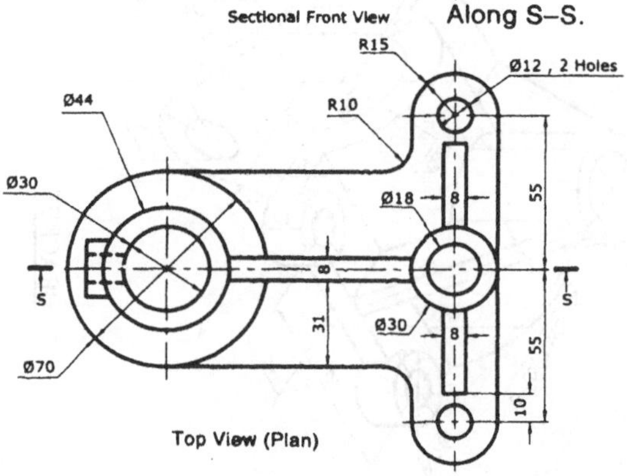

Top View (Plan)

Fig. shows a pictorial view of A C. I. BRACKET. Draw to full size the following views, by using first angle method of projection.

(a) An elevation looking in the direction of arrow X.

(b) Sectional end view along A-A, looking in the direction of arrow Y.

(c) Plan

Show important dimensions. [Note : Solution not given] [B.T.E.M.S.]

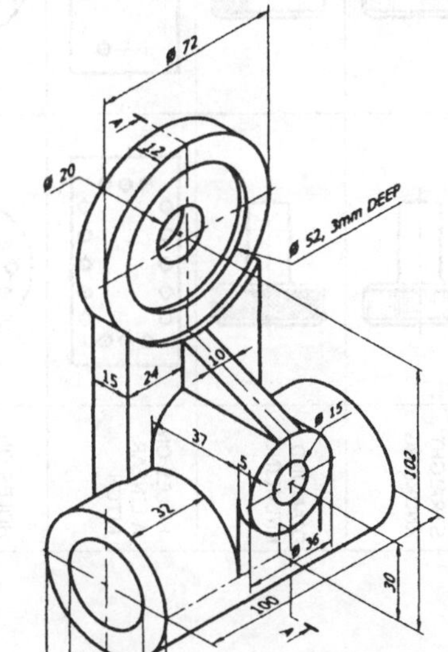

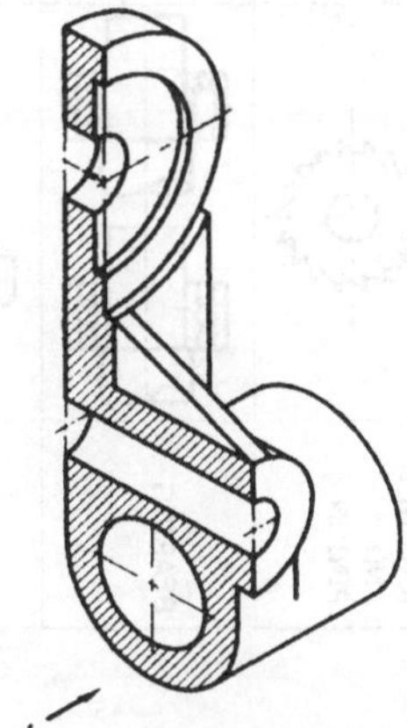

SECTIONAL VIEW ALONG A-A

C.I.BRACKET

A pictorial view of an object is shown in Fig. below, using first angle method of projection, draw the following views.

(a) Sectional elevation in the direction of arrow X.

(b) Plan

(c) End view in the direction of arrow Y.

Give all dimensions. [Note : Solution not given] [Pune University]

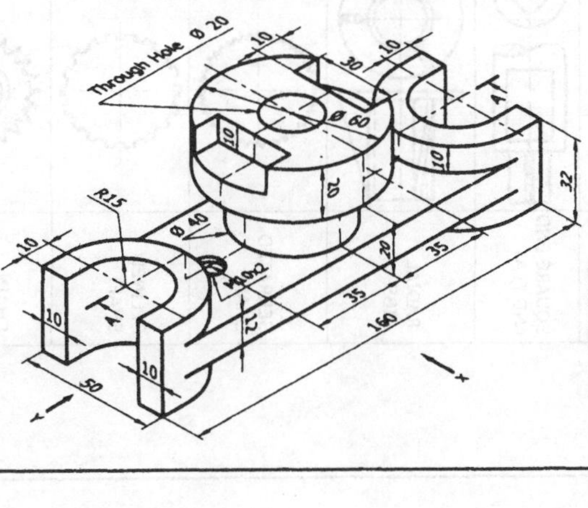

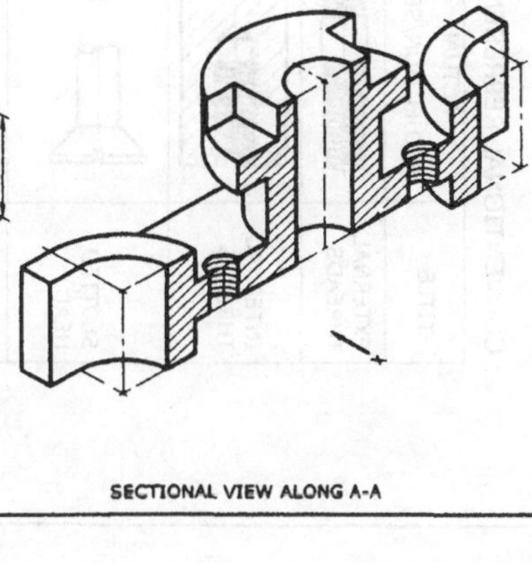

SECTIONAL VIEW ALONG A-A

Some common features, which have complicated and / or difficult to draw geometry, are presented in drawing by conventional representation instead of their actual drawings. They are given in 2-Tables below.

CONVENTIONAL REPRESENTATION OF COMMON FEATURES

TITLE	ACTUAL PROJECTION/SECTION	CONVENTION
EXTERNAL THREADS	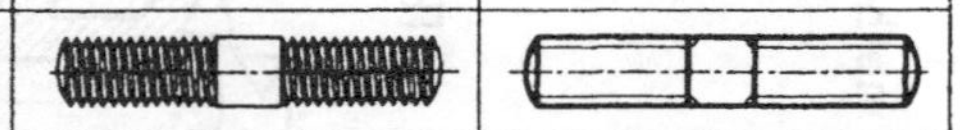	
INTERNAL THREADS	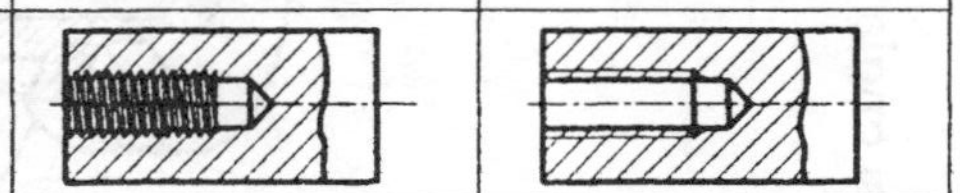	
SLOTTED HEAD	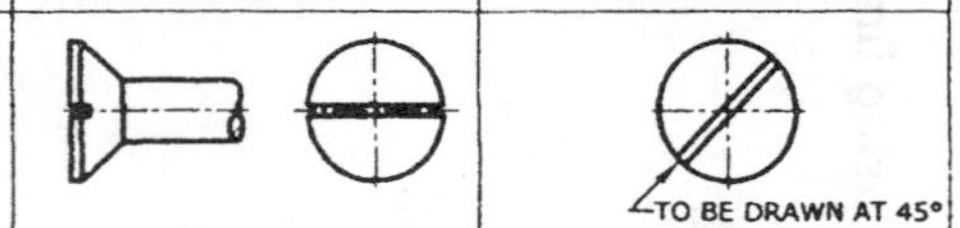	
SQUARE END AND FLAT	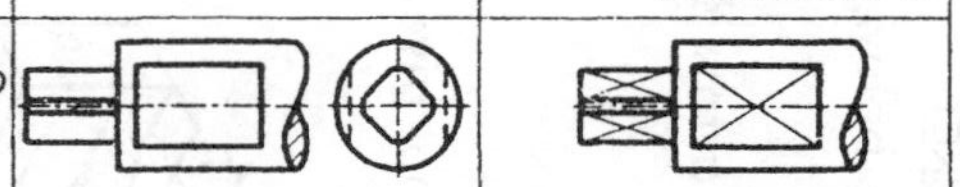	
RADIAL RIBS	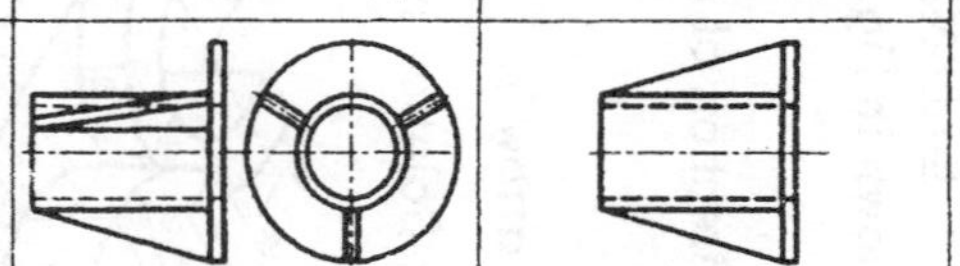	
SERRATED SHAFT	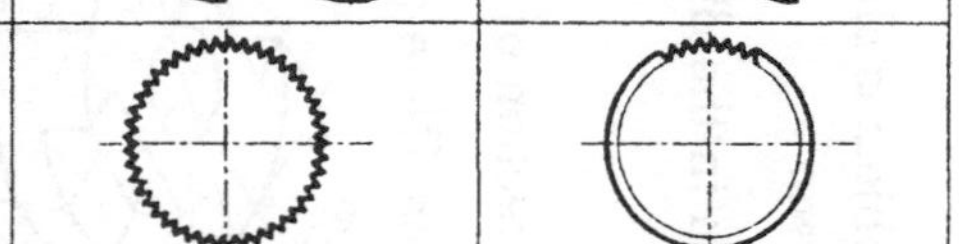	
SPLINED SHAFT	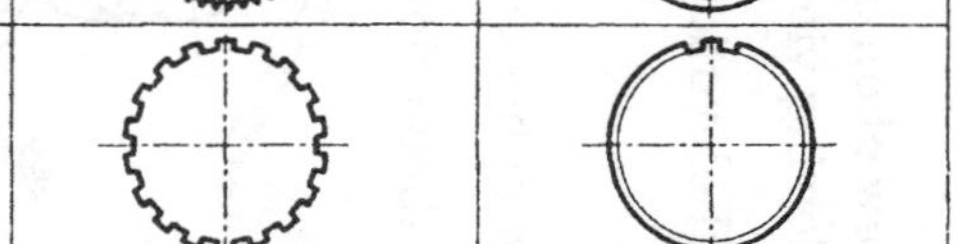	
CHAIN WHEEL	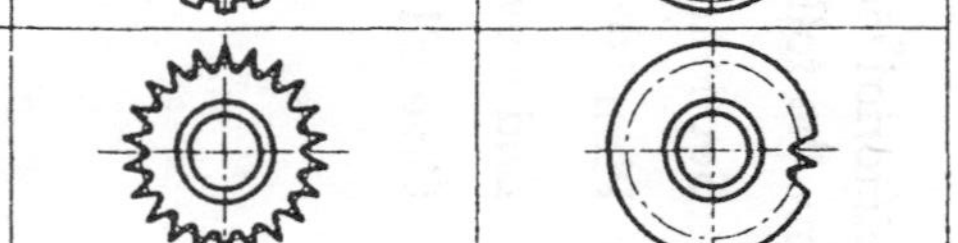	

CONVENTIONAL REPRESENTATION OF COMMON FEATURES

TITLE	ACTUAL PROJECTION/SECTION	CONVENTION
RATCHET AND PINION		
BEARING	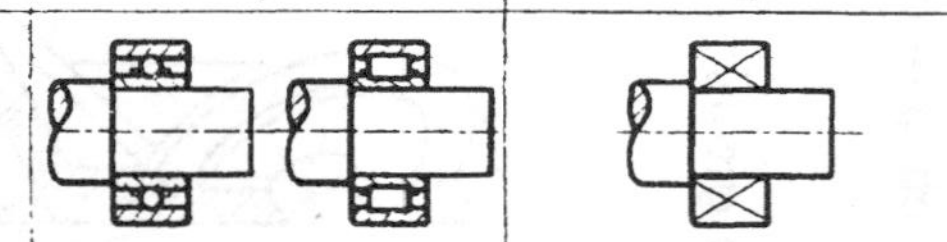	
STRAIGHT KNURLING	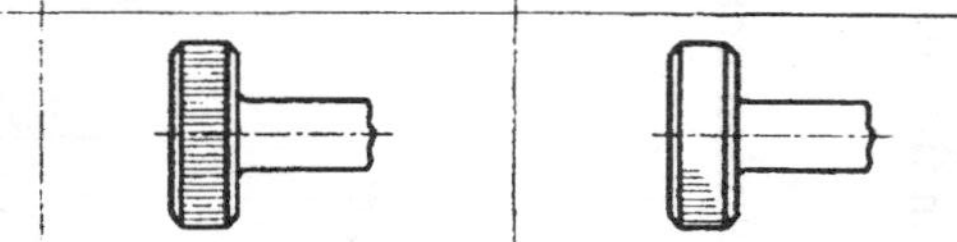	
DIAMOND KNURLING	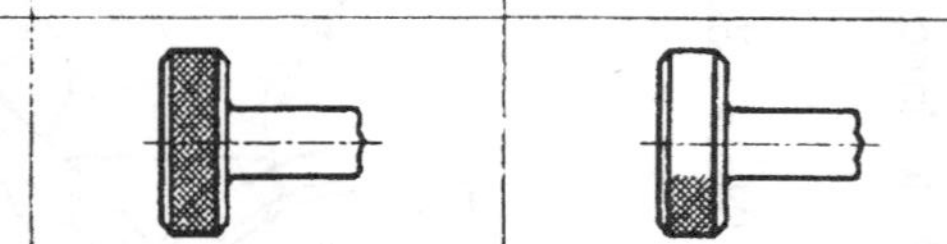	
HOLES ON A LINEAR PITCH	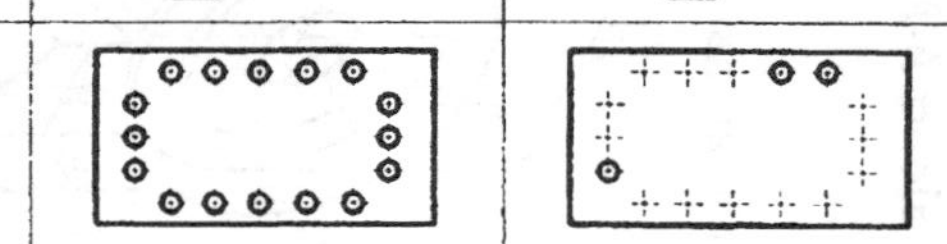	
HOLES ON CIRCULAR PITCH	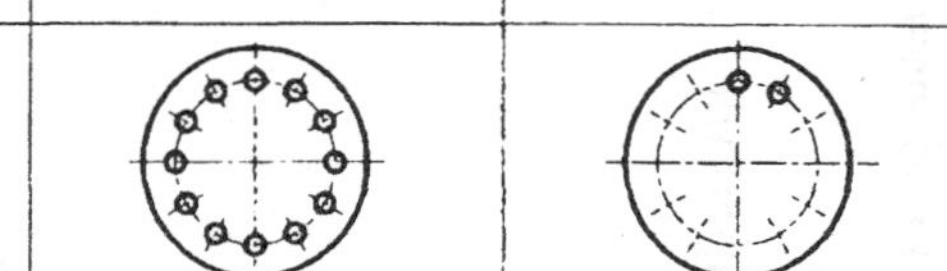	
REPEATED PARTS	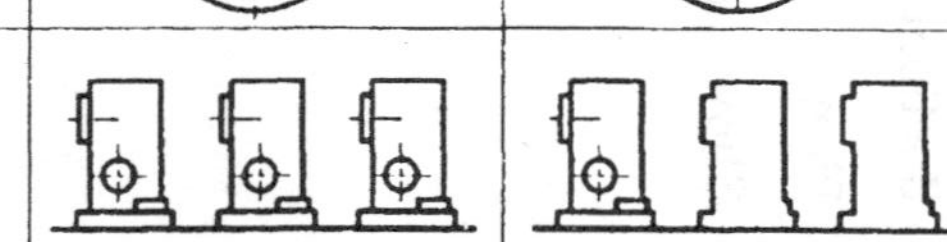	

CHAPTER 17
ORTHOGRAPHIC READING

1. Introduction :

Engineering drawing is a language of all persons connected with technical activities. Language has writing and reading practices. To draw orthographic projections from pictorial view is a writing practice. Orthographic reading means to understand or to interpret or to visualise the shape and size of the object from the orthographic projections. In fact it is a reverse process to that of making orthographic drawing.

Reading orthographic projections is a bit difficult task to that of preparing orthographic projections from pictorial views. One can obtain expertise in reading practice by having thorough knowledge of the rules and conventional practices followed in preparing orthographic projections and by having sufficient amount of practice in preparing orthographic drawings.

Reading here does not mean speaking by the use of tongue but it is a mental work of correlating different lines, different shapes and different projections and thereby evolving in mind the three dimensional picture of the object with correct size. Most of the engineering objects are combination of simple objects like cylinder, cone, prism, pyramid, sphere, flat plates and their frustum or truncated frustum joined together in different positions and locations. One should not try to visualise the whole object at a time or at a glance but should try to visualise piece by piece and should keep in memory the shape, size and location of each piece separately, To help reading, the reader can prepare three dimensional pictorial sketch of all pieces, After visualising all pieces (simple objects) combine them together in respective positions and locations to get a picture of the whole complicated object in mind. Reader should do this systematically and patiently. Correlating respective lines of two views, three views or many views is of utmost importance in reading.

2. Important Findings

The following findings will be useful in reading orthographic projections.

(1) If one view, of any edge of an object, is seen as a point view and two other views of the same edge are seen parallel to ground lines (i. e. vertical or horizontal) then that edge must be perpendicular to the principal plane on which it shows point view and parallel to two other principal planes.

(2) If one view of any edge of an object is seen as inclined line and two other views of the same edge seen parallel to ground lines (i.e. horizontal or vertical) then that edge must be parallel to the principal plane on which it shows inclined line.

(3) If all views of any edge of an object are inclined lines then that edge must be inclined to all principal planes. In no view it will show the true length.

(4) If any flat surface of the object is seen as straight line views in two projections then that plane must give the true shape in the third projection. It will be parallel to the principal plane on which it shows projection in the true shape.

(5) If any flat surface of the object is seen as an inclined straight line in one view then that flat surface must be perpendicular to the principal plane on which it gives inclined straight line projection. Other two views will show the area less than the true shape.

(6) If any flat surface of the object is not giving in any projection straight line view means that plane will be inclined to all three principal planes and hence in all projections the area will be seen, which will not be the true shape.

(7) Area enclosed by one boundary in projection represents one surface either flat or curved.

(8) Intersection of two surfaces forms an edge of an object.

(9) Corner of a solid is formed by minimum three edges. Three surfaces of solid meet at a corner of a solid.

(10) Surface limit of the curved boundary is seen as a line in a drawing.

(11) A point in projection may represent a corner or an edge of the object.

(12) A line in projection may represent a line or surface of the object.

Moreover, students are requested to practise by imagining different possible objects for a given plan, for a given elevation and for given two views. This is necessary to train the mind in the direction of orthographic reading. Method of projecting other views from the given views must be known to the reader.

Now, we shall practise reading of orthographic projections by some illustrative problems:

3. Illustrative Problems

Problem 1 : In Fig. 17.1(a) front view is given. Imagine different possible objects for the same F.V. and draw corresponding top views.

In Fig. 17.1(b) five possible objects are shown whose front view will be the same as the given front view but whose top views will be different. Students are requested to imagine many more objects for the same front view and draw their corresponding top views. Three dimensional drawings of the objects shown in Fig. 17.1(b) are for the purpose of easy understanding. You are not required to draw these three dimensional drawings. You have to draw simply elevation and their corresponding top views as shown in Fig. 17.1(c). Pictorial drawings of the objects are kept in memory. Drawings are given in the 3rd angle projection system.

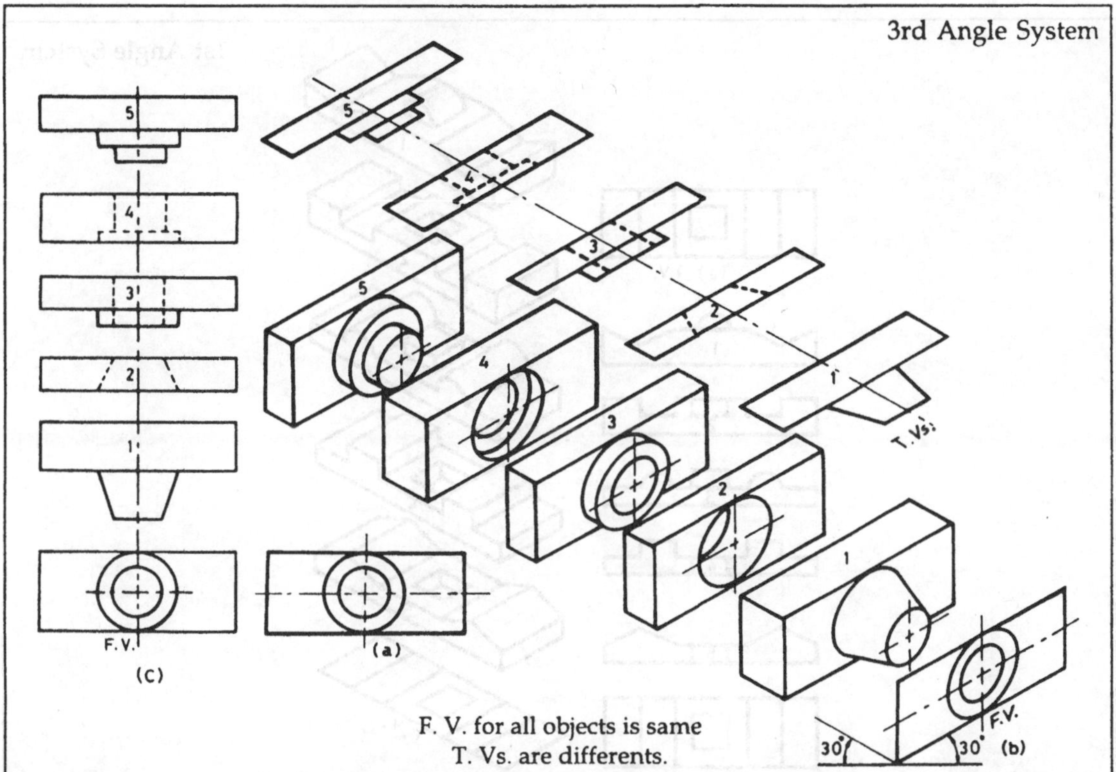

Fig. 17.1 Same F. V. - Different T. Vs.

Problem 2 : In Fig. 17.2(a) top view is given. Imagine different possible objects for the same T.V. and draw corresponding front views.

In Fig. 17.2(b) five possible objects are shown whose top view will be the same as the given T.V. but whose front views will be different. Imagine many more objects and draw corresponding front views. Solution is given in Fig. 17.2(c) in the 1st angle projection system.

Problem 3 : In Fig. 17.3(a) front view and top view are given. Imagine different possible objects for the same F.V. and T.V. and draw corresponding different side views.

In Fig. 17.3(b) six possible objects are shown whose F.V. and T.V. will be the same as the given but whose side views will be different. Imagine many more. Side views are shown in Fig. 17.3(c). Solution is in 3rd angle projection system.

Students are requested to solve many more such exercises to train the mind in mastering orthographic reading.

Now, we shall study the projection method to get other missing views when two or three views are given by taking same exercise for both the systems of projections i.e. 1st angle and 3rd angle systems.

1st Angle System

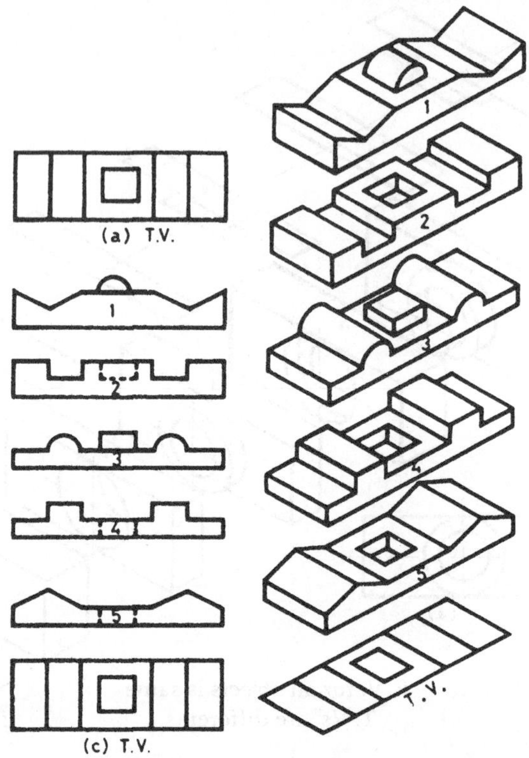

Fig. 17.2 Same T. V. - Different F. Vs.

3rd Angle System

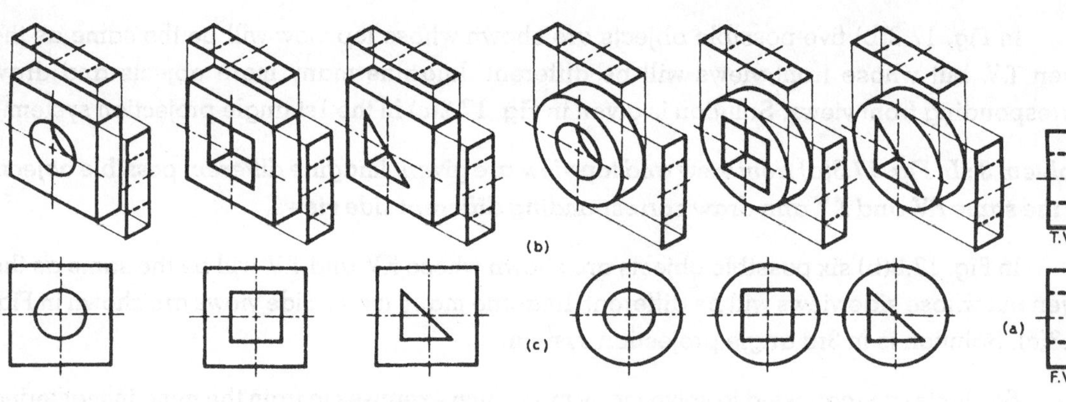

Fig. 17.3 Same F. V. & T. V. - Different S. Vs.

Problem 4 : In Fig. 17.4 front view and left hand side view of an object are given. Draw four other missing views i.e. top view, bottom view, R.H. side view and rear view. Use 1st angle projection system.

Before proceeding for the solution of the problem, let us study some of the important aspects of first angle projection system.

1st Angle System

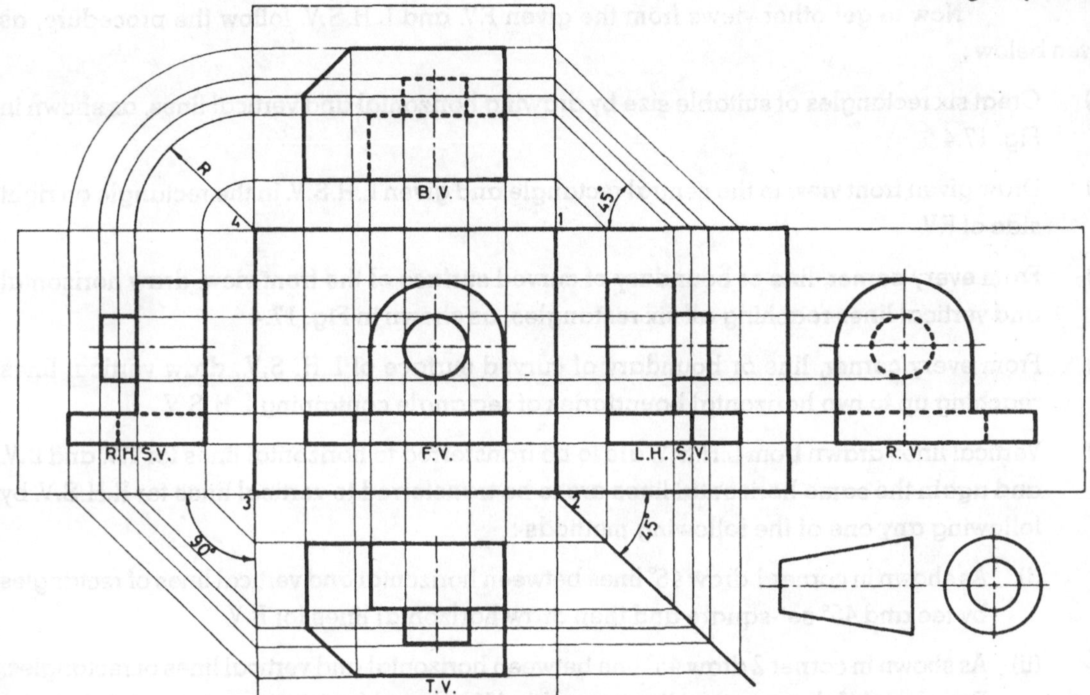

Fig. 17.4 Method of Projection for Missing Views.

(1) King view, front view is always situated at the centre.

(2) T.V. is situated below F.V.

(3) B.V. is situated above F.V.

(4) R.H.S.V. is situated on left side of F.V.

(5) L.H.S.V. is situated on right side of F.V.

(6) Rear view is situated on extreme right beside L.H.S.V.

(7) Portion of the object which is located on the front side of the object is seen away from front view, i.e. not facing front view, in all four views surrounding F.V. e.g. in Fig. 17.4 cylindrical portion, which is on front side of the object, is seen away from front view.

(8) Portion of the object, which is located on the rear side of the object, is seen nearer to front view, i.e. facing F.V. in all four views, surrounding F.V. e.g. in Fig. 17.4, plate (rectangular at the bottom side and semicylindrical on the upper side) situated on the rear side of the object is seen nearer (facing) to the front view in all four views surrounding the F.V.

In fact (7) and (8) aspects are the tests to be carried out to find out whether the given projections are in the 1st angle system or in the 3rd angle system. As the front view of cylindrical portion is full/continuous circle it can be concluded that it is on the front side of the object. In all four views surrounding F.V. it is seen not facing F.V. or away from F.V.

Now to get other views from the given F.V. and L.H.S.V. follow the procedure, as given below :

(1) Creat six rectangles of suitable size by drawing horizontal and vertical lines, as shown in Fig. 17.4

(2) Draw given front view in the central rectangle and given L.H.S.V. in the rectangle on right side of F.V.

(3) From every corner, line or boundary of curved surface of the front view, draw horizontal and vertical lines reaching all six rectangles, as shown in Fig. 17.4

(4) From every corner, line or boundary of curved surface of L.H. S.V., draw vertical lines reaching up to two horizontal boundaries of rectangle containing L.H.S.V.

(5) Vertical lines drawn from L.H.S.V. are to be transferred to horizontal lines for T.V. and B.V. and again the same horizontal lines are to be transferred to vertical lines for R.H.S.V. by following any one of the following methods :

   (i) As shown in corner 1 draw 45° lines between horizontal and vertical lines of rectangles by tee and 45° set-square and then draw horizontal lines for B.V.

   (ii) As shown in corner 2 draw 45° line between horizontal and vertical lines of rectangles. From L.H.S.V. draw vertical lines upto this 45° line and from intersecting points transfer them to horizontal lines for T.V. Use tee and 45° set-square for the same.

   (iii) As shown in corner 4 draw arcs of circles to transfer horizontal lines from B.V. to get vertical lines for R.H.S.V.

(6) In all four rectangles surrounding F.V., by intersection of lines drawn as above, boundary of projection is created. Imagine shapes of various parts of the object and accordingly make necessary lines thicker and darker to evolve projections.

(7) Rear view is drawn by taking dimensions on horizontal lines from F.V. In rear view, portion of the object of left side will be seen on right side and portion of the object of right side will be seen on left side.

(8) For drawing full and dotted lines one will have to imagine the object, Prepare freehand pictorial view of the object if you can.

Normally front portion of the object is seen full in front view and rear portion as dotted in front view. Rear portion is seen full in R.V. and front portion as dotted in R.V. Left side portion is seen full in L.H.S.V. and right side portion is seen dotted in L.H.S.V. Similarly, right side portion is seen full in R.H.S.V. and left side portion is seen dotted in R.H.S.V. Upper portion of the object is seen full in T.V. and lower portion of the object is seen dotted in T.V. Similarly, lower portion of the object is seen full in B.V. and upper portion of the object is seen dotted in B.V. These are guide lines only. They are to be applied with the imagination. Imagination plays important role in obtaining correct solution. Develop your mind for imagination.

Problem 5 : In Fig. 17.5 front view and top view of an object are given. Draw (4) four missing views i.e. B.V.; R.H.S.V.; L.H.S.V. and Rear view. Use 3rd angle projection system.

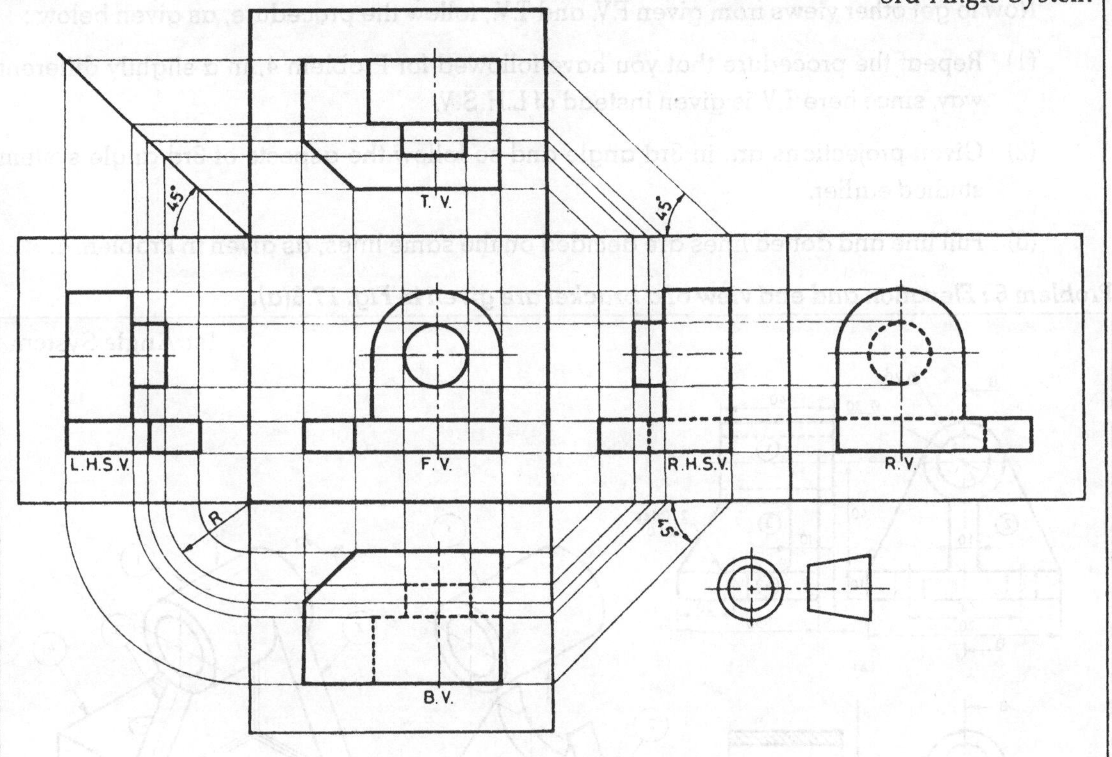

Fig. 17.5 Method of Projection for Missing Views.

Let us study first some of the aspects of 3rd angle projection system.

(1) King view, F.V. is always situated at the centre.

(2) T.V. is situated above F.V.

(3) B.V. is situated below F.V.

(4) R.H.S.V. is situated on right of F.V.

(5) L.H.S.V. is situated on left of F.V.

(6) Rear view is situated on the extreme right beside R.H.S.V.

(7) Portion of the object which is located on the front side of the object is seen nearer to F.V. in all four views surrounding F.V. e.g. in Fig. 17.5 cylindrical portion which is on front side of the object is seen nearer to F.V. in all four views.

(8) Portion of the object which is on rear side is seen away from the front view in all four views surrounding F.V. e.g.in Fig. 17.5 plate (rectangular at the bottom side and semicylindrical on the upper side), situated on rear side of the object, is seen away from front view in all four views surrounding the F.V.

In fact the above aspects at No.7 and 8 are the real tests to be carried out to decide whether the given projections are in 1st angle or 3rd angle system.

Now to get other views from given F.V. and T.V., follow the procedure, as given below :

(1) Repeat the procedure that you have followed for Problem 4, in a slightly different way, since here T.V. is given instead of L.H.S.V.

(2) Given projections are in 3rd angle and so follow the aspects of 3rd angle system studied earlier.

(3) Full line and dotted lines are decided on the same lines, as given in Problem 4.

Problem 6 : Elevation and end view of a bracket are given in Fig. 17.6(a).

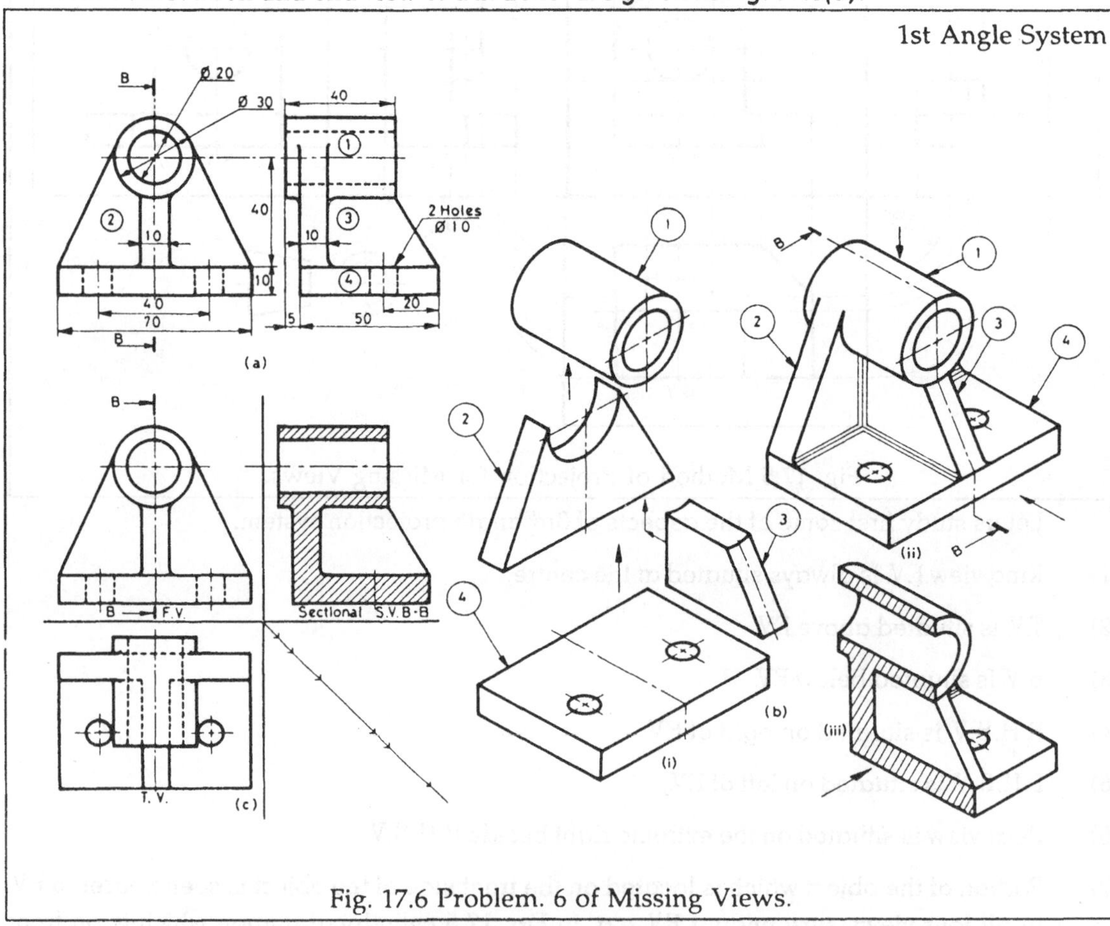

Fig. 17.6 Problem. 6 of Missing Views.

(i) Find in which system the projections are given ?

(ii) Draw front elevation, left hand sectional side view, take section at B-B and top plan in the same system of projections.

(i) Part No.3 i.e. rib must be in front of the bracket since in the front view it is shown by full lines. In side view rib is seen not nearer to F.V. but away from it and hence we can conclude that the projections given are in the Ist angle system.

(ii) To solve part (ii) first we have to imagine the object from the given two views.

As said earlier let us imagine the object piece by piece. Object can be conveniently divided into four pieces.

Correlating two concentric circles of F.V. with full and dotted rectangles of the side view we get hollow cylinder as piece No. 1. Similarly, pieces No. 2, 3 and 4 are imagined by correlating respective shapes of elevation and end view.

Having imagined four pieces No. 1, 2, 3 and 4, they are drawn in pictorial views as shown in Fig. 17.6(b)(i). Later on these pieces are joined by moving them in the direction shown in Fig. 17.6(b)(i), to get the pictorial drawing of complete bracket, as shown in Fig. 17.6(b)(ii).

Imagine now the cut object after cutting it along central plane B-B. Cut object is also shown in the Fig. 17.6(b)(iii).

Do not worry if you cannot draw pictorial views as shown in Fig. 17.6(b). Students are required to imagine such pictorial views in their minds.

Now having imagined the full object and the half cut object we can draw (i) Front Elevation (same as given), (ii) Full Sectional Side View and (iii) Top Plan. Solution is shown/given in Fig. 17.6(c) in 1st angle system.

Problem 7 : Elevation and side view are given in Fig. 17.7(i). (i) Find system of projection, (ii) Draw (a) Sectional Elevation along A-A, (b) Offset Sectional Side View along B-B and (c) Sectional Top Plan along C-C. Draw projections in the same system of projection.

(i) Given projections are in the 3rd angle system.

(ii) Shape of the component may be visualised by imagining the different pieces of the component separately and then by combining them into one piece. After visualising the complete component imagine its cutting along section A-A, along offset section B-B and along section C-C.

For your understanding three cut components after cutting it along A-A, B-B and C-C are shown in Fig. 17.7 (ii) in pictorial views.

Having imagined three different cut components, as shown in Fig. 17.7 (ii) draw (i) Sectional Elevation along A-A, (ii) Offset Sectional Side View along B-B and (iii) Sectional Plan along C-C, as shown in Fig. 17.7(iii).

Material that is cut by cutting plane is shown by means of section (hatching) lines. In normal circumstances dotted lines, to show hidden features, are avoided in the sectional views.

Having practised seven illustrative problems, students are required to practise many more problems of different varieties given as solved & unsolved exercise problems to grasp orthographic reading and develop self confidence in reading orthographic projection.

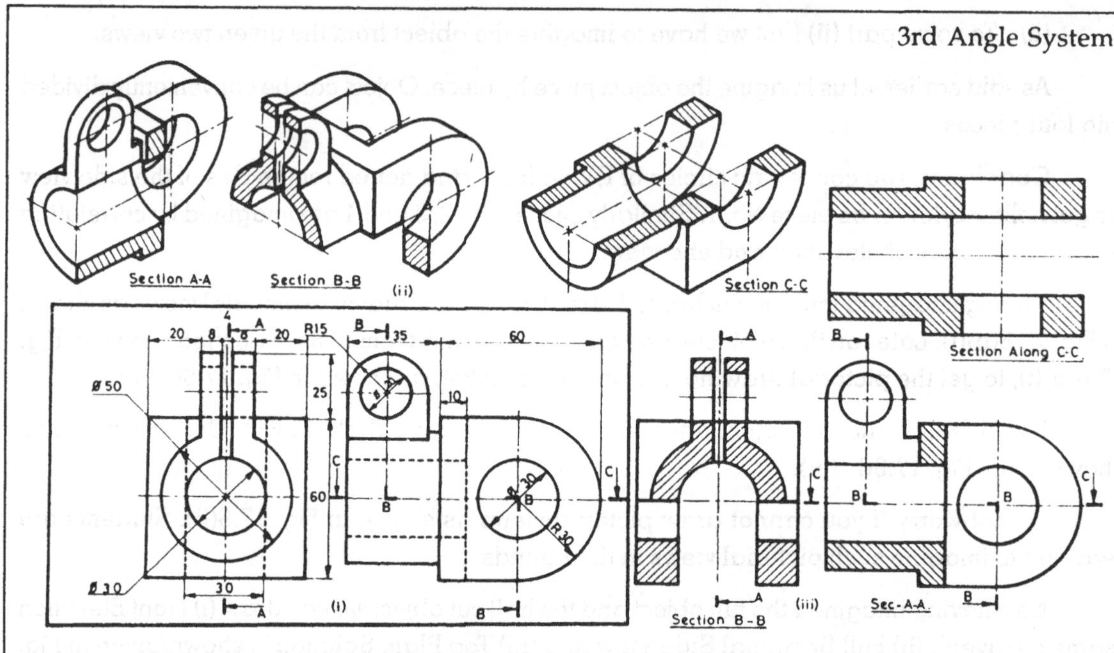

Fig. 17.7 Missing Views [Sectional]

Solved Exercise

S.E.1 The projections of a **'Bearing Cap'** are given in Fig. (A). Draw to full scale using 1st Angle Method of Projection the following views.

(i) H.S. Elevation (ii) H.S.L.H.S.V. Solution is given in Fig. (B)

S.E.2 Two views of **C.I. Bracket** are given in Fig. (C). Using the same method of projection draw the following views.

(i) Sectional F.V. (ii) Sectional T.V. (iii) S.V. from right Solution is given in Fig. (D)

S.E.3 T.V. and F.V. of **A Block** are given in Fig. (E). Draw, in the same method of projection, the following views.

(i) Sectional F.V. (ii) R.H.S.V. (iii) L.H.S.V. Solution is given in Fig. (F).

S.E.4 **A C.I. machine part** is shown in Fig. (G). Using the same method of projection, draw the following views.

(i) Sectional F.V. (ii) Sectional T.V. (iii) S.V. Solution is given in Fig. (H)

S.E.5 Two views of **C.I. bracket** are given in Fig. (J). Draw to Full Size Scale and according to First Angle Projection Method the following views.

(i) Sectional F.V. (Sect. A-A)

(ii) T.V. Solution is given in Fig. (K)

(iii) Both Side Views

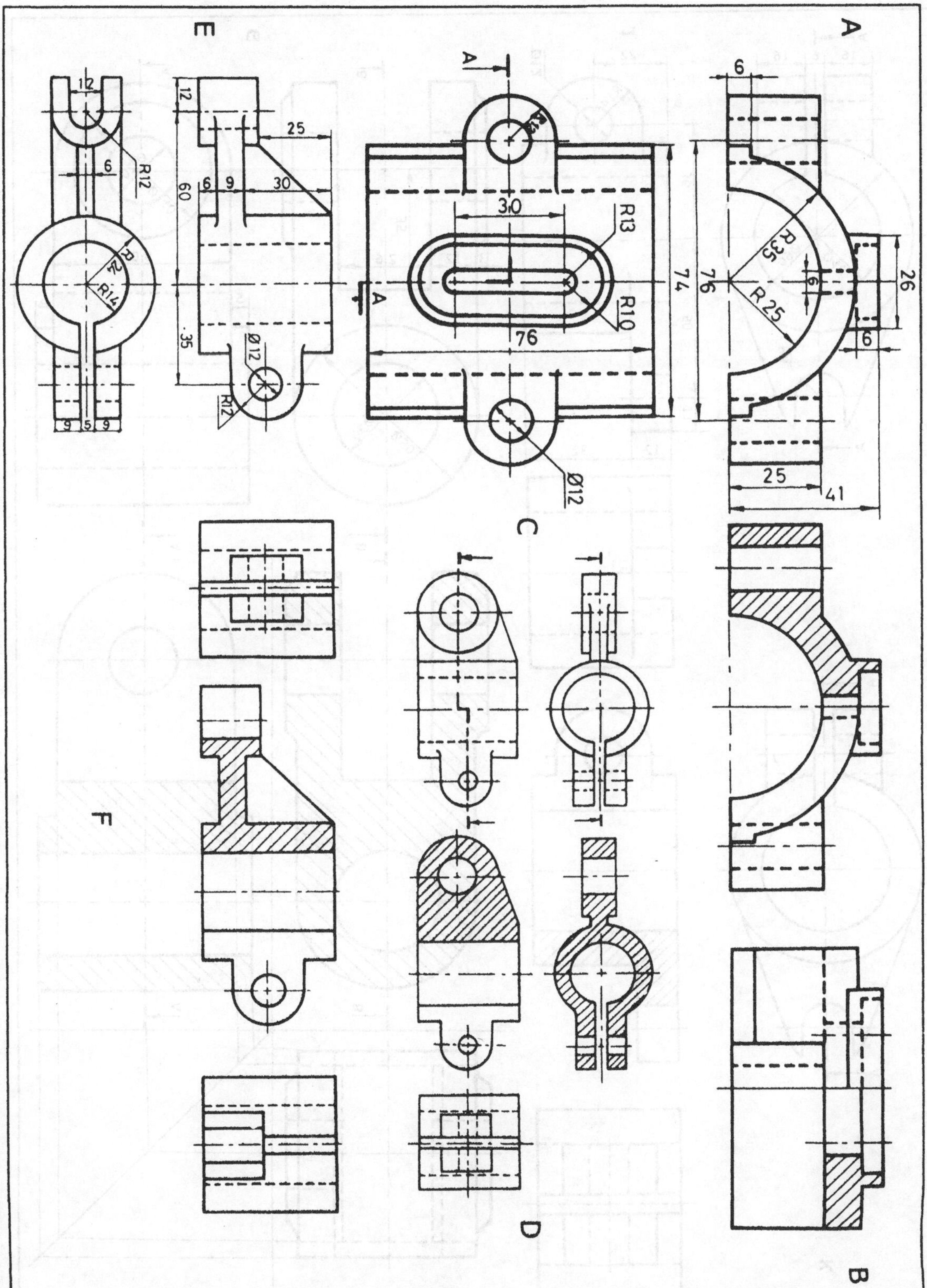

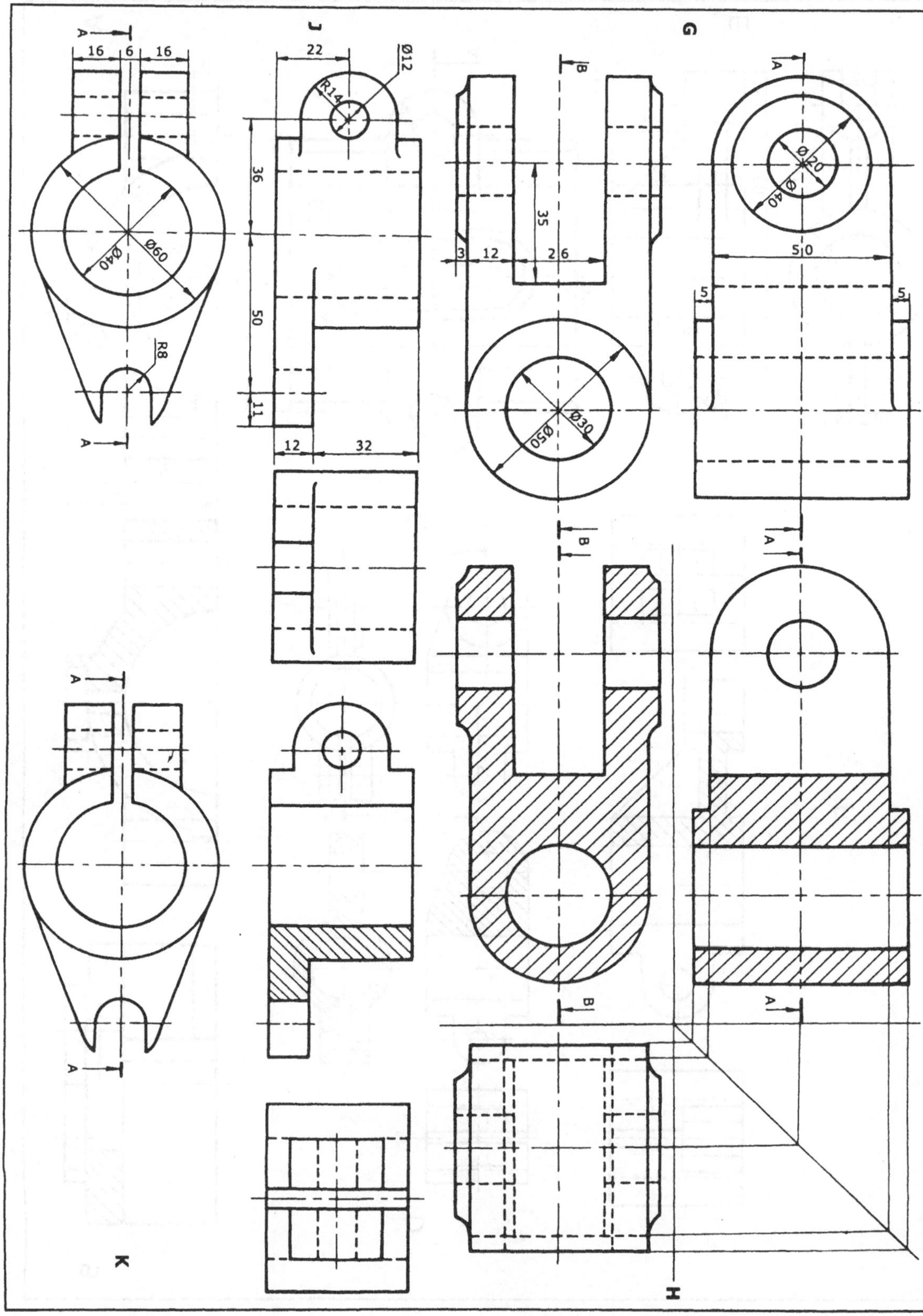
A
16
6
16
Ø40
Ø60
R8
J
22
Ø12
R14
36
50
11
12
32
G
B
35
3
12
2 6
Ø50
Ø30
Ø20
Ø40
50
5
5
K
H

Exercise

Ex. 1 In Fig 17.8 T. V. is given. Draw according to the 1st angle projection system elevation of different objects whose Top View will be same as the given T. V.

Ex. 2 In Fig. 17.9 T. V is given. Imagine different possible objects whose top view will be the same as given and then draw their elevations in the 3rd angle system of projection.

Ex. 3. In Fig. 17.10 F. V. is given. Draw different plans and side views which satisfy the given F. V., in 1st angle system.

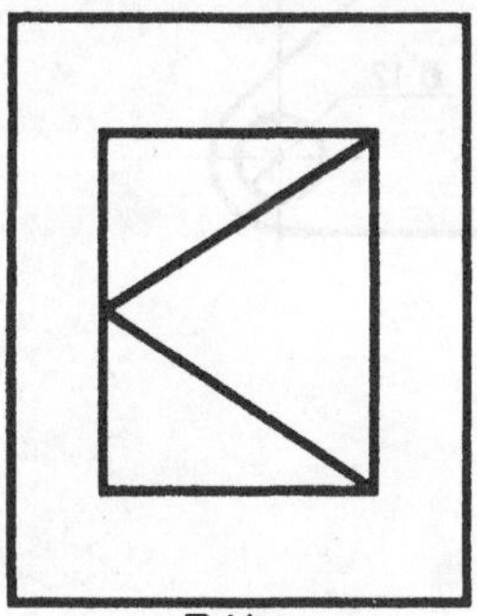

Fig. 17.8

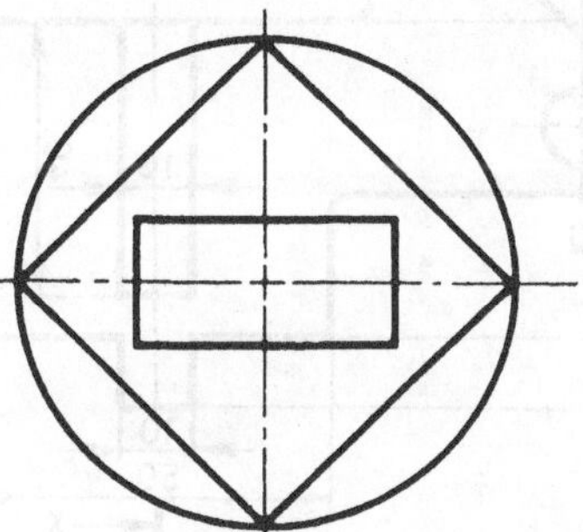

Fig. 17.9

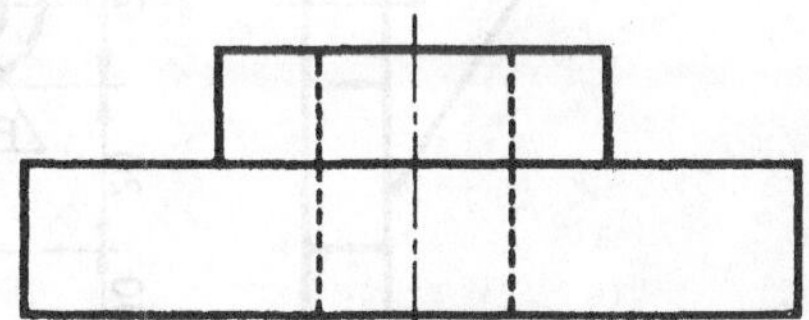

Fig. 17.10

Ex. 4. In Fig. 17.11 F. V. is given. Draw free hand pictorial sketches of the different objects whose front view will be the same as given. Draw also possible T. Vs. and S. Vs. in any system of projection.

Ex. 5. In Fig. 17.12 F. V. and T. V. are given. Imagine different objects whose F. V. and T. V. remain the same but whose side views will be different and then draw different possible side views.

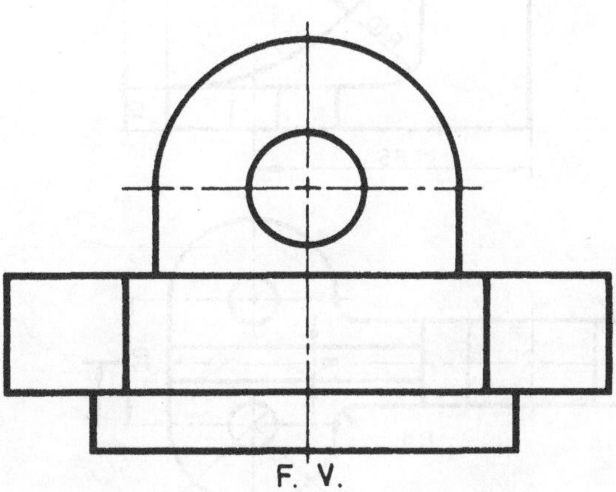

Fig. 17.11

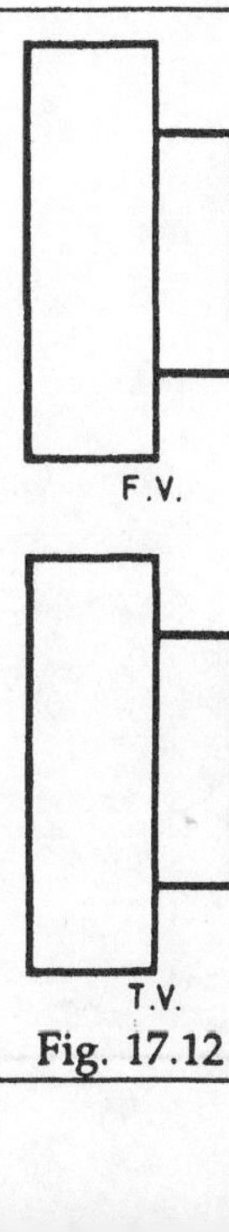

Fig. 17.12

Ex.6. Fig. 17.13. Draw by the same method of projection, following views :

(i) Front View

(ii) Sectional Side View, take section along X-X

(iii) Top Plan

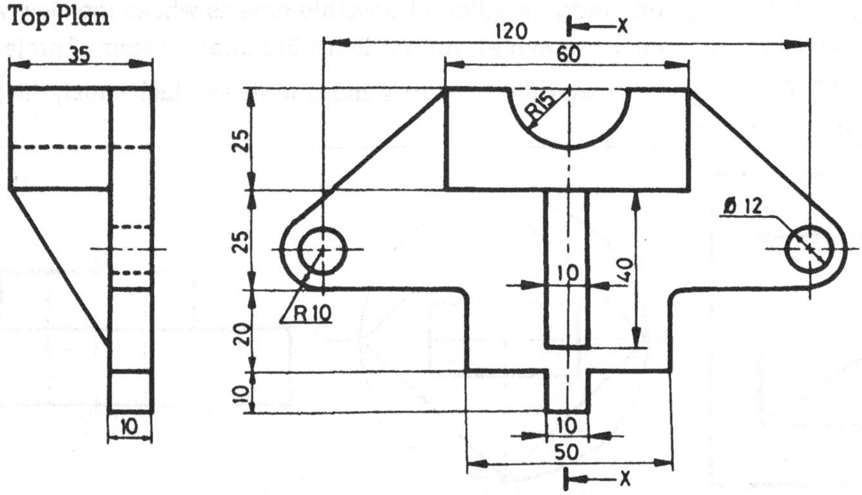

Fig. 17.13

Ex.7. Fig. 17.14. Draw by the 1st angle projection method :

(i) Sectional Front View, take section along A - A

(ii) Top Plan

(iii) R. H. S. V. and L. H. S. V.

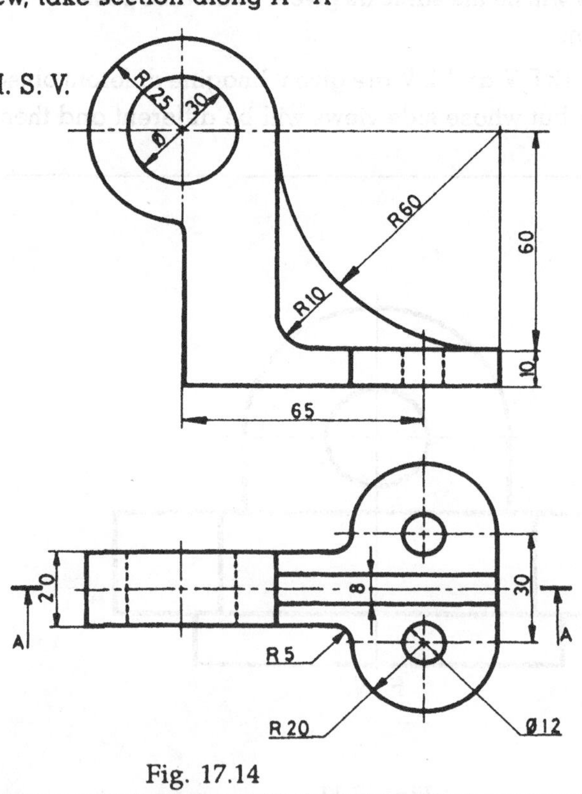

Fig. 17.14

Ex.8. Fig. 17.15. Draw by the same method of projection :

(i) Sectional Elevation, take section along A - A

(ii) Bottom Plan

(iii) L. H. Side View

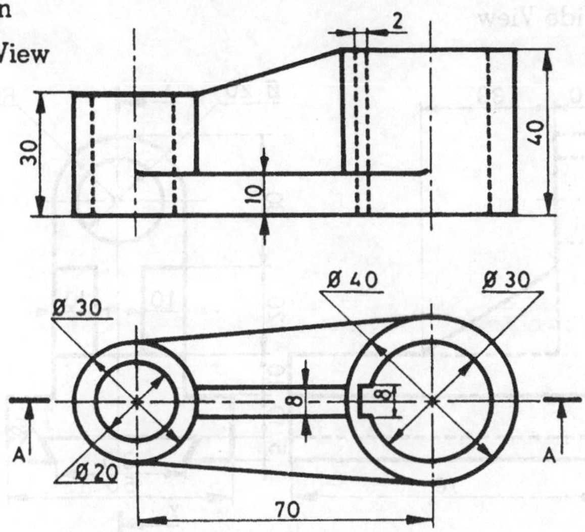

Fig. 17.15

Ex.9. Fig. 17.16. Draw by the same method of projection :

(i) Sectional elevation, take section along X - X

(ii) R. H. Side View

(iii) Top Plan

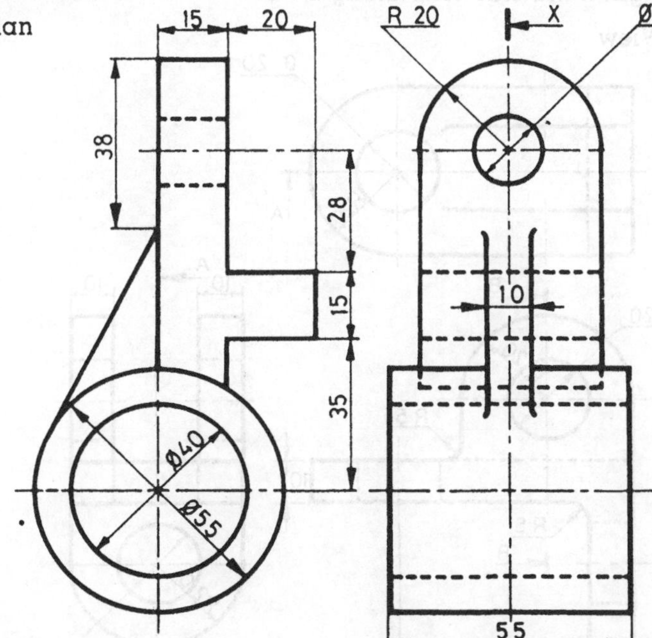

Fig. 17.16

Ex.10. Fig. 17.17. Draw by the same method of projection :

(i) Sectional Elevation, take section along X - X

(ii) Left Hand Side View

(iii) Top Plan

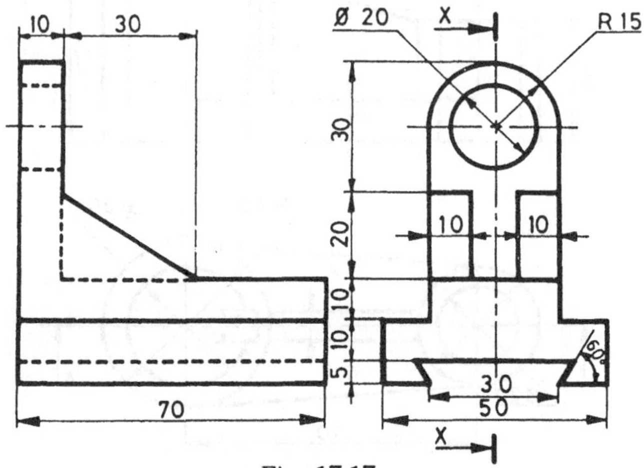

Fig. 17.17

Ex.11. Fig. 17.18. Draw by the same method of projection :

(i) Sectional Elevation along A - A

(ii) Sectional right hand side view along B - B

(iii) L. H. Side View

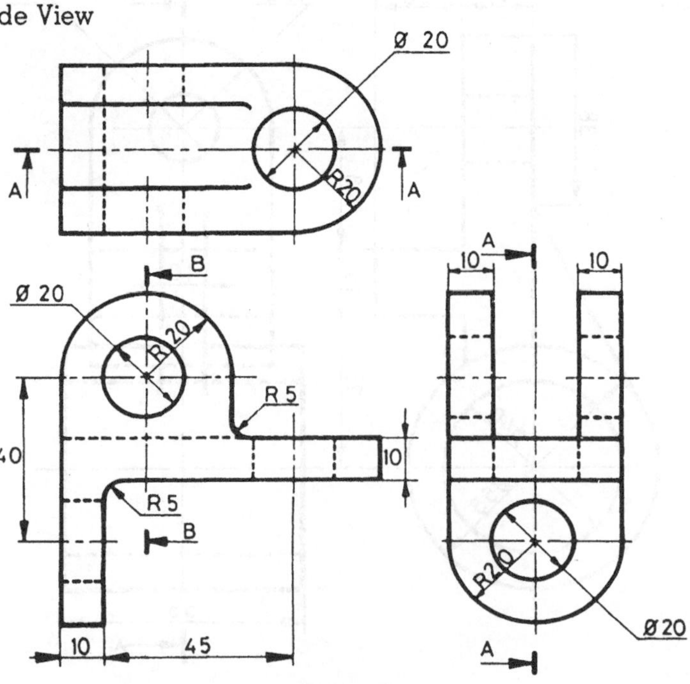

Fig. 17.18

Ex.12. Fig. 17.19. Draw by other than the system of projection used :

(i) Sectional Elevation (Section X - X)

(ii) Bottom Plan

(iii) R. H. Side View

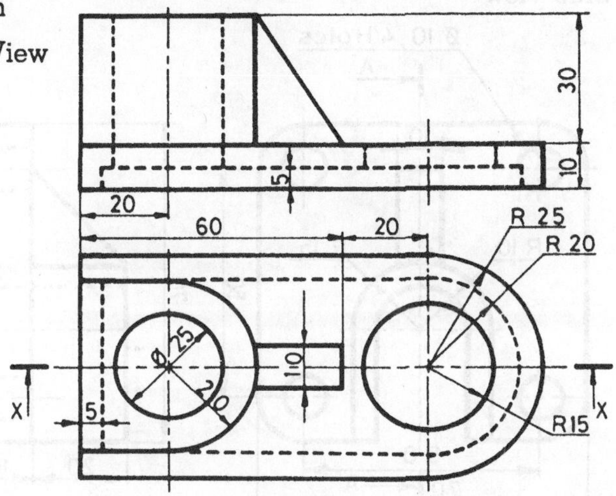

Fig. 17.19

Ex.13. Fig. 17.20. Draw by the 3rd angle projection method :

(i) Sectional Front View - Section along A - A

(ii) Sectional Side View - Section along B - B

(iii) Left Hand Side View

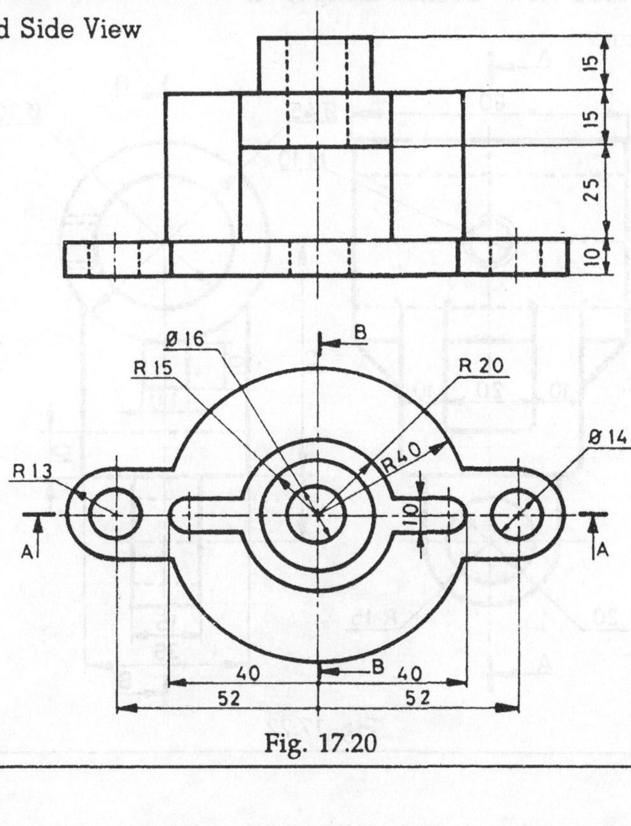

Fig. 17.20

Ex.14. Fig. 17.21. Draw according to the 1st angle projection system :

(i) Front View

(ii) Sectional Side View

(iii) Top Plan

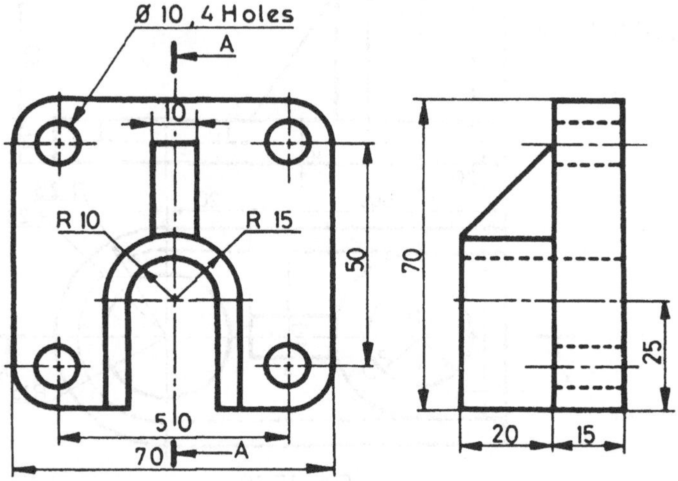

Fig. 17.21

Ex.15. Fig. 17.22. Draw by the same system of projection :

(i) Sectional Elevation - Section along B - B

(ii) Sectional Side View - Section along A - A

(iii) Bottom Plan

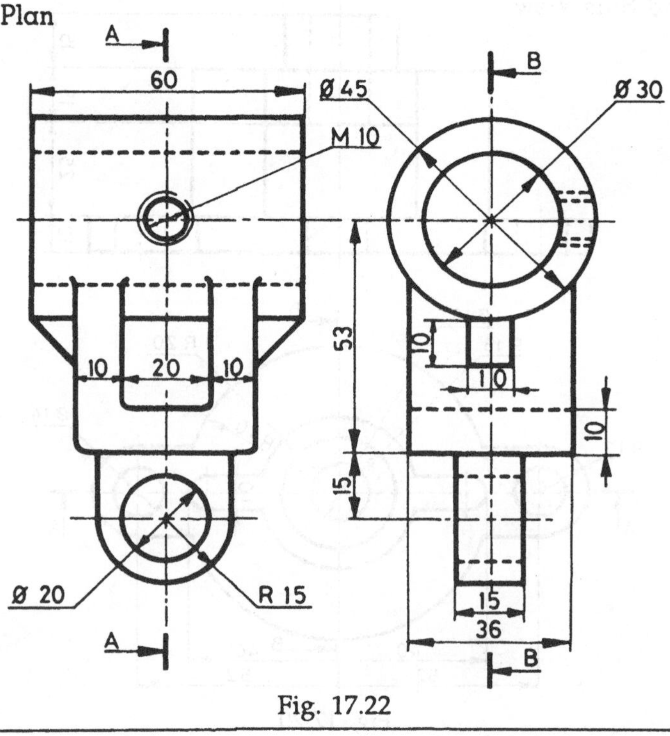

Fig. 17.22

Ex.16. Fig. 17.23. Draw by the 1st angle system of projection :

(i) Sectional Elevation Section along A - A

(ii) Sectional Side View Section along B - B

(iii) Top Plan

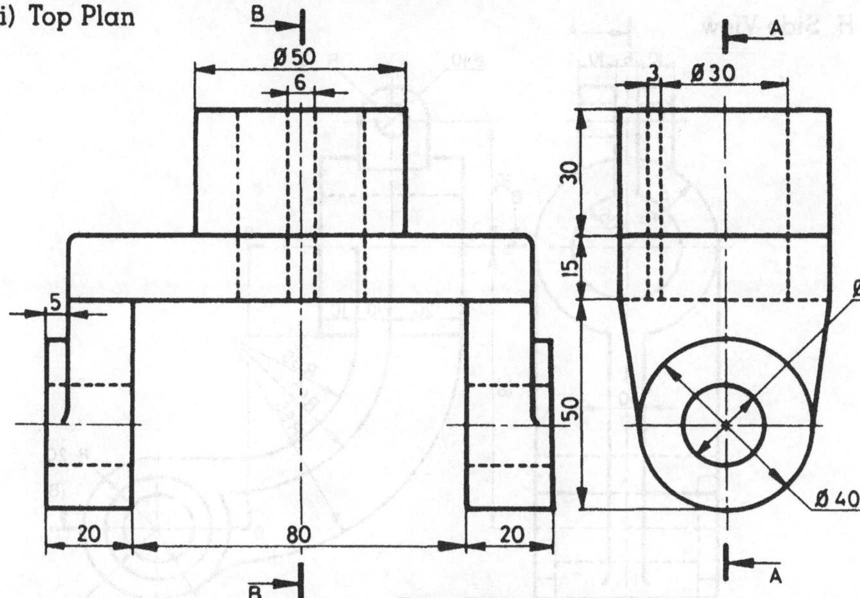

Fig. 17.23

Ex.17. Fig. 17.24. Draw by the 1st angle system of projection :

(i) Sectional Elevation - Section along A - A

(ii) Sectional Side View - Section along B - B

(iii) Left Hand Side View

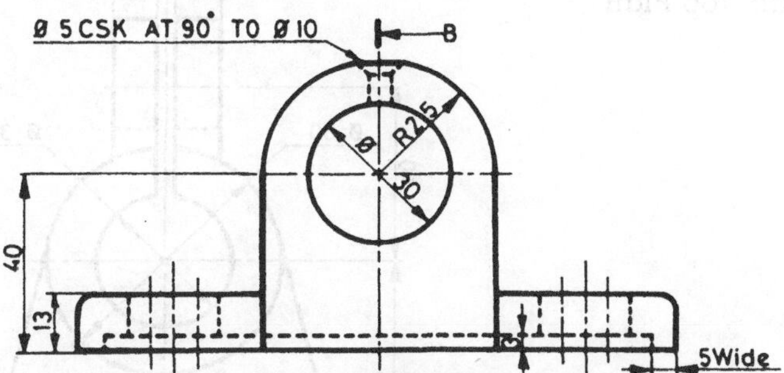

Fig. 17.24

Ex.18. Fig. 17.25. Draw by other than the system of projection used :

(i) Sectional Elevation - Section along A - A

(ii) Sectional Top Plan - Section along B - B

(iii) R. H. Side View

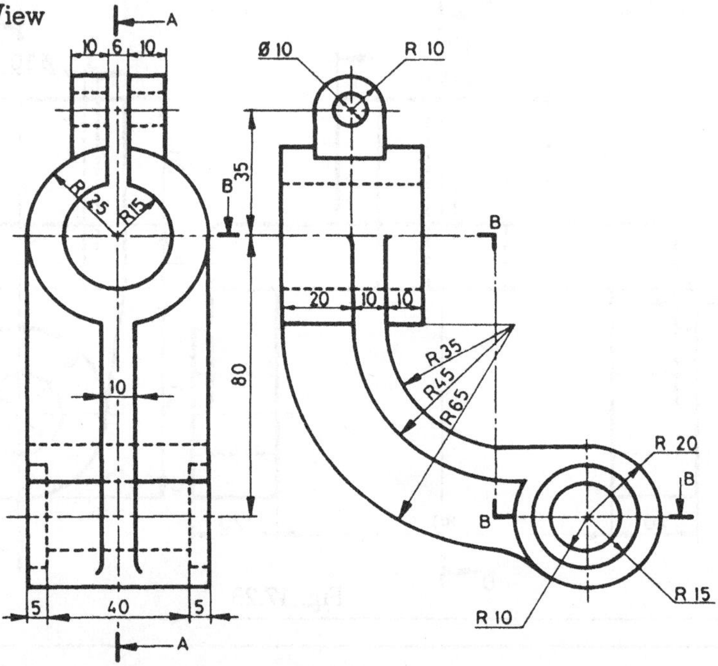

Fig. 17.25

Ex.19. Fig. 17.26. Draw according to the system of projection used :

(i) Sectional Elevation along A - A

(ii) Sectional Side View along B - B

(iii) Top Plan

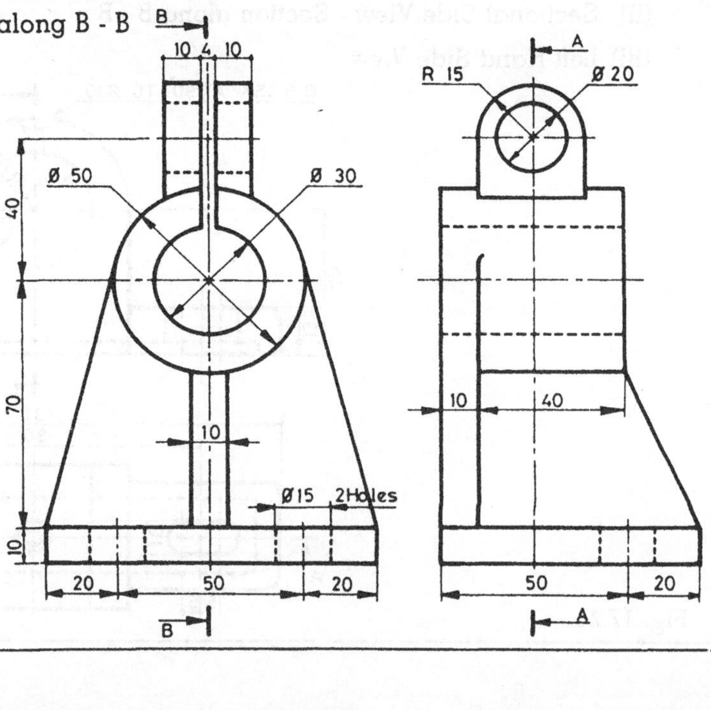

Fig. 17.26

Ex.20. Fig. 17.27. Draw by the 3rd Angle Projection System :

(i) Sectional F. V. take section along X - X

(ii) R. H. Side View

(iii) L. H. Side View

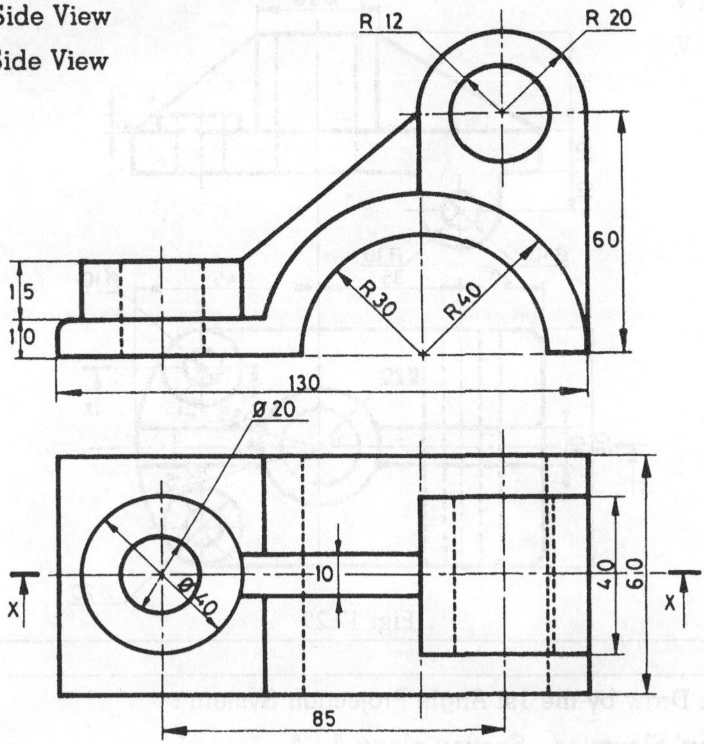

Fig. 17.27

Ex.21. Fig. 17.28. Draw by the 1st Angle System of Projection :

(i) Sectional Elevation along A - A

(ii) R. H. S. V.

(iii) L. H. S. V.

(iv) Sectional Plan along B - B

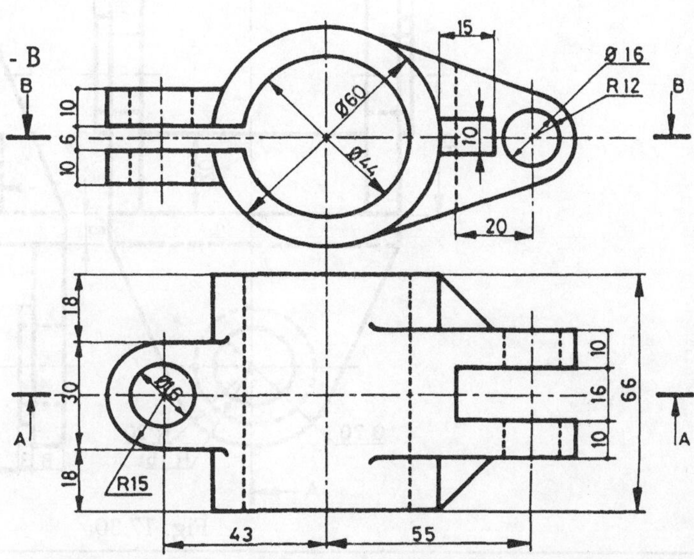

Fig. 17.28

Ex.22. Fig. 17.29. Draw by the same system of Projection as used :

(i) Offset Sectional Elevation along X - X

(ii) R. H. S. V.

(iii) L. H. S. V.

(iv) T. V.

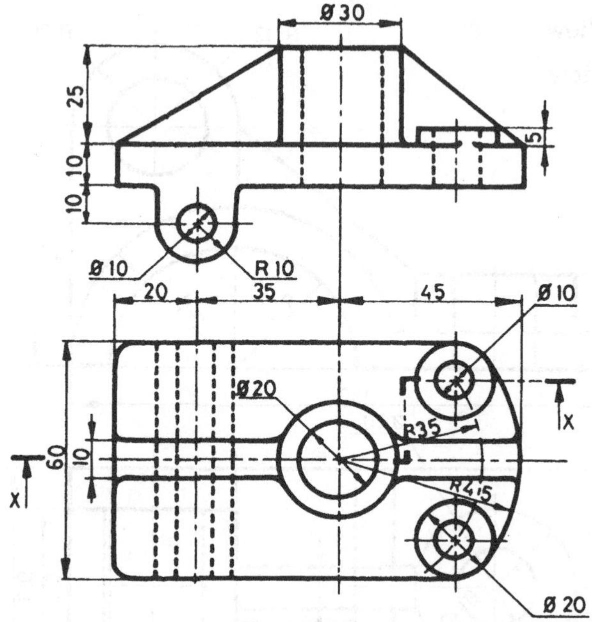

Fig. 17.29

Ex.23. Fig. 17.30. Draw by the 1st Angle Projection System :

(i) Sectional Elevation - Section along A - A

(ii) Sectional Side View - Section along B - B

(iii) Top Plan

(iv) Bottom Plan

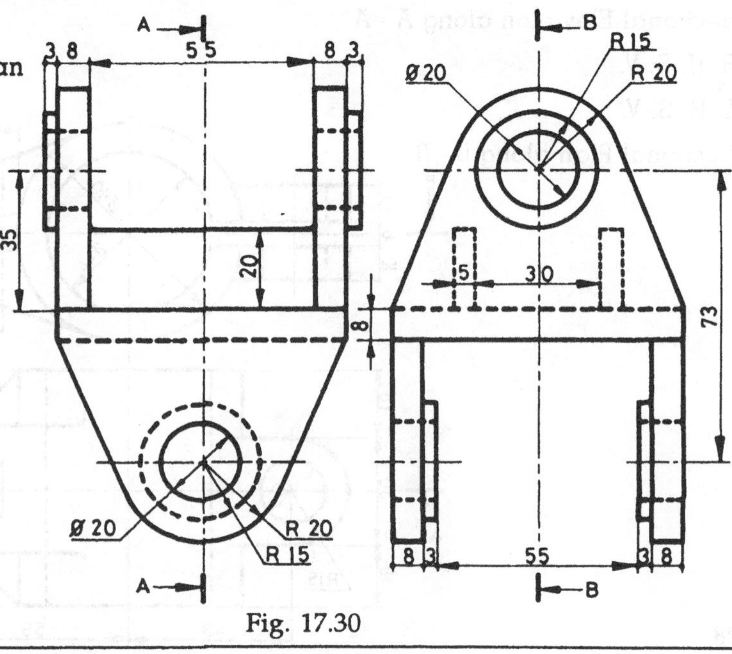

Fig. 17.30

Ex.24. Fig. 17.31. Draw by the same System of Projection :

(i) Full Sectional Elevation along A - A

(ii) Top Plan

(iii) Left Hand Side View

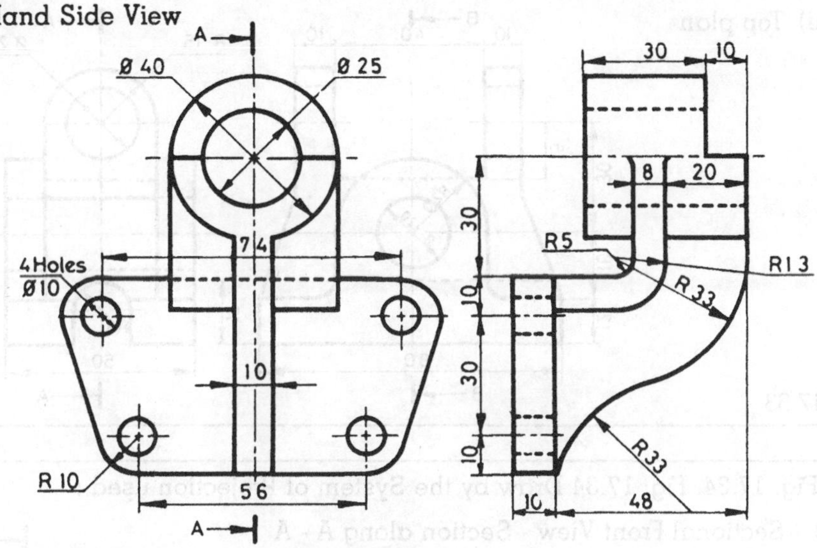

Fig. 17.31

Ex.25. Elevation and Top Plan of an object are given in Fig. 17.32 Draw in the same system of projection :

(i) Half Sectional Elevation, take section along A - B

(ii) Half Sectional R. H. Side View, take section along A - B

(iii) Top Plan

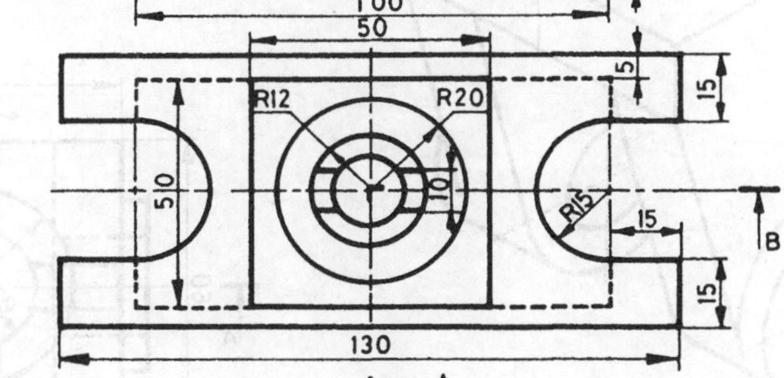

Fig. 17.32

Ex.26. Fig. 17.33. Draw in the 1st Angle Projection System :

(i) Sectional Elevation along A - A

(ii) Sectional Side View along B - B

(iii) Top plan

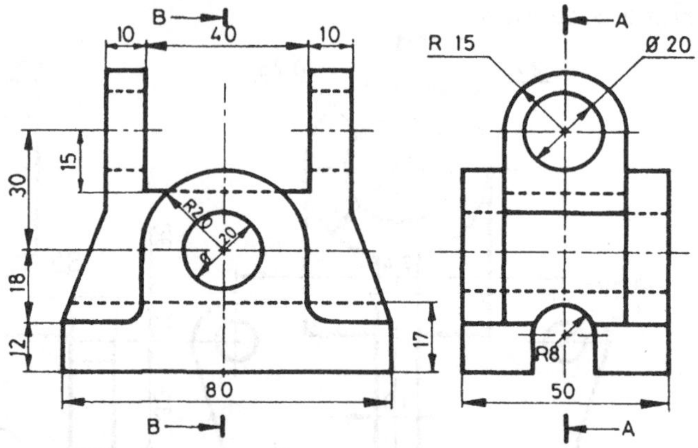

Fig. 17.33

Ex.27. Fig. 17.34. Fig. 17.34 Draw by the System of Projection used :

(i) Sectional Front View - Section along A - A

(ii) Sectional Side View - Section along B - B

(iii) L. H. S. V.

Hints : As the component is difficult to visualise, pictorial drawing is given for understanding.

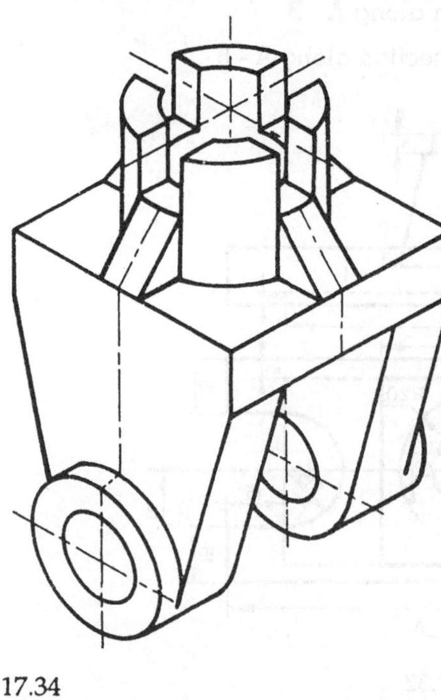

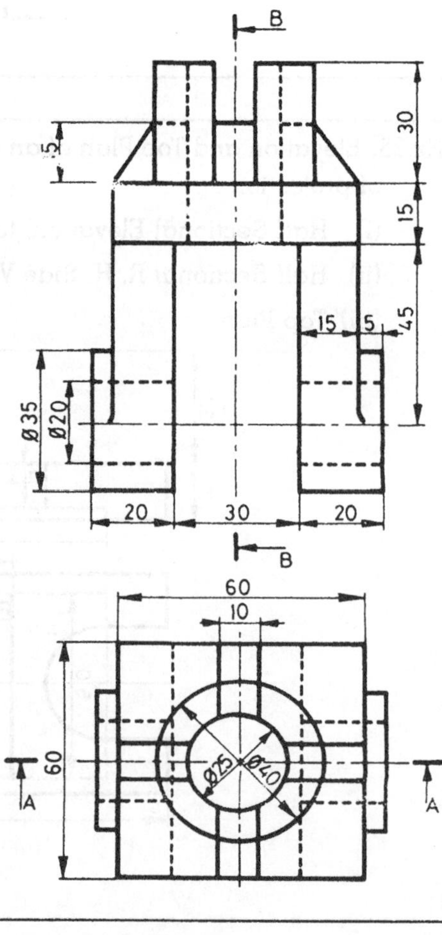

Fig. 17.34

Ex.28. Fig. 17.35. Draw by the same System of Projection :

(i) Sectional Elevation along A - A

(ii) Sectional Plan along B - B

(iii) Sectional Side View along C - C

(iv) Side View from left

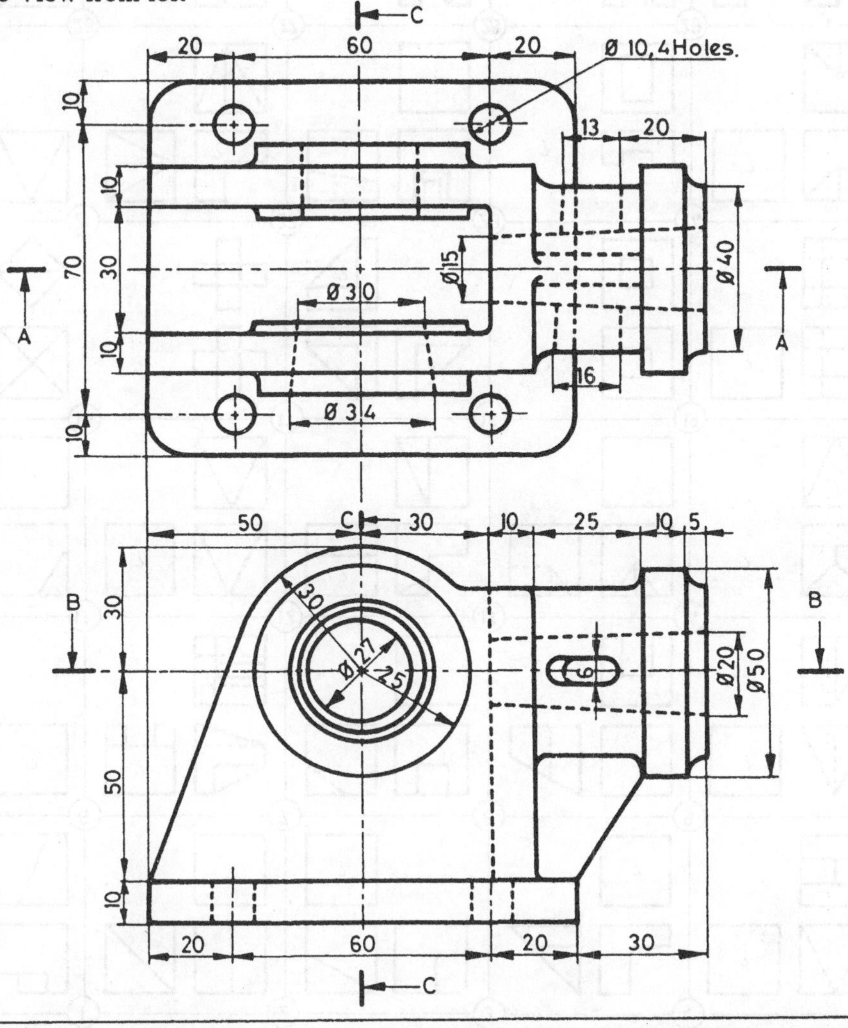

Fig. 17.35

Ex.29. Two or three orthographic views of 35 different objects, with either (i) a line or (ii) lines or (iii) a view missing in view marked by sign (?), are given in Fig. 17.36.

Imagine three dimensional view of each object and prepare a sketch of the same. Complete the view in which line or lines missing. Draw third missing view wherever not given.

(Answers are given in Fig. 17.37 in the form of three dimensional drawings. One can easily find the missing line or lines or view from the pictorial drawings given. The given Fig. 17.37 is not to scale.)

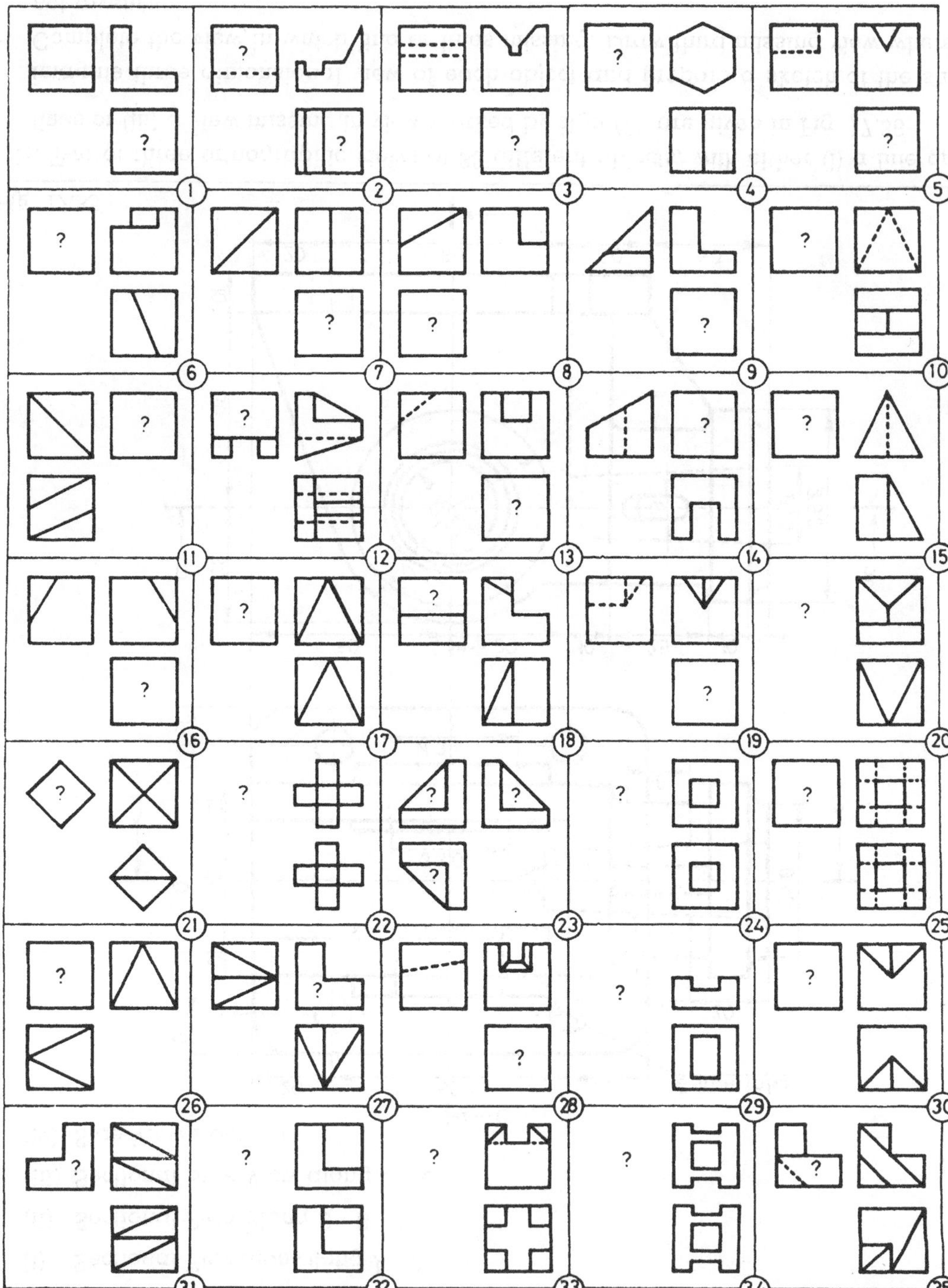

Fig. 17.36 Ex. 29

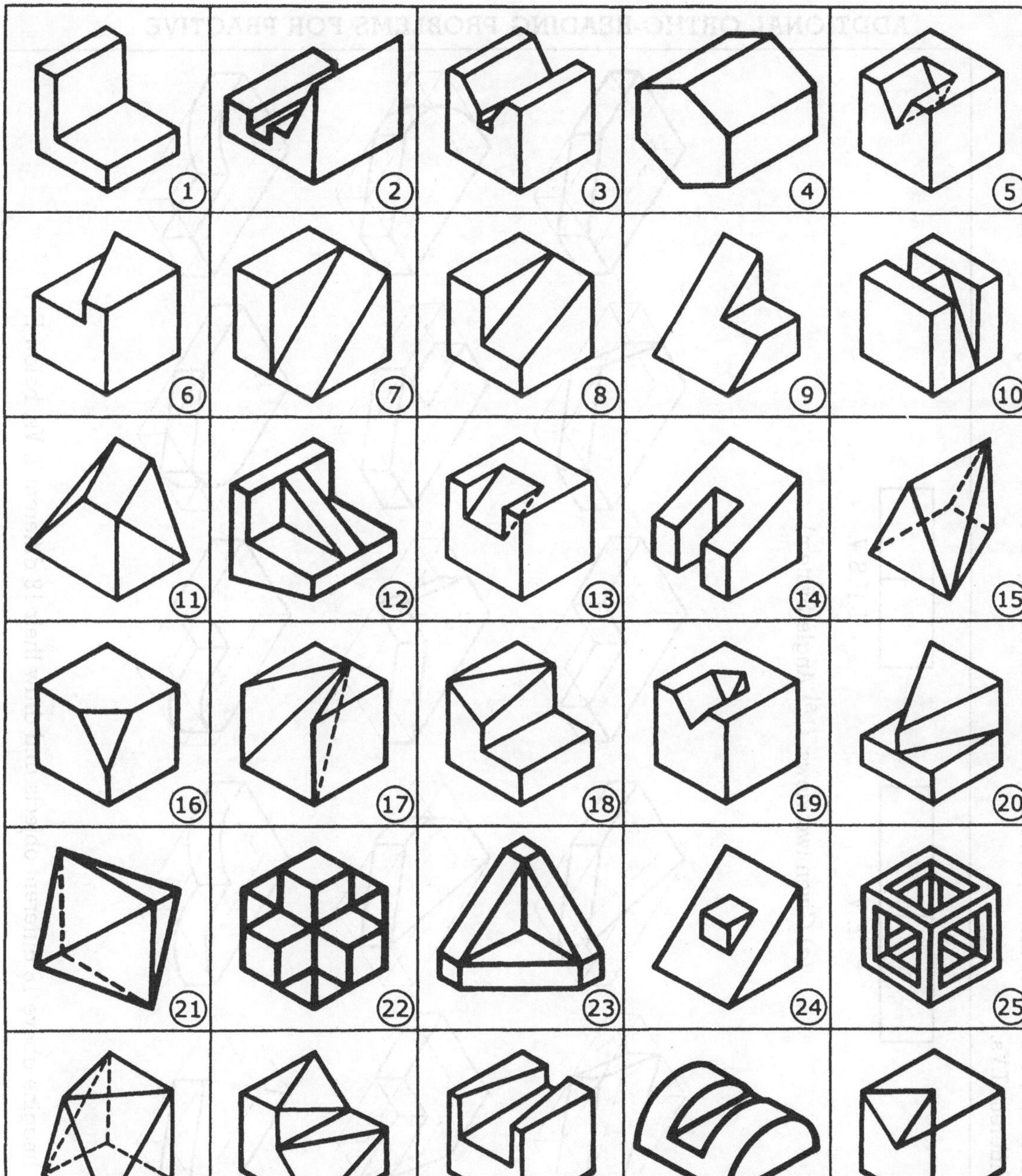

Fig. 17.37 Answers of Ex. 29

ADDTIONAL ORTHO-READING PROBLEMS FOR PRACTIVE

Problem : Figure (a) shows F. V. and L.H.S.V. of the object. Imagine 18 different objects, which satisfy given. F. V. and L.H.S.V. and draw their corresponding different T.Vs.

(a) Given two views. (1st Angle System)

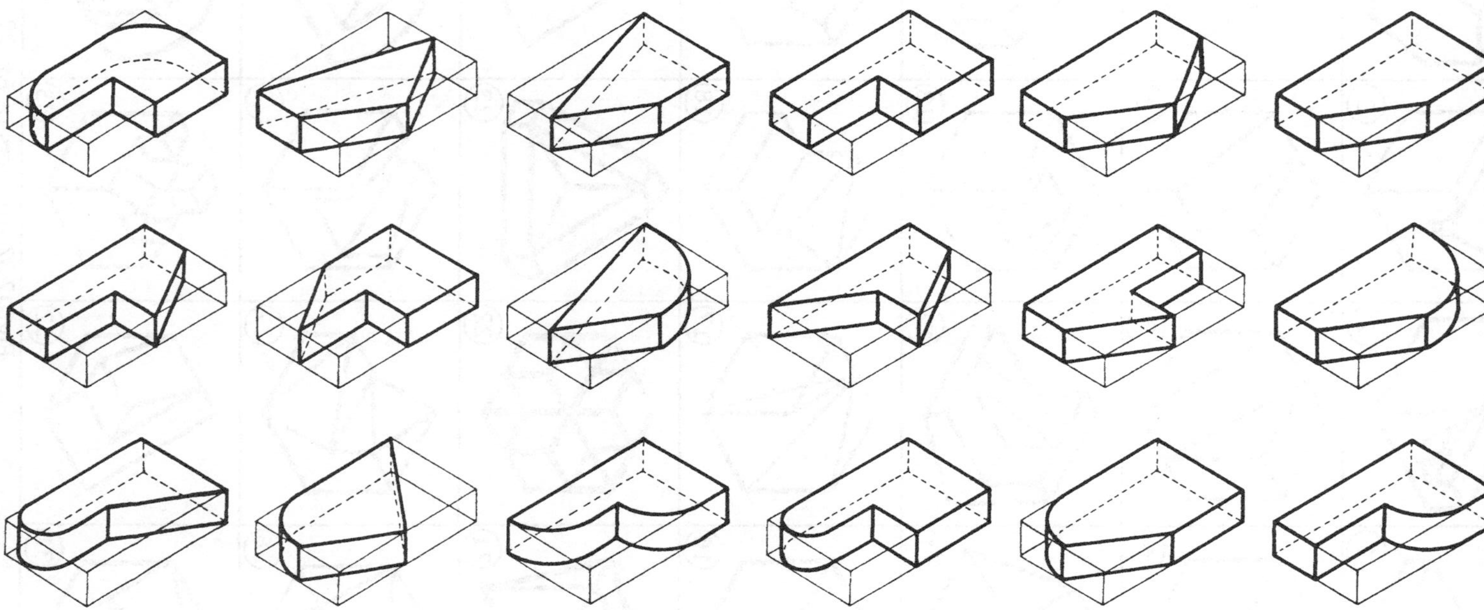

(b) Imagine above 18 different objects and draw their 18 different T. Vs. below F. V.

Problem : Figure (a) shows two views of an object. Draw the following views.

(i) Elevation (Sectional) OR (i) Sectional F. V.

(ii) Plan (T. V.) (ii) T. V.

(iii) R. H. S. V. (iii) R.H.S.V.

[Pune University, May 1994]

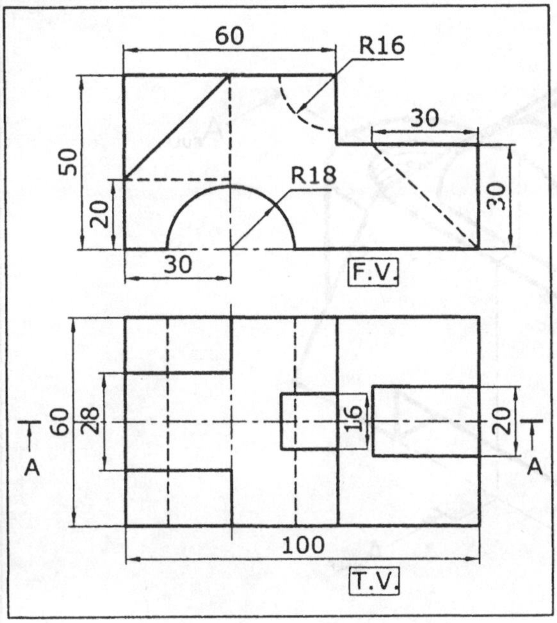

(a) Given Data

(b) Imagine 3D of the object.

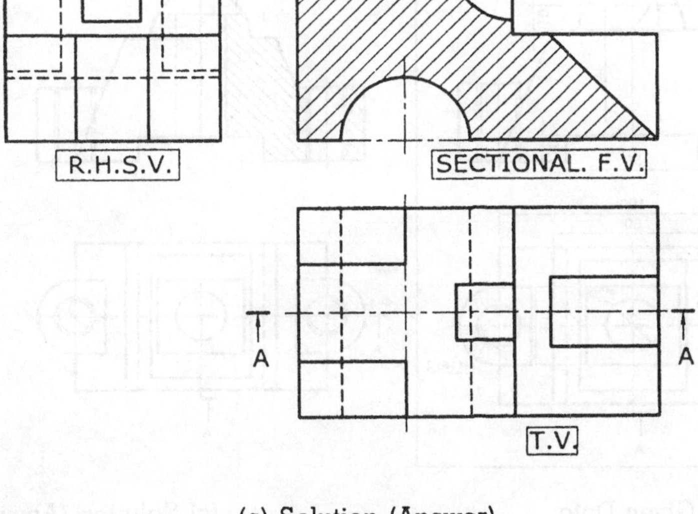

(c) Solution (Answer)

Problem : Figure (a) shows the elevation and plan of an object. Draw the following views by 1st angle method of projections

(i) Half Sectional F. V. [Section along A-A $_{\text{Half}}$]

(ii) T. V.

(iii) L. H. S. V. [Pune University]

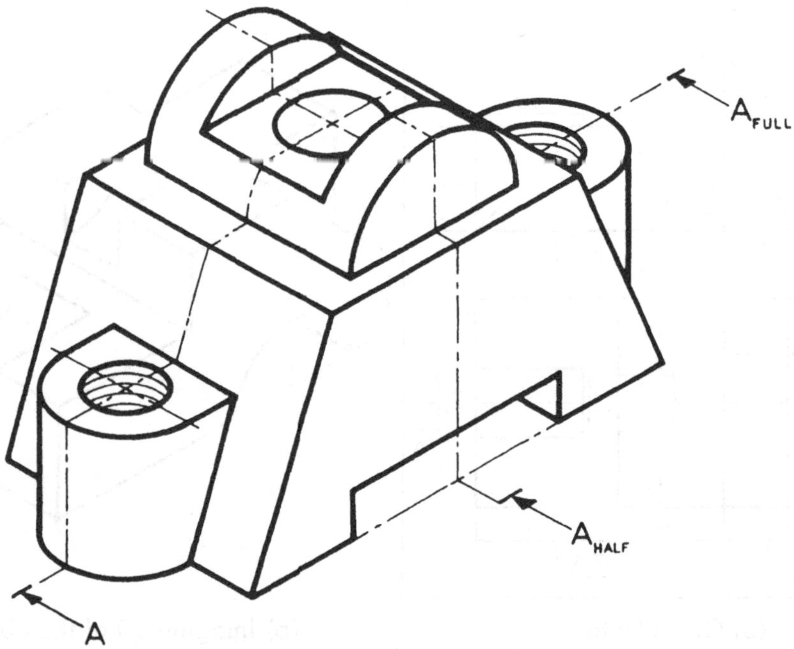

(b) Imagine 3D of the object

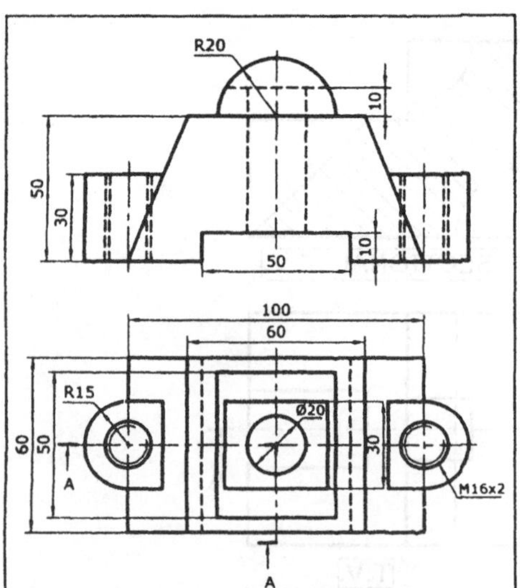

(a) Given Data

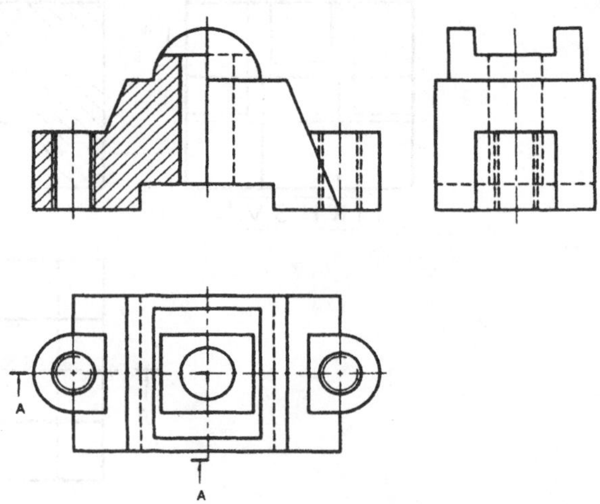

(c) Solution (Answer)

Problem : Figure (a) shows F. V. and T. V. of an object. Using the same method of projections, draw

(i) Sectional F. V. (Section along A-A)

(ii) T. V.

(iii) L. H. S. V. [Pune University, May 1996]

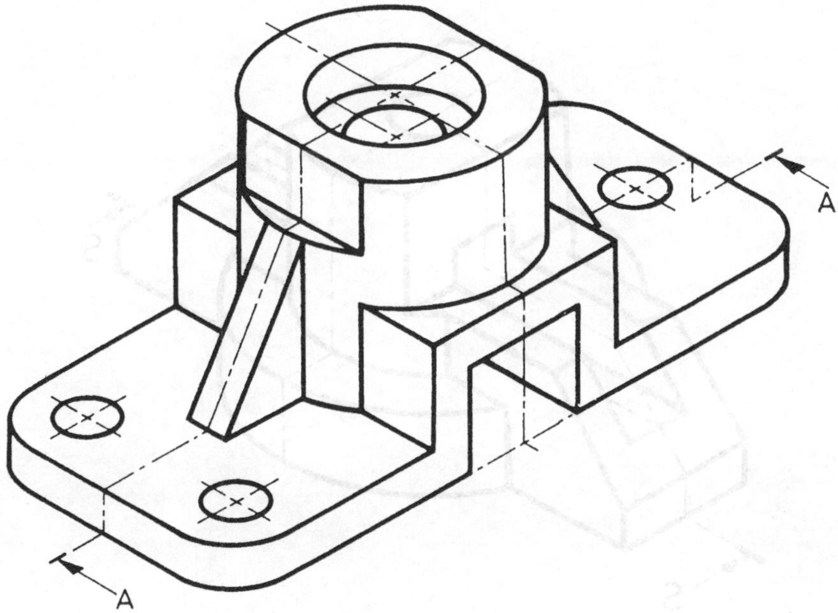

(b) Imagine 3D of the object.

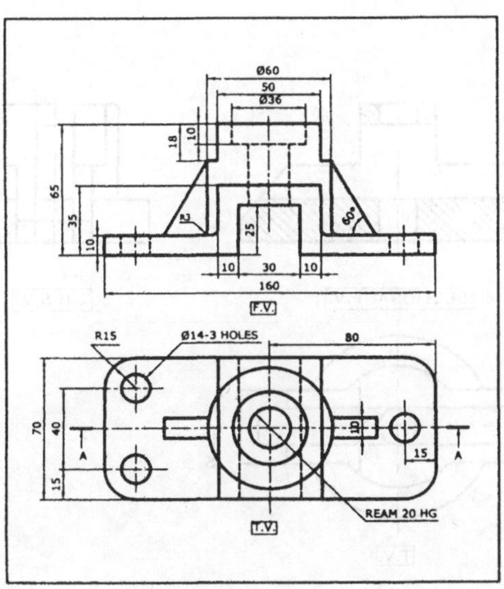

(a) Given Data

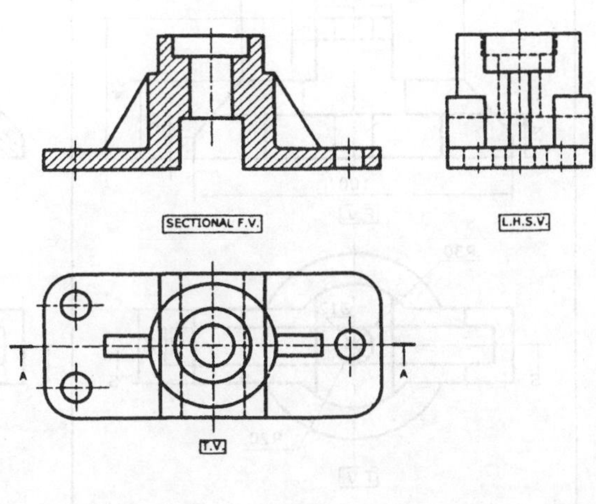

(c) Solution (Answer)

Problem : Figure (a) shows F. V. and T. V. of an object. Draw

(i) Sectional F. V. [Section along S-S]

(ii) T. V.

(iii) L. H. S. V.

Using same method of projections [Pune University]

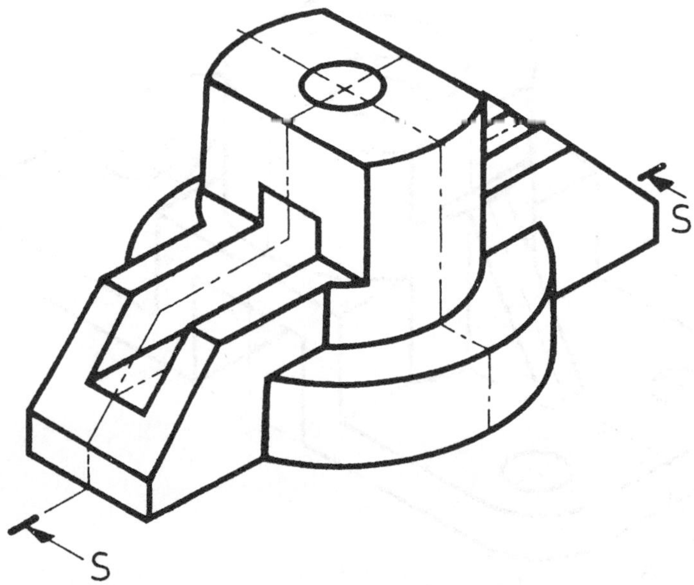

(b) Imagine above 3D of the object

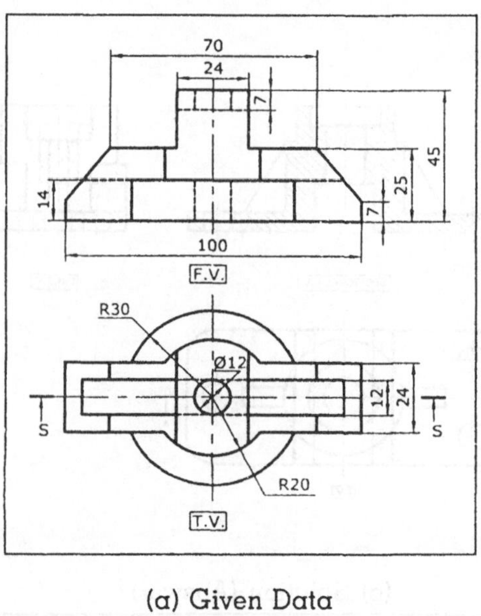

(a) Given Data

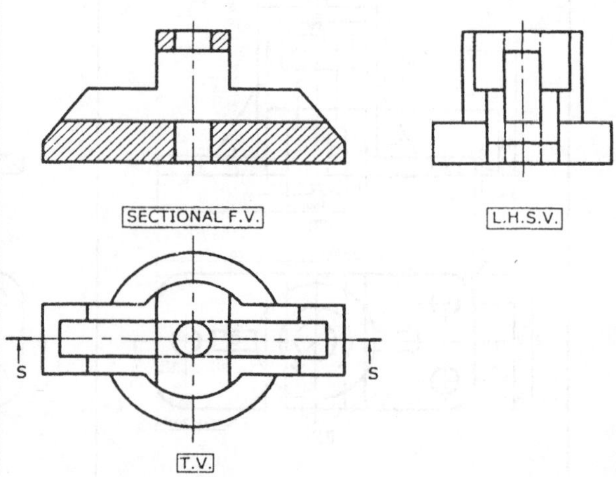

(c) Solution (Answer)

Problem : Figure (a) shows F. V. and L. H. S. V. of A MACHINE PART. Draw the following views using 1:1 scale

(i) Sectional F. V. across section plane A-A.

(ii) Sectional L. H. S. V. across section plane B-B.

(iii) Missing T. V. showing all hidden details.

(iv) Insert atleast 10 important dimensions

[Do dimensioning as done in data] [Mumbai University, November 1997]

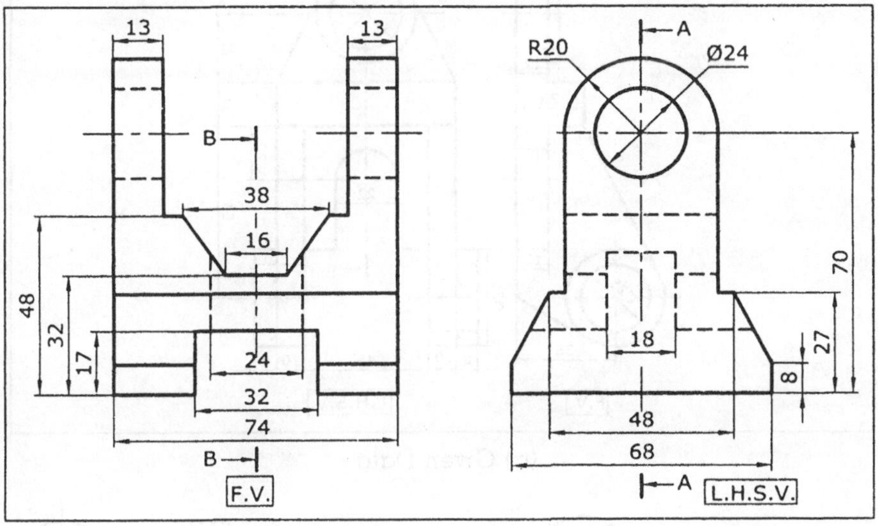

(a) Given Data

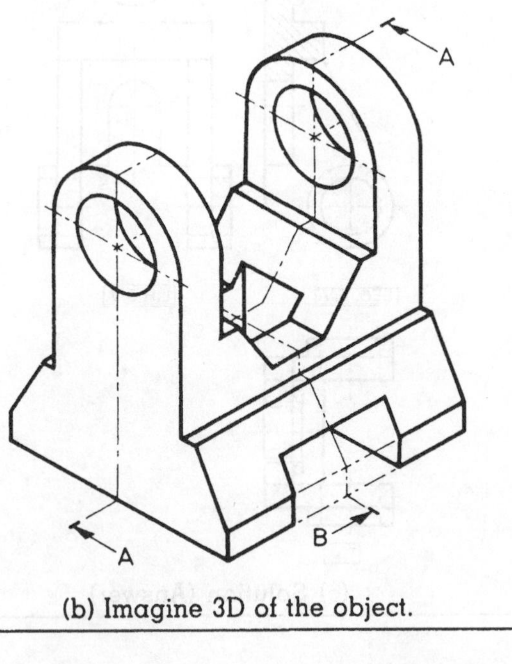

(b) Imagine 3D of the object.

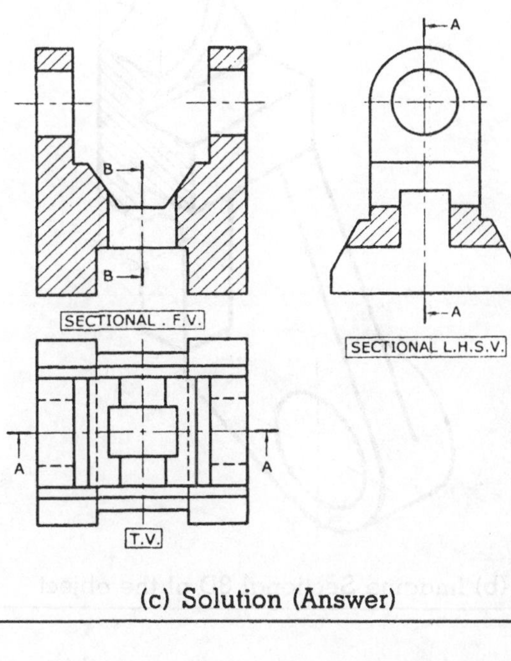

(c) Solution (Answer)

Problem : Figure (a) shows F. V. and L. H. S. V. of A Bracket. Draw

(i) Sectional F. V. along S-S

(ii) L. H. S. V.

(iii) T. V.

Using same method of projections [Pune University, November 1997]

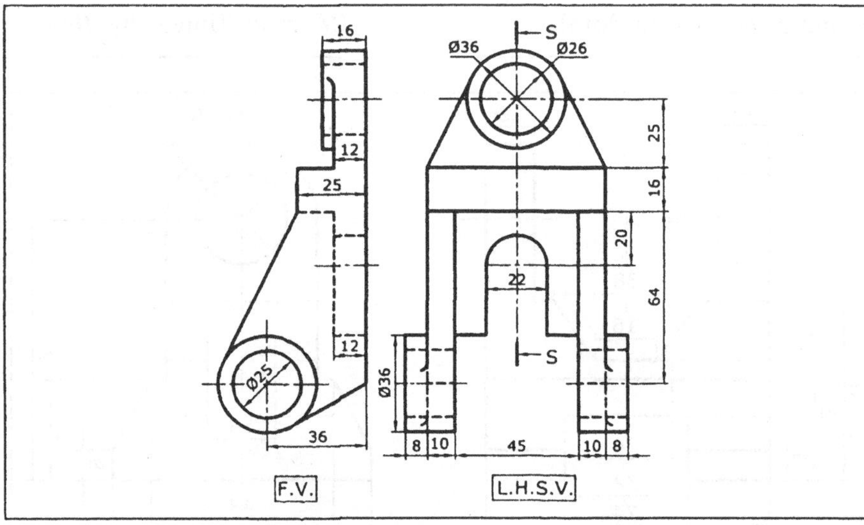

(a) Given Data

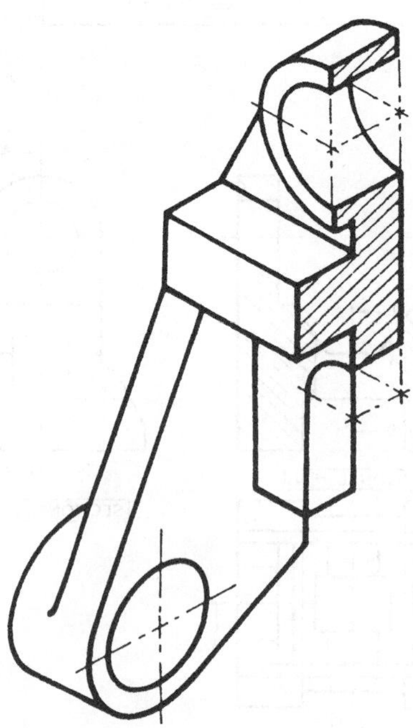

(b) Imagine Sectional 3D of the object

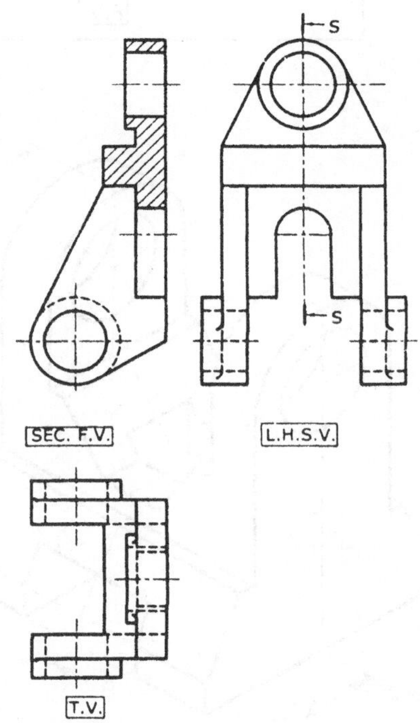

(c) Solution (Answer)

Problem : Figure (a) shows two views of an object. Draw the following views by 1st angle-method.

(i) Sectional Elevation. [Section along A-A]

(ii) Plan (T. V.)

(iii) L. H. S. V. [Pune University, November 1993]

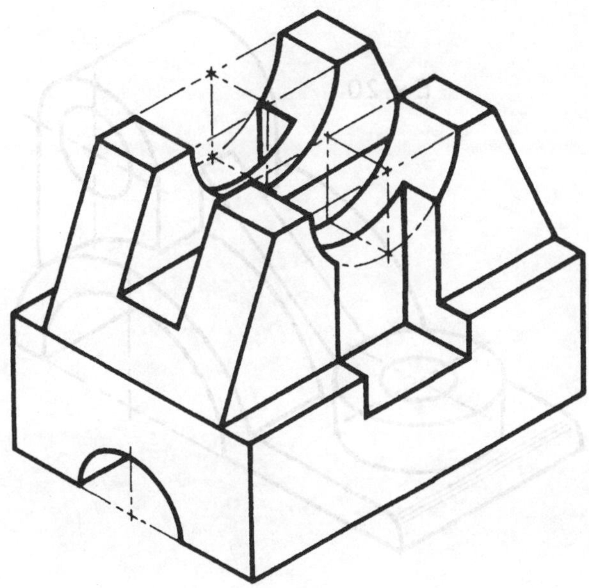

(b) Imagine 3D of the object.

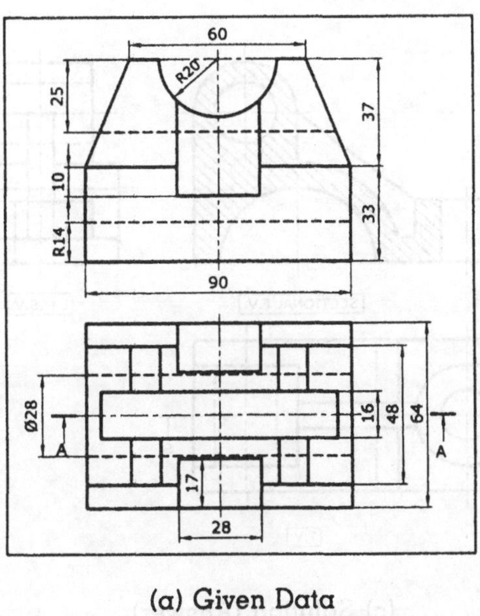

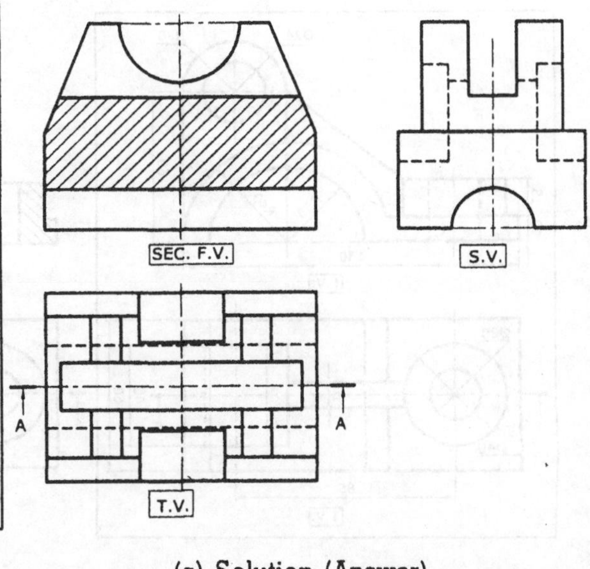

(a) Given Data

(c) Solution (Answer)

Problem : Figure (a) shows two views of an object. Using the same method of projection, draw

(i) Sectional F. V. [Section along S-S]

(ii) Given T. V.

(iii) Left Hand Side View. [Pune University, May 1996]

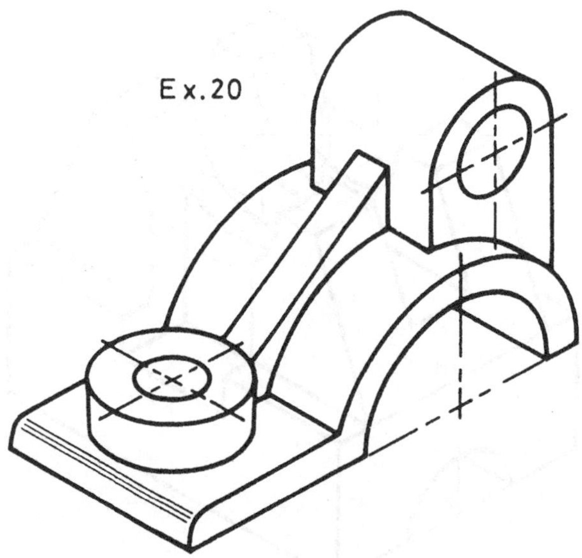

(b) Imagine 3D of the object

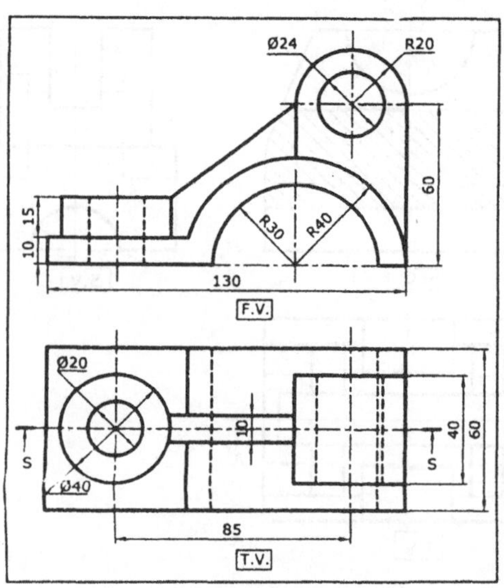

(a) Given Data

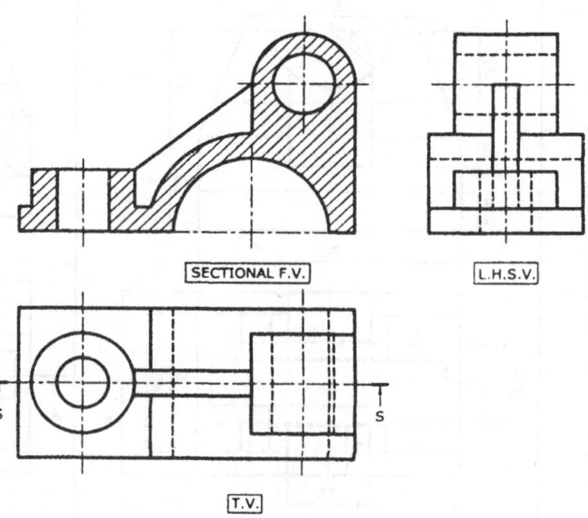

(c) Solution (Answer)

Problem : Two views of an object are shown in Figure (a).

Using 1st angle projection method, draw the following views.

(i) Sectional Elevation [Section along A-A]

(ii) Sectional Plan [Section along B-B]

(iii) Right Hand Side View. [Mumbai University, November 1994]

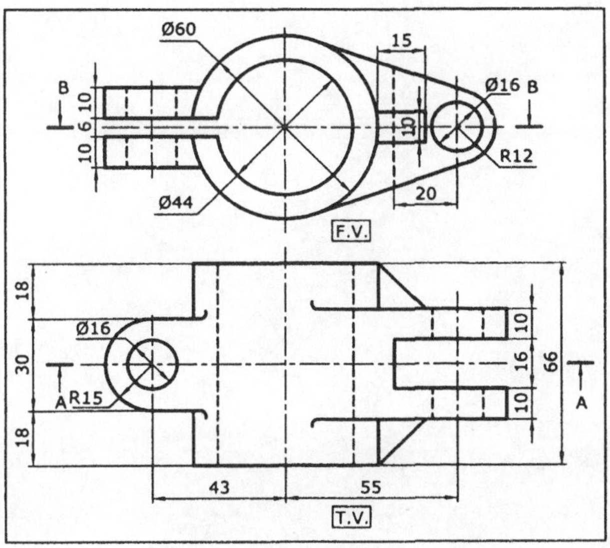

(a) Given Data

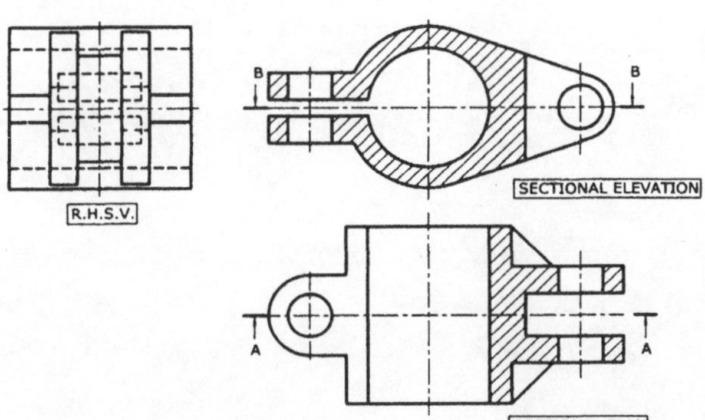

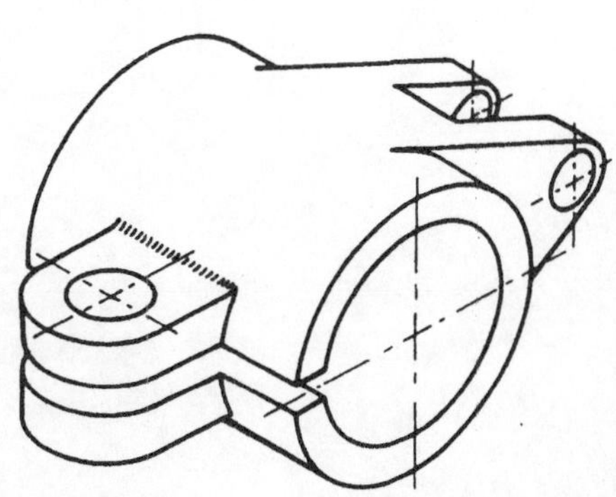

(c) Solution (Answer)

(b) Imagine 3D of the object

CHAPTER 18

ISOMETRIC PROJECTION/VIEW/DRAWING

(Axonometric Projection)

1. Definition

We have studied earlier that **Axonometric Projection** is one kind of orthographic projection. We have also seen that a plane inclined to all the three principal planes is known as **Axonometric Plane** and orthographic projection of an object on that plane is known as **Axonometric Projection**. It is a pictorial projection.

If Axonometric Plane is equally inclined to all the three principal planes then that plane is known as the isometric plane and orthographic projection on it is known as isometric projection. Similarly, if Axonometric Plane is equally inclined to two of the three principal planes then that is known as the diametric plane and orthographic projection on it is known as diametric projection. Similarly, if Axonometric Plane is having different inclination with all the three principal planes, then that plane is known as the trimetric plane and orthographic projection on it is known as trimetric projection.

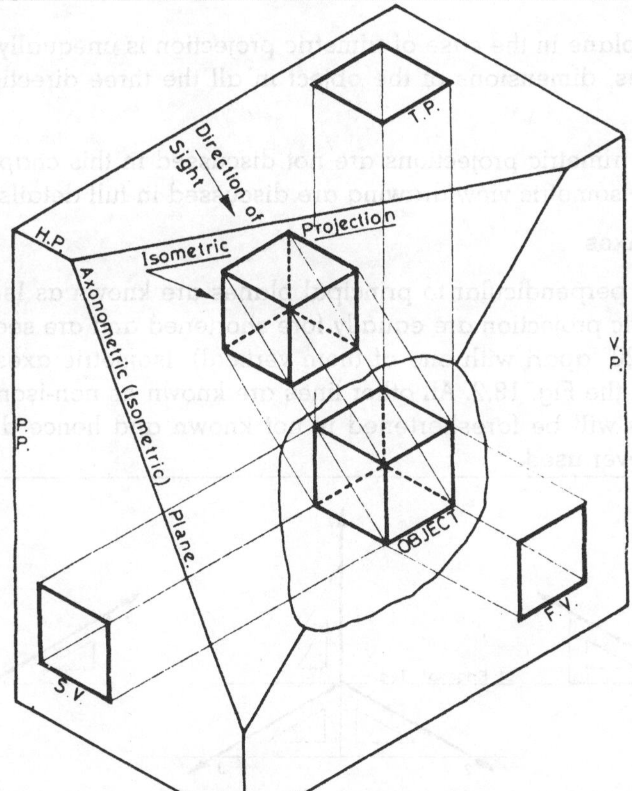

Fig. 18.1 Theory of Isometric Projection

In Fig. 18.1, object (Cube) is kept inside the box with its faces parallel to principal planes i.e. (V.P., H.P. and P.P.). Principal orthographic projections of an object, that is F.V., T.V. and S.V., are shown on V.P., H.P. and P.P. respectively. In Fig. 18.1, Axonometric plane is shown and projection on it of the object is also shown. Here projection is isometric projection.

Direction of sight is perpendicular to the plane of projection. i.e. Axonometric plane.

2. Isometric

As the picture plane in the case of isometric projection is equally inclined to all the three principal planes, dimensions of the object in all the three directions are equally foreshortened. And hence same scale can be used on all the three axes. Isometric means equal measurement. 4 edges of the cube are perpendicular to V.P., 4 edges are perpendicular to H.P. and 4 edges are perpendicular to P.P. These edges in a set of four are seen along parallel lines in three directions 120° apart, out of which one is vertical.

3. Diametric

As the picture plane, in the case of diametric projection, is equally inclined to two of the three principal planes dimensions of the object in two directions are equally foreshortened and in the third direction it is foreshortened by some other amount.

4. Trimetric

As the picture plane in the case of trimetric projection is unequally inclined to all the three principal planes, dimensions of the object in all the three directions are unequally foreshortened.

Diametric and trimetric projections are not discussed in this chapter. Only isometric projection as well as isometric view/drawing are discussed in full details.

5. Isometric Lines/Axes

Lines in space perpendicular to principal planes are known as **Isometric Lines** and these lines in Isometric projection are equally fore shortened and are seen along **Isometric Axes**. (Three lines 120° apart with one of them vertical). Isometric axes in three different formats are shown in the Fig. 18.2. All other lines are known as non-isometric lines and by what proportion they will be foreshortened is not known and hence dimensions of non-isometric lines are never used.

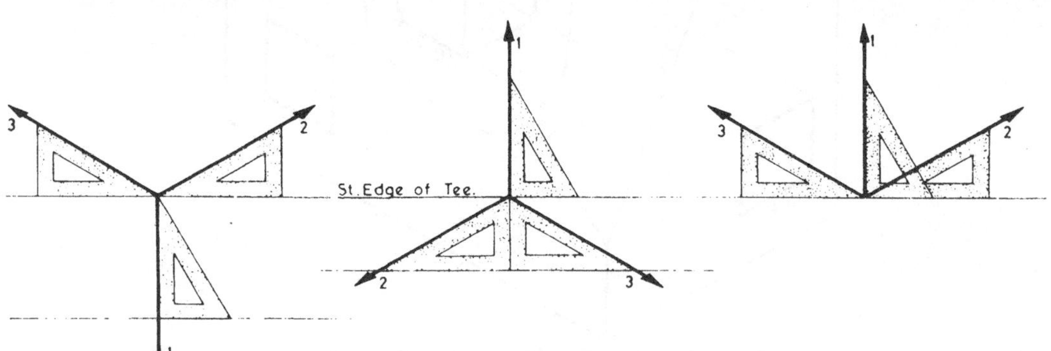

Fig. 18.2 Isometric Axes

6. Isometric Planes

Any plane in space parallel to one of the principal planes and obviously perpendicular to two other P. Ps. is known as the **Isometric Plane** and its isometric projection can be drawn by the use of corresponding two axes as shown in Fig. 18.3 All the three isometric planes are shown in Fig. 18.3.

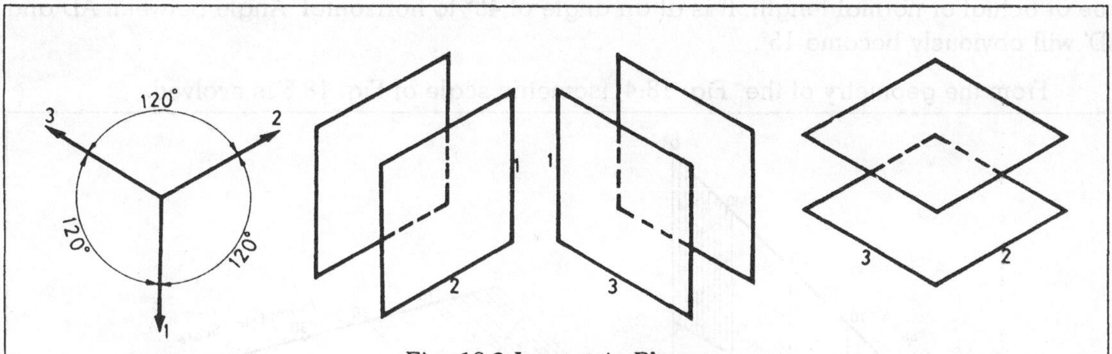

Fig. 18.3 Isometric Planes

7. Isometric Scale

Axonometric plane, which is equally inclined to all the three principal planes, is parallel to one of the diagonals of a cube's top square surface. Try to visualise this from Fig. 18.1

In the Fig. 18.4 diagonal AC, since it happens to be parallel to axonometric (isometric) plane, is shown in its true length while diagonal BD is shown of shortened length. Three sets of 4 edges of cube are seen at 120° out of which 1 set is vertical. As all the 12 edges of the cube are equally inclined to axonometric (isometric) plane, they are reduced by the same proportion and hence seen equal. The square faces of the cube are seen as rhombuses.

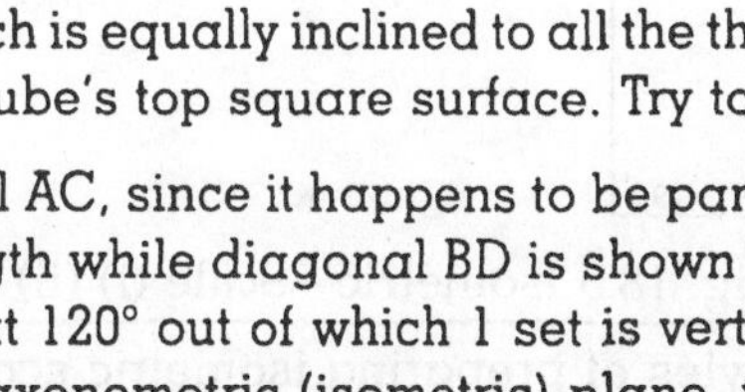

Fig. 18.4 Iso-scale formation

Now rhombus ABCD, the isometric of top surface of the cube, is having one diagonal (AC) of true length and another diagonal BD of shortened length. They intersect at right angles at point O.

Draw OD' and OB' equal to AO or OC. AB'CD' will become a square of exact size of cube's top surface, since in construction we have taken plane diagonals AC and B'D' of same size and they are mutually at right angles. AB', B'C, CD' and D'A represent true lengths. AB, BC, CD and DA represent lengths in isometric projection.

To understand the scale formation, we take a set of lines AD and AD'.

AD represents the isometric length. It is at an angle of 30° to horizontal. AD' represents true or actual or normal length. It is at an angle of 45° to horizontal. Angle between AD and AD′ will obviously become 15°.

From the geometry of the Fig. 18.4, isometric scale of Fig. 18.5 is evolved.

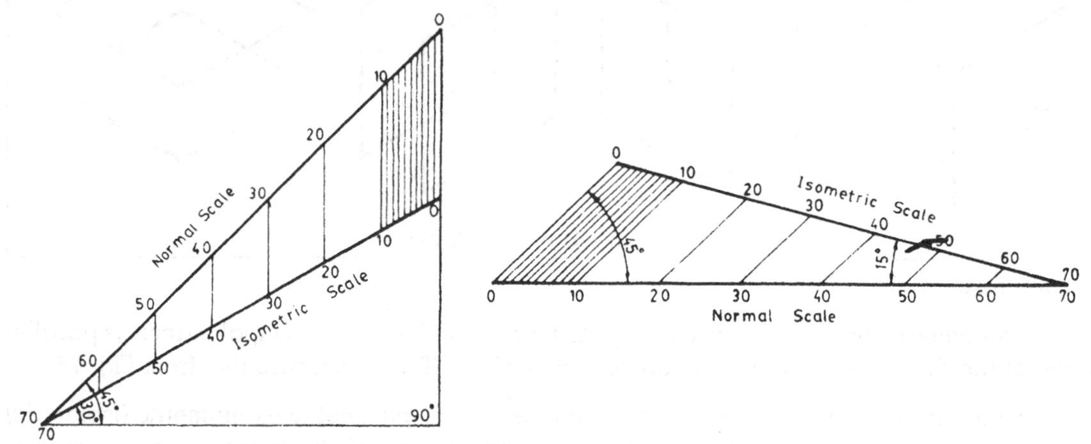

Fig. 18.5 Isometric - Scale (i) (ii)

There are two different styles of preparing isometric scale.

(1) In the style first, normal, true or actual scale is drawn on a line at 45° to the horizontal line. By drawing vertical lines from normal scale line up to a line at 30° to the horizontal, we get isometric scale on a 30° line as shown in the Fig. 18.5(i)

(2) In the style second, normal, true or actual scale is drawn on a horizontal line. By drawing parallel lines at 45° to the horizontal line from normal scale line upto a line at 15° to the horizontal, we get isometric scale on a 15° line as shown in the Fig. 18.5 (ii).

Reduction of isometric line in isometric projection is

$$\frac{\text{Isometric length}}{\text{Normal, true or actual length}} = \frac{\sqrt{2}}{\sqrt{3}} = 0.815 \ \underline{\Omega} \ \frac{9}{11}$$

In short, figure in isometric projection is reduced lengthwise to approximately 81.5%. Volumewise it is reduced to (0.815 x 0.815 x 0.815 = 0.5413) or 54.13%.

8. Isometric View or Drawing

We are more interested in the shape of isometric of an object rather than its size and hence normally isometric view or drawing is drawn without reducing dimensions by isometric scale. When projection is drawn dimensions are reduced by the scale and then used. When iso-view or iso-drawing is drawn, then the actual dimensions are used i.e. without reducing them by the scale.

In Fig. 18.6, comparison of isometric view or drawing with isometric projection is shown. View or drawing of a cube is shown drawn by the normal scale while projection of a cube is shown drawn by the isometric scale. Method of drawing is absolutely same, there is only difference in the size. View or drawing is bigger in size (of actual size) while projection is smaller in size (of shortened size.)

$$\frac{\text{Isometric length}}{\text{True length}} = \frac{\sqrt{2}}{\sqrt{3}} = 0.815 \;\underline{\Omega}\; \frac{9}{11}$$

9 Divisions of NormalScale $\underline{\Omega}$11 Divisions of Isometric Scale

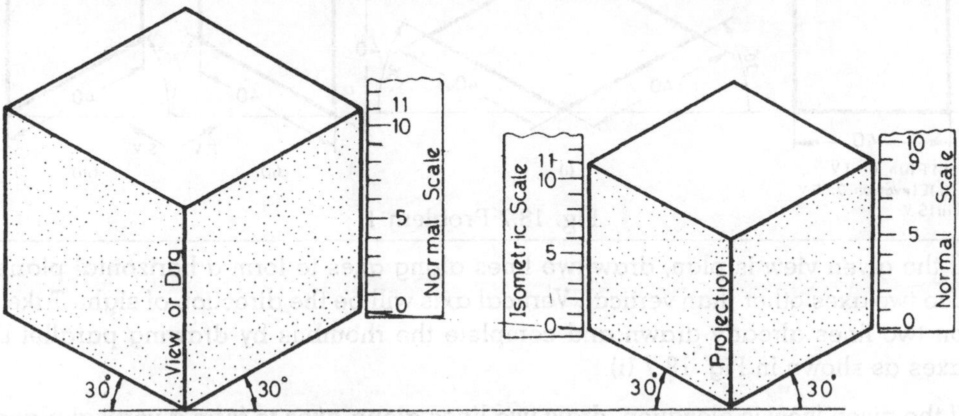

Fig. 18.6 Comparison of Iso-projection with Iso-view

9. Methods of Construction of Isometric

There are two methods of construction :

(i) Conventional method

(ii) Exact method

(i) Conventional Method :

This is a practical method. In this method for getting projection we have to use one scale for isometric, two scales for diametric and three scales for trimetric projections. Usually isometric drawing is drawn instead of projection by making measurements with a normal scale. While constructing isometric, measurements should be made only on or parallel to the three isometric axes.

(ii) Exact Method :

This method is theoretically correct and exact. It is a simple and superior method. In this method scales are not required to be used. Scales are taken care of automatically and exactly. Diametrics and trimetrics can be drawn very easily with perfect flexibility as to choice of angle of axes and position of object.

First of all we shall study the conventional method and solve problems also by the same method. Later on in the chapter we shall study the exact method and solve few problems also by the same method.

10. Isometric Projection of Plane Figures

We shall study this by taking different problems.

Problem 1 : The (i) Plan or (ii) Elevation or (iii) S. V. of a 40 mm square plate is given in Fig. 18.7. Draw its isometric view.

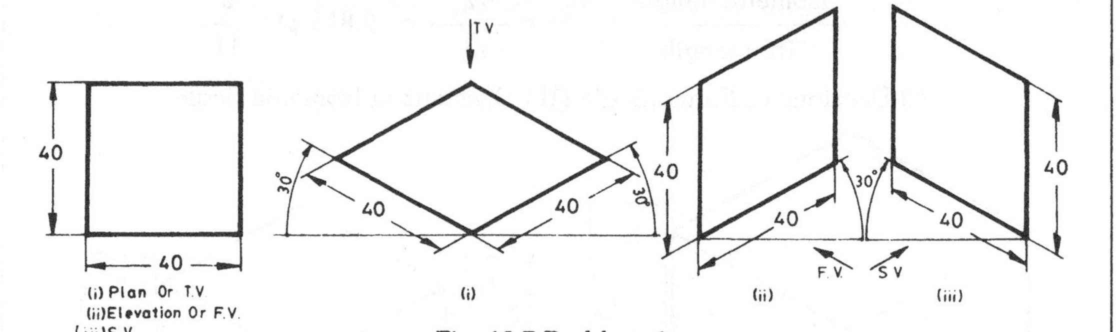

Fig. 18.7 Problem 1

(i) If the given view is plan, draw two lines along axes to form a horizontal plane. Here use two axes other than vertical. Vertical axis will be the direction of sight. Take 40 mm on two lines already drawn and complete the rhombus by drawing parallel lines to axes as shown in Fig. 18.7 (i).

(ii) If the given view is elevation, draw two lines along axes to form a vertical plane. Here use vertical axis and one suitable axis, remaining third axis will be the direction of sight. Take 40 mm on two lines already drawn. Complete the rhombus by drawing parallel lines to the axes as shown in Fig. 18.7 (ii).

(iii) If the given view is a side view, draw two lines along axes to form a plane parallel to the profile plane. Use vertical axis and one suitable axis. Direction of sight for S. V. will be the third remaining axis. On both the axes, already drawn, take 40 mm and then draw parallel lines to get rhombus as shown in Fig. 18.7 (iii).

Problem 2 : The (i) F. V. or (ii) S. V. or (iii) T. V. of a rectangular plate having its surface parallel to (i) V, P. or (ii) P. P. or (iii) H. P. respectively is given in the Fig. 18.8 Draw its isometric projection.

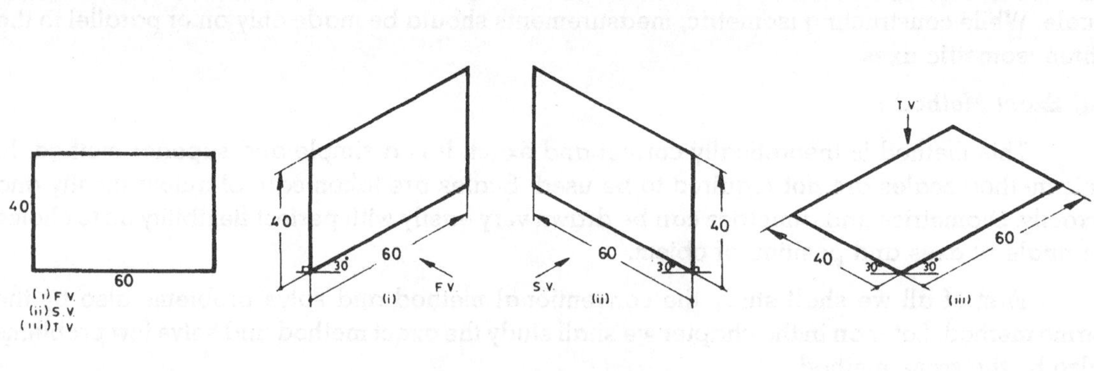

Fig. 18.8 Problem 2

Here take a set of two suitable axes according to the given view and take 60 mm and 40 mm (Iso-lengths) on the two axes and draw parallelograms to get isometric projections. See Fig. 18.8 (i), (ii) & (iii).

Here as the projection is required, reduce the dimensions 60 mm and 40 mm by the isometric scale.

Problem 3 : The (i) Elevation or (ii) Plan of a pentagonal plate having its surface parallel to (i) V. P. or (ii) H.P. respectively is given. Draw its isometric view.

As the edges 12, 34, 45 and 51 are non isometric lines, inscribe the pentagon 12345 in a rectangle PQRS as shown in Fig. 18.9 Here PQ, QR, RS and SP are isometric lines.

(i) Draw the isometric view of a rectangle PQRS on a vertical plane as studied in problem 2.

Now to plot point 1, take dimension Y_1 on a divider from Q on the line QP and plot accordingly in isometric view on iso-line QP. Similarly, by taking co-ordinates X_2,(y_4 , x_4) and x_5, points 2, 4 and 5 respectively can be plotted. Point 3 coincides with R and hence mark it. Join all points 1, 2, 3, 4 and 5 to get isometric view of the pentagonal plate 12345 as shown in Fig. 18.9 (i)

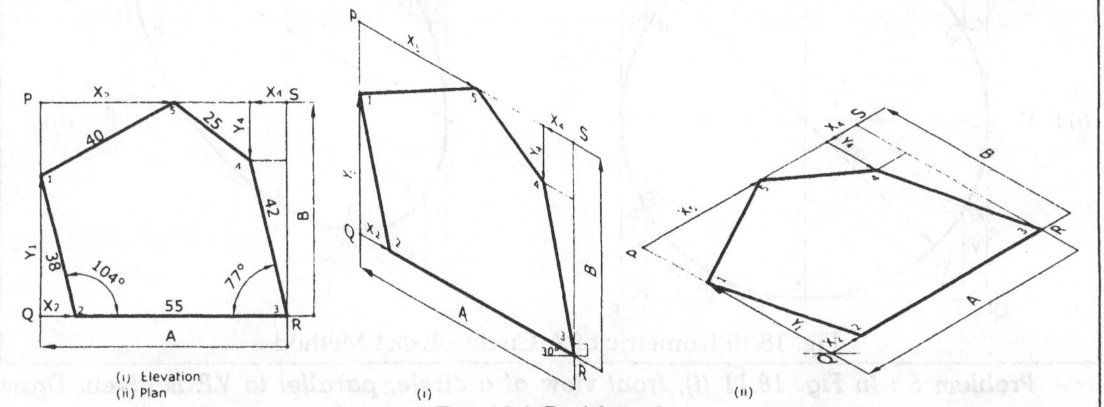

Fig. 18.9 Problem 3

(ii) Here isometric view of the rectangle PQRS should be drawn on a horizontal plane instead of vertical as shown in Fig. 18.9 (ii). Other procedure remains the same.

Students should note that during construction of the above isometric views (i) and (ii) we have not used anywhere given dimensions 42, 25, 40 and 38 of non isometric lines 34, 45 51 and 12 respectively, directly or indirectly. Angles given 77° and 104° are also not used. This is done because non-isometric lines reduce by indefinite proportion. Also the angles do not increase or decrease by any fixed proportion.

Whenever, non-isometric lines are required to be drawn in isometric projections or views, their end points are taken on isometric lines and first the end points are plotted and then joined.

Problem 4 : In Fig. 18.10 (i), front view of a circle of Ø 100 mm, parallel to V.P. is given. Draw isometric view of the circle. Use the exact method.

Inscribe the circle in a square PQRS. Divide the circle into 12 equal parts and mark points 1, 2,.....12 as shown in Fig. 18.10(i). From points 2, 3, 5, 6, 8, 9, 11 and 12 draw horizontal and vertical lines and find out co-ordinates (x_2, y_2), (x_3, y_3), (x_5, y_5), (x_6, y_6), (x_8, y_8), (x_9, y_9), (x_{11}, y_{11}) and (x_{12}, y_{12}) as shown in figure. The points 1, 4, 7 and 10 will become the mid points of PQ, QR, RS and SP respectively.

Now draw rhombus PQRS, the isometric view of the square PQRS as shown in Fig. 18.10(ii). Now with the known co-ordinates (x_2y_2), for the point 2, plot the point 2 taking distances $x_2$and y_2 along isometric axes from Q as shown in the Fig. 18.10(ii). Similarly, find

points 3, 5, 6, 8, 9, 11 and 12. Find also the points 1, 4, 7 and 10 by marking the mid points of PQ, QR, RS and SP respectively.

Join all points 1, 2, 3,.....12 and 1 by a smooth curve to get the isometric view of the circle. It is an ellipse.

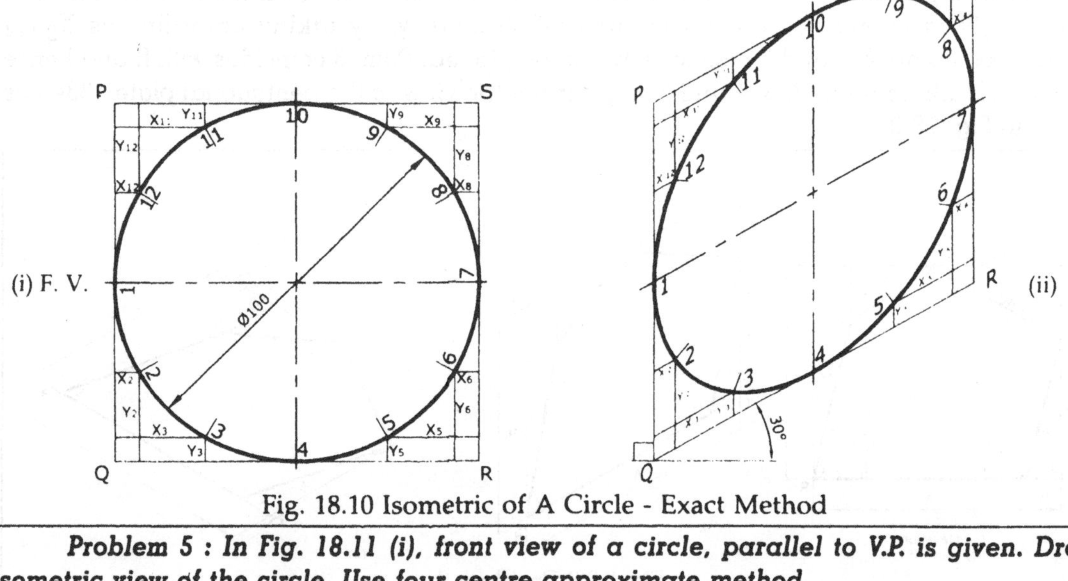

Fig. 18.10 Isometric of A Circle - Exact Method

Problem 5 : In Fig. 18.11 (i), front view of a circle, parallel to V.P. is given. Draw isometric view of the circle. Use four centre approximate method.

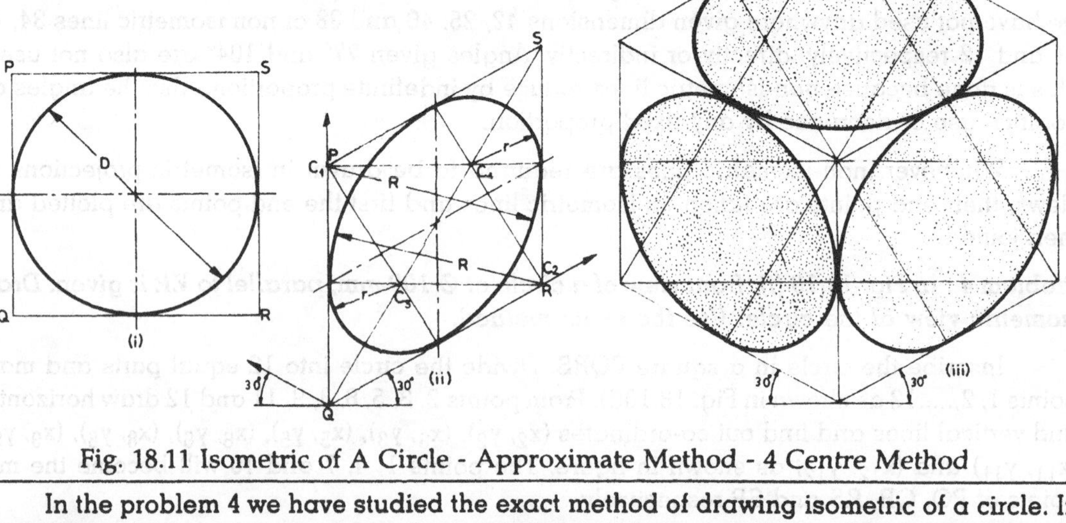

Fig. 18.11 Isometric of A Circle - Approximate Method - 4 Centre Method

In the problem 4 we have studied the exact method of drawing isometric of a circle. In the exact method, student is required to join 12 points by a smooth free hand curve. To join 12 points by a perfectly smooth curve is a difficult task. Eventhough it is an exact method, it is rarely used.

Ellipse, the isometric projection/view of a circle, is constructed by a combination of 4 arcs of circles drawn with 4 different centres. The method of construction is explained as under. The method shown is for the case when the given circle is the F. V.

1. First inscribe the given circle in the square PQRS as shown in Fig. 18.11 (i).
2. Draw the rhombus PQRS, the isometric of the square PQRS. See Fig. 18.11 (ii).
3. Join the bigger diagonal of the rhombus. Here join QS.
4. Mark two centres C_1 and C_2 at the points P and R, the end points of the smaller diagonal, respectively.
5. From P as well as from R draw two lines meeting at the mid points of the opposite sides of the rhombus.
6. Intersection of above two lines with the bigger diagonal gives us two more centres C_3 and C_4 on the bigger diagonal QS. See Fig. 18.11 (ii).
7. Now with C_1 and C_2 as the centres and the radius equal to (R) draw two arcs of a circle between the mid points of the opposite sides.
8. Similarly, with C_3 and C_4 as the centres and the radius equal to (r) draw two arcs of a circle meeting smoothly, the arcs already drawn, at the mid points.

Now in Fig. 18.11 (iii) three ellipses are shown, one for F.V. another for S.V. and the third for T.V.

To get the isometric of a circle, given in the plan, rhombus is constructed by two axes which are at 30° to the horizontal. For the given circle in F. V. and S. V. along with vertical axis another axis is suitably taken for the construction of the rhombus.

Problem 6 : In Fig. 18.12 (i) F. V. of a semi circle parallel to V. P. is given. Draw its isometric view.

Follow the procedure as given below :

1. Inscribe the semi circle into the rectangle PQRS. See Fig. 18.12 (i)
2. Draw the parallelogram PQRS with PQ on the vertical iso-axis and QR on the iso-axis at 30°. This parallelogram PQRS is an isometric of the rectangle PQRS. See Fig. 18.12 (ii).
3. Complete the circle and inscribe it in the square PQ_1R_1S. See Fig. 18.12(i).
4. Extend PQ by same amount upto Q_1 and SR by the same amount upto R_1 to get the rhombus PQ_1R_1S, the isometric for square PQ_1R_1S. See Fig. 18.12(ii).
5. Join the bigger diagonal Q_1 S. Join P with the mid-point R and get by intersection centre C_2 on Q_1 S. End point R_1 will be the centre C_1.
6. Now with centres as C_1 and C_2 and radii as R and r draw two arcs of circles to get isometric projection of a semi circle. See Fig. 18.12 (ii).

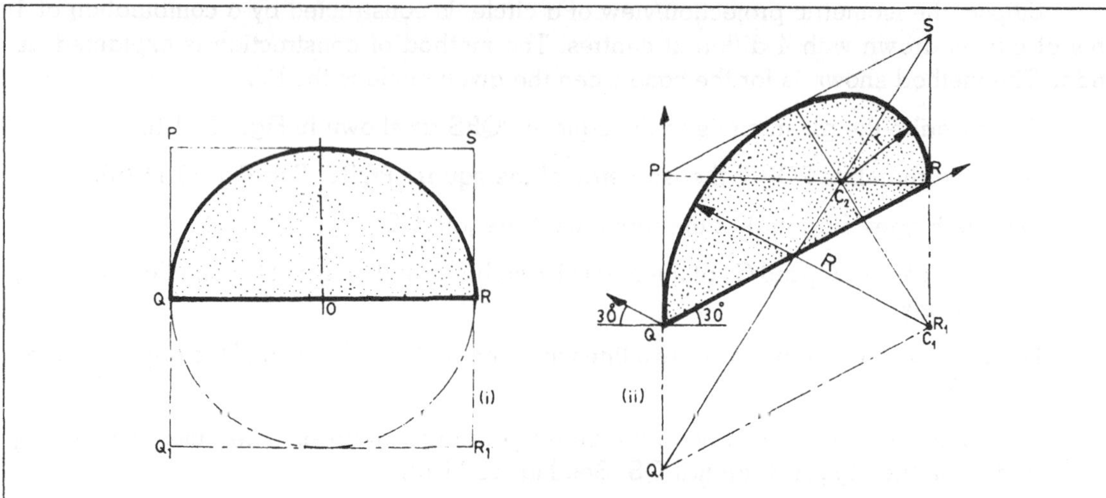

Fig. 18.12 Isometric of A Semi - Circle

Problem 7 : In Fig. 18.13 (i) Elevation and L. H. S. V., of a rectangular plate with top two corners curved are given. Draw the isometric view of a plate.

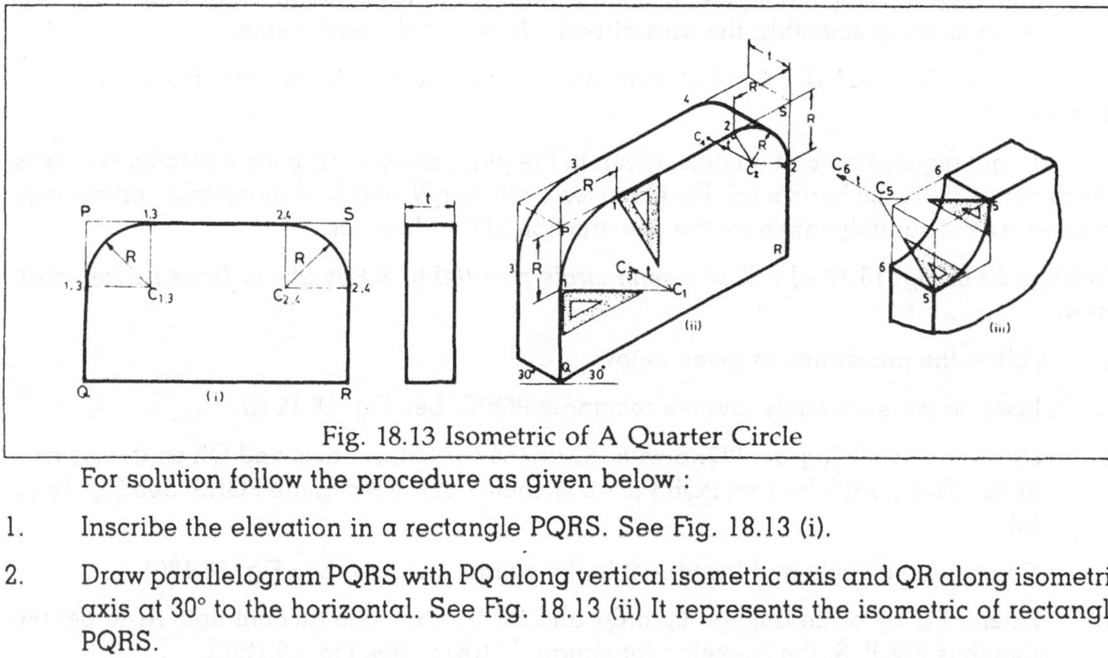

Fig. 18.13 Isometric of A Quarter Circle

For solution follow the procedure as given below :

1. Inscribe the elevation in a rectangle PQRS. See Fig. 18.13 (i).
2. Draw parallelogram PQRS with PQ along vertical isometric axis and QR along isometric axis at 30° to the horizontal. See Fig. 18.13 (ii) It represents the isometric of rectangle PQRS.
3. Along PS and PQ take distances R (the radius of curvature R of the plate) and mark two points 1 as shown in Fig. 18.13 (ii).
4. At two points 1 draw perpendicular lines to PS and PQ intersecting at the point C_1. Two setsquares indicate perpendicularity in the drawing. See Fig. 18.13 (ii).
5. Now with C_1 as the centre and radius equal to $C_1$1 draw an arc touching PQ and PS smoothly at points 1.
6. At the corner S repeat the same construction and get the centre C_2. Now with C_2 as the centre and radius equal to $C_2$2 draw an arc touching SR and SP smoothly at points 2.

7. Draw thickness 't' of the plate along the third unutilised iso-axis at the points P, Q, R and S as shown in Fig. 18.13 (ii) and get isometric of a rectangular slab.

8. Either repeat the same construction in the rear plane of the plate or from C_1 and C_2 draw parallel lines to the third axis or lines in the direction of thickness and take on them measurement 't' to get centres C_3 and C_4 respectively.

9. With C_3 as centre draw arc 3-3 and with C_4 as the centre draw arc 4-4 using same radii $C_1 1$ and $C_2 2$ respectively.

10. Draw common tangent line, parallel to the third axis, showing the thickness, to two arcs already drawn with centres C_2 and C_4 to complete the isometric view. See Fig. 18.13 (ii).

11. If the radius is in the concave form draw as per Fig. (iii).

Problem 8 : In Fig. 18.14(i), F. V. of a "Gujarati Alphabet ટ" having irregular curve is given. Draw the isometric view of the alphabet.

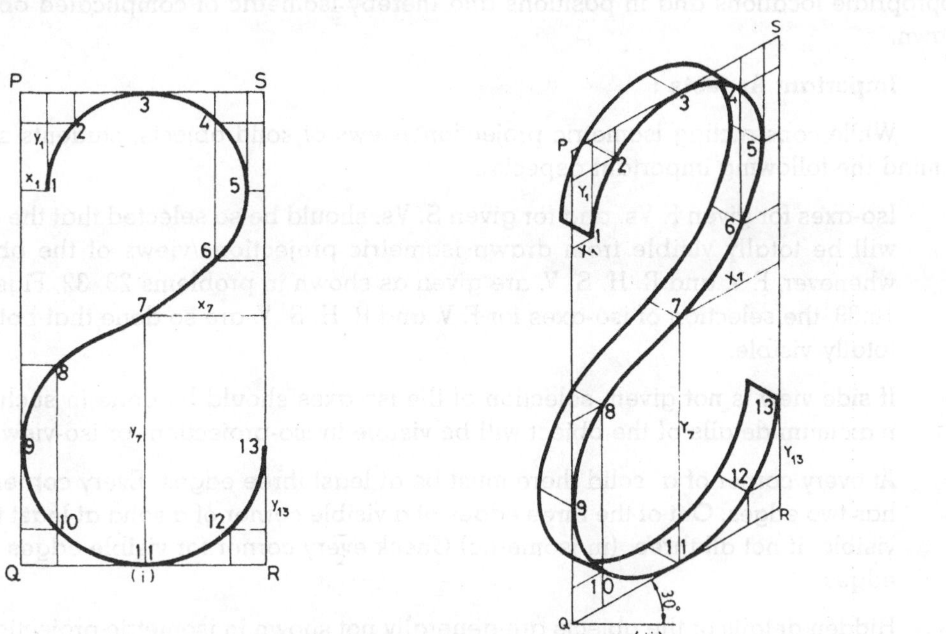

Fig. 18.14 Isometric of An Irregular Curve

For solution, follow the procedure as given below :

1. Inscribe the F. V. of alphabet in a rectangle PQRS.

2. Take points 1, 2,... 13 on the irregular curve of alphabet at nearly equal distances. From points. 1, 2,... 13 draw horizontal and vertical lines to get co-ordinates (x_1, y_1), (x_2, y_2),......$(x_{13}\ y_{13})$ respectively w. r. t. the corner points of the rectangle.

3. Now draw PQRS parallelogram with PQ on vertical iso-axis and QR on iso-axis at 30° to the horizontal.

4. Take x_1 and y_1 distances from reference point P along the direction of iso-axes to get point 1 as shown in Fig. 18.14 (ii). Similarly, find points 2, 3... 13 by taking (x_2, y_2), $(x_3,$

Y_3)....($X_{13}Y_{13}$) from corresponding reference corner points. Join points 1, 2, 3,... 13 by a smooth curve to get the isometric projection/view of the alphabet.

5. Draw equal thickness lines in the direction of 3rd axis to provide thickness to the alphabet as shown.

11. Isometric of Cylinders, Prisms, Pyramids, Cones, their Frustums and Truncated Frustums

Isometric projections/views of above solids are drawn mostly by box method. We shall study first the box method and then some important aspects relevant to the **Box Method**. In this method of isometric projection/view, solid object is assumed to be inside the box. The shape of the box, in which object is enclosed, is mostly square or rectangular prism. First, isometric of the square or rectangular prism is drawn and then by using isometric lines and iso-planes of the box, isometric projection/view of the object is constructed. By joing suitably the points, lines and planes thus obtained we can get isometric of the object.

This box-method can also be used for isometric projections/views of many complicated objects. Sometimes separate boxes for different portions of an object can be constructed at appropriate locations and in positions and thereby isometric of complicated object can be drawn.

Important Aspects :

While constructing isometric projections/views of solid objects, students should keep in mind the following important aspects :

1. Iso-axes for given F. Vs. and for given S. Vs. should be so selected that the given views will be totally visible from drawn isometric projections/views of the objects, e.g. whenever, F. V. and R. H. S. V. are given as shown in problems 29, 32, Figs. 18.36 and 18.39, the selection of iso-axes for F. V. and R. H. S. V. are so done that both views are totally visible.

   If side view is not given, selection of the iso-axes should be done in such a way that maximum details of the object will be visible in iso-projection or iso-view.

2. At every corner of a solid there must be at least three edges. Every corner of a plane has two edges. Out of the three edges of a visible corner of a solid at least two must be visible, if not all three. (in isometric) Check every corner for visible edges out of three edges.

3. Hidden details of the objects are generally not shown in isometric projections or views even with dashed or dotted lines. If hidden details of the objects are of such a nature that one cannot imagine them on the basis of visible portion then only those hidden details are shown by dotted lines.

4. In isometric, two edges (visible-full-continous) will never cross each other i.e. out of the two crossing edges one will be full and another will be dotted.

Now having studied the box method and important aspects for isometric projections of solids we shall solve few problems of cylinders, cones, prisms, pyramids, etc.

Problem 9 : In Fig. 18.15 (i) elevation and R. H. S. V. of an object are given. Draw the isometric view.

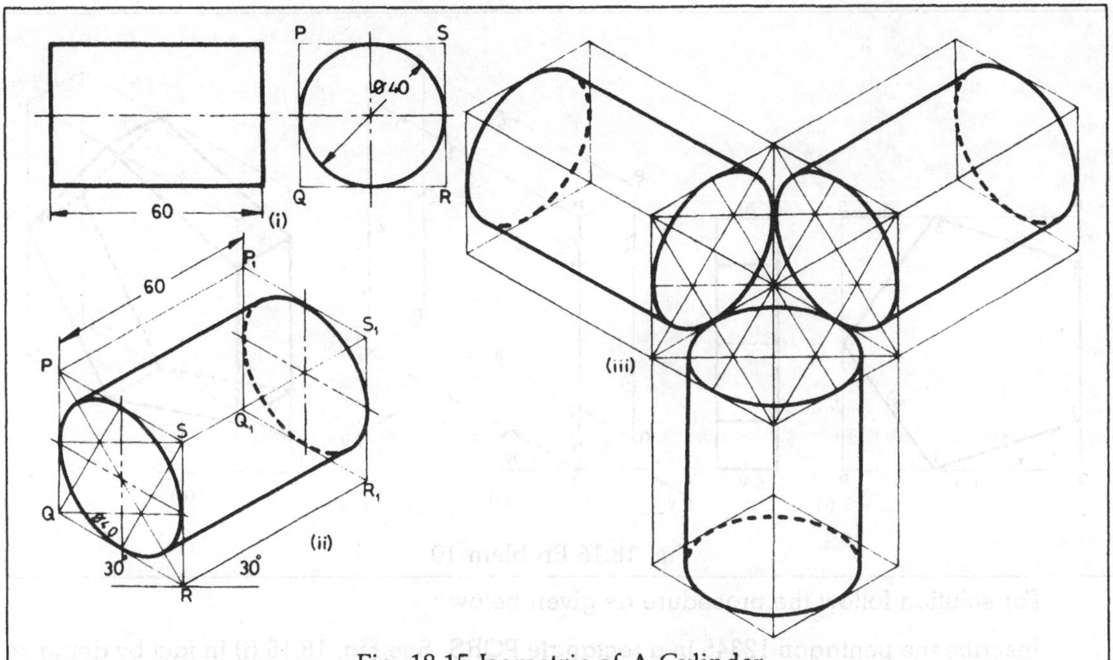

Fig. 18.15 Isometric of A Cylinder

For solution, follow the procedure as given below :

1. Inscribe the circle of elevation in a square PQRS. In fact by doing so we have enclosed the object (cylinder) in a square prism.

2. For elevation select vertical axis and left hand side 30° axis so that both F. V. and R. H.S.V. will be totally visible.

3. Draw rhombus PQRS with SR on vertical axis and QR on left hand side 30° axis. By four centre method construct the ellipse, the isometric of a circle.

4. On the third, right hand side 30°, axis take length available from side view and get P_1, Q_1, R_1 and S_1. Join them to get rhombus $P_1Q_1R_1S_1$ as shown in Fig. 18.15 (ii).

5. By four centre method construct the ellipse in rhombus $P_1Q_1R_1S_1$.

6. Join both the ellipses by two common tangent lines parallel to the third, right hand side 30°, axis.

7. Make visible lines of the object thicker and darker and retain all other construction lines thinner and lighter.

8. In Fig. 18.15 (iii) isometric drawings of cylinders in three different positions are shown. Depending upon the given views iso-axes are selected and correspondingly we can get isometric drawings of cylinders in three different positions as shown in Fig. 18.15 (iii).

Problem 10 : In Fig. 18.16 (i) F. V. and L. H. S. V. of an object are given. Draw the isometric drawing.

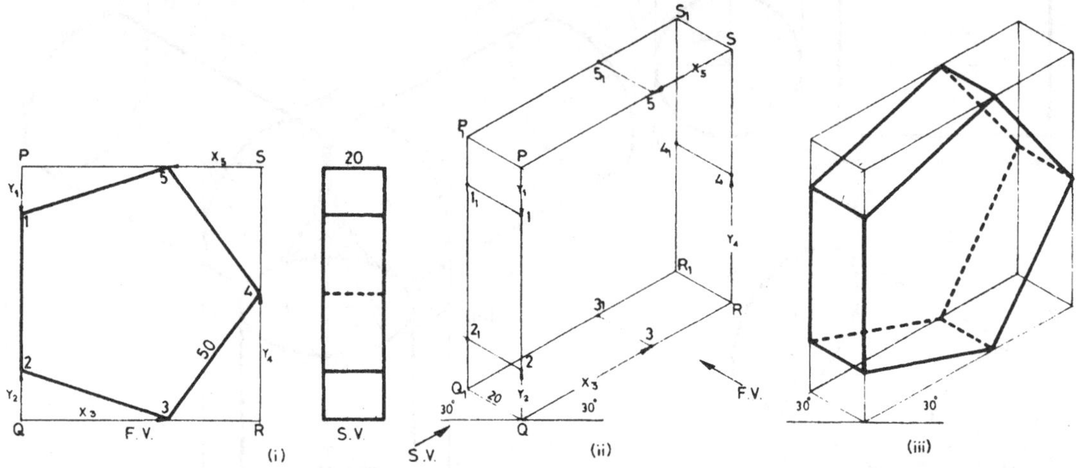

Fig. 18.16 Problem 10

For solution follow the procedure as given below :

1. Inscribe the pentagon 12345 in a rectangle PQRS. See Fig. 18.16 (i) In fact by doing so we have enclosed the object in a rectangular slab.
2. For F. V. select vertical axis and right hand side 30° axis so that both given F. V. and L. H. S. V. will be totally visible.
3. Draw parallelogram PQRS with PQ on vertical axis and QR on R.H. side 30° axis. By taking y_2 distance from F. V. and plotting it from Q on QP line in isometric drawing we get point 2 marked. Similarly, mark points 3, 4, 5, and 1 on lines QR, RS, SP and PQ by taking dimensions x_3, y_4, x_5 and y_1 respectively. Join them to get isometric drawing of a pentagon 12345 see Fig. (ii).
4. By taking thickness of rectangular slab from S. V. and plotting it along third iso-axis we get parallelogram $P_1Q_1R_1S_1$ as shown in Fig. 18.16 (ii).
5. Plot points 1_1, 2_1, 3_1, 4_1 and 5_1 by the same method or by drawing parallel lines to third axis equal to thickness from 1, 2, 3, 4, and 5 respectively.
6. Complete the figure as shown in Fig. 18.16 (iii).

Problem 11 : In Fig. 18.17(i) F. V. and T. V. of an object are given. Draw the isometric drawing

OR

Problem 11: Hexagonal prism is resting on H. P. on its base with two edges of the base perpendicular to V. P. Draw its isometric drawing.

For solution, follow the procedure as given below :

1. Inscribe the hexagon of plan in a rectangle PQRS. See Fig. 18.17 (i). In fact by doing so we have enclosed the hexagonal prism in a vertical rectangular prism.
2. First draw the iso-drawing of rectangular prism as shown in Fig. 18.17 (ii) and then on lines PQ, QR, RS and SP of bottom rectangular surface plot points (1, 2), (3), (4, 5) and (6) respectively by taking co-ordinates. On similar lines plot points (A, B), (C), (D E)

and (F) on top rectangular surface PQRS.

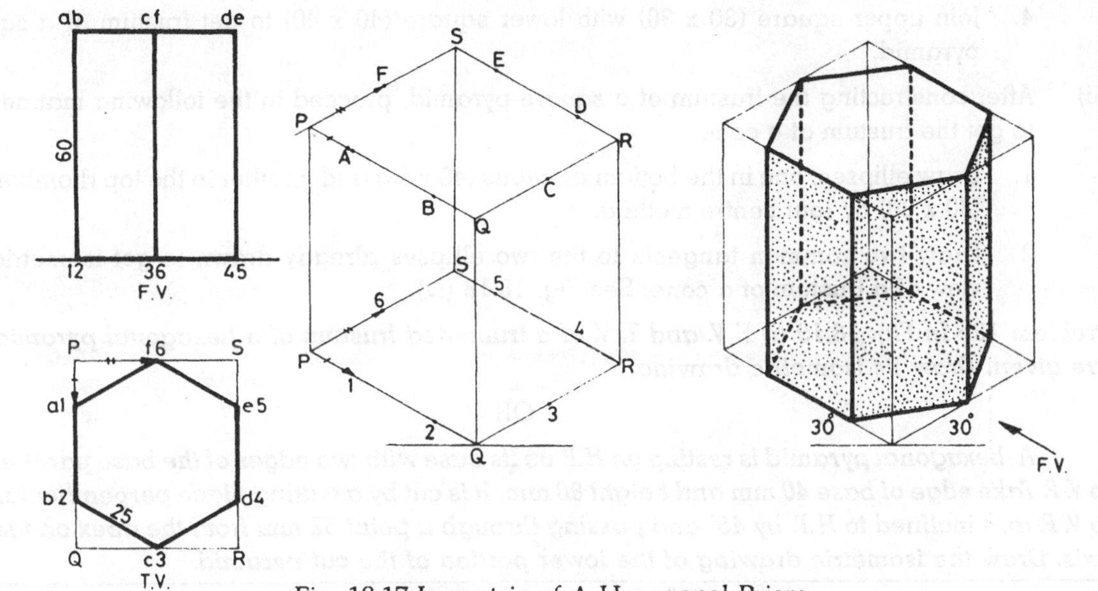

Fig. 18.17 Isometric of A Hexagonal Prism

3. Join these points. Visible lines should be drawn darker and thicker. Hidden lines may be avoided or drawn by medium thick dashed lines. Retain construction lines thinner and lighter. Final isometric is given in Fig. 18.17 (iii).

Problem 12 : In Fig. 18.18 (i) F. V. and T. V. of a frustum of a square pyramid are given in 3rd angle system. Draw the isometric view. (ii) In Fig. 18.18 (ii) F. V. and T. V. of a frustum of a cone are given in 3rd angle system. Draw the isometric view.

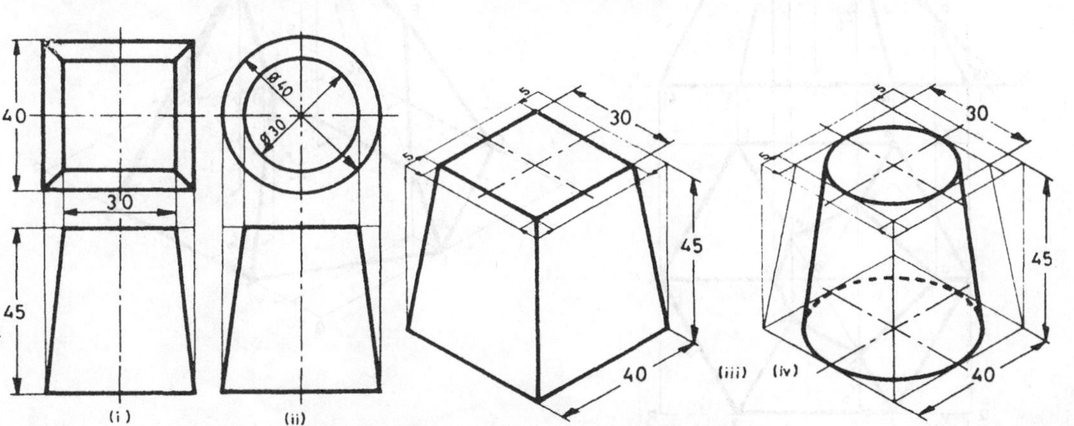

Fig. 18.18 Isometric of A Frustum

For solution follow the procedure as given below :

(i) 1. Inscribe F. V. in a rectangle. In fact by doing so, frustum of a sq-pyramid is enclosed in a sq-prism. See Fig. 18.18 (i).

2. First draw isometric view of a sq-prism with 40 mm edge of base and 45 mm height. See Fig. 18.18 (iii).

3. On the top sq-surface (40 x 40) of prism cut down 5 mm on all sides at corners to get square of 30 mm. See Fig. 18.18 (iii).
4. Join upper square (30 x 30) with lower square (40 x 40) to get frustum of a sq-pyramid.

(ii) After constructing the frustum of a square-pyramid, proceed in the following manner to get the frustum of a cone.

1. Draw ellipses, one in the bottom rhombus (40 x 40) and another in the top rhombus (30 x 30) by four centre method.
2. Draw two common tangents to the two ellipses already drawn to get isometric view of a frustum of a cone. See Fig. 18.18 (iv).

Problem 13 : In Fig. 18.19 (i) F. V. and T. V. of a truncated frustum of a hexagonal pyramid are given. Draw its isometric drawing.

OR

A hexagonal pyramid is resting on H.P. on its base with two edges of the base parallel to V. P. Take edge of base 40 mm and height 80 mm. It is cut by a cutting plane perpendicular to V. P. and inclined to H.P. by 45° and passing through a point 32 mm from the apex on the axis. Draw the isometric drawing of the lower portion of the cut pyramid.

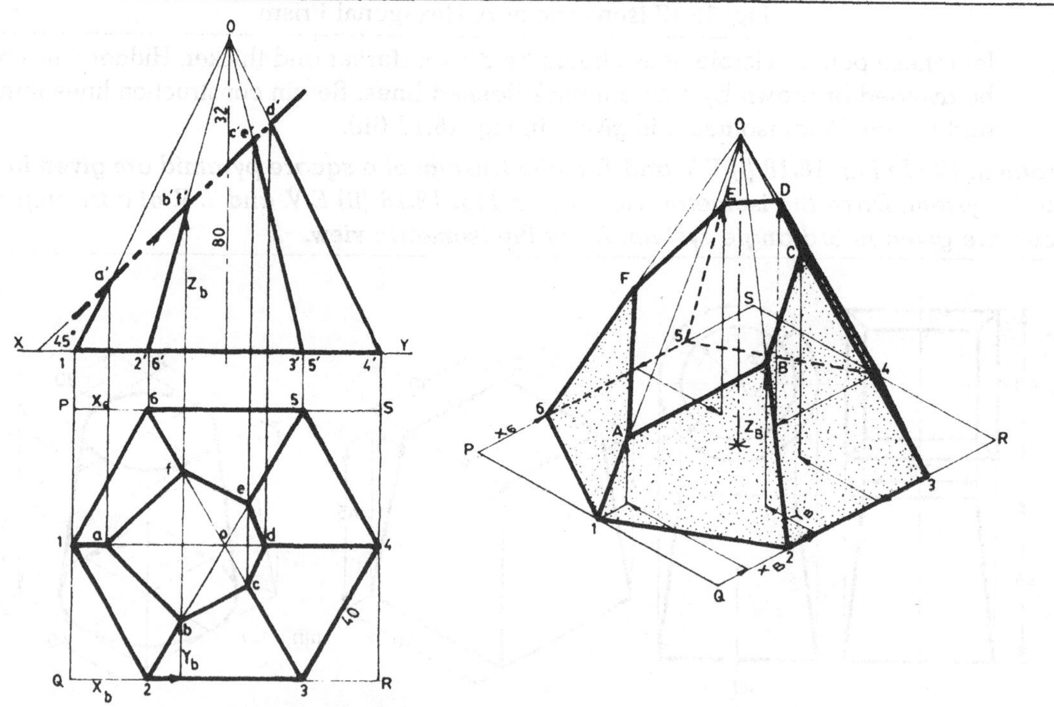

Fig. 18.19 Isometric of A Truncated Frustum of Hex-Pyramid

For solution see Fig. 18.19(ii) and follow the procedure as given below :

1. Inscribe hexagon 123456 in a rectangle PQRS.
2. Construct rhombus PQRS by taking two iso-axes other than vertical. Plot points 1, (2, 3), 4, (5 and 6) on lines PQ, QR, RS and SP respectively by taking co-ordinates. Similar to x_6 shown for point 6.

3. Now to plot points A, B, C, D, E and F, which are on an inclined section plane, take 3 co-ordinates for each point w.r.t. the known reference point.
4. For point B take reference point Q and find co-ordinates x_b, y_b (from plan) and x_b, z_b (from elevation) and plot them in isometric from the reference point Q in the direction of corresponding iso-axes as shown in Fig. 18.19(ii). Similarly, plot points C, D, E, F and A.
5. Join all points and get iso-drawing of the object.
6. If lines 1A, 2B,...6F are extended they will meet at O, the apex of the pyramid.
7. The other method of construction of truncated frustum is to first construct the full pyramid by plotting the point O by the use of co-ordinates (x_o, y_o, z_o) and then get point A, B,...F on lines 1O, 2O, 3O,......6O by taking only two co-ordinates (x_a, y_a), (x_b, y_b),....(x_f, y_f). Here heights of the vertical co-ordinates (z) are not required.

12. Isometric Projections/Views of A Sphere and Spherical Object

All types of orthographic projections of a sphere are circles of the true radius, i.e. F.V., T.V., S.V. Aux-Plan, Aux-Elev. Iso-projection, diametric projection and trimetric projection all are circles of the true radius.

In the chapter on section of solids, we have studied that a section of a sphere is a circle. If the section plane passes through the centre of the sphere then the section of the sphere will be circle of diameter equal to diameter of the sphere. We shall take three mutually perpendicular circles, as sections of a sphere by three section planes parallel to three principal planes and then we will plot three ellipses, the isometrics of above circles, in three mutually perpendicular principal planes. We shall find the touching point of the sphere with the surface on which it is resting and radius of the circle representing the isometric projection of the sphere.

For solution see Figs. 18.20(i) and (ii) and follow the procedure as given below :

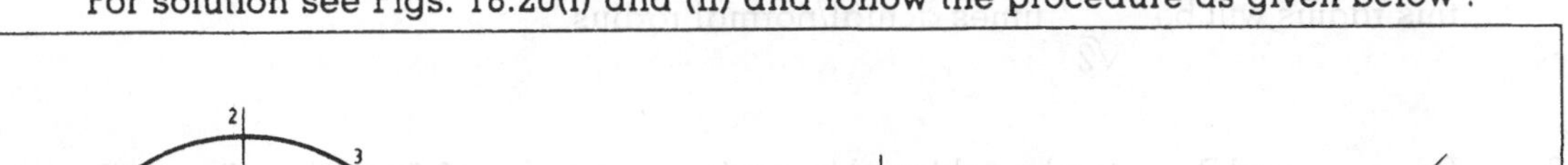

Fig. 18.20 Isometric of A Sphere

1. Imagine the sphere cut by following three section planes through centre :

   (1) Horizontal section plane-1 (H.P.)

   (2) Section plane perpendicular to H.P. and V.P.-2 (P.P.) and

   (3) Vertical section plane parallel to V.P.-3 (V.P.)

2. Horizontal section plane 1 will give in section of sphere a horizontal circle of diameter D of sphere.

   Isometric projection of that circle 1 is constructed in H.P. plane (rhombus D x D) by four centre method. See Fig. 18.20(ii).

3. The isometric projection of circle 2 of diameter D, the section of sphere by section plane-2 is constructed in P. P. plane (rhombus D x D) also by four centre method. See Fig. 18.20(ii).

4. Similarly, the isometric projection of circle-3 is constructed in V.P. plane (rhombus D x D), as shown in Fig. 18.20(ii).

5. Keep centres of all the rhombuses H.P; V.P. and P.P. at one point C which will eventally become the centre of the sphere in isometric.

6. Point of intersection P of two ellipses at the bottom most side will give touching point of sphere with the ground. Measured length CP will be equal to isometric length of the actual radius.

7. With centre C, draw a circle touching and enclosing all the three ellipses 1-2-3 drawn in mutually perpendicular planes. Note that radius of this circle will be equal to the actual radius of the sphere i.e. R_{normal}. In case of isometric view/drawing of sphere this radius will be $\frac{\sqrt{3}}{\sqrt{2}}$ times actual/normal radius.

8. $R_{isometric}$ and R_{view} can be achieved by scale shown in Fig 18.20 (iii)from known R_{normal}.

Important Note :

In construction of **Isometric Projection** of object containing sphere or spherical portion all dimensions are taken after reducing it by iso-scale except the radius of the circle, representing isometric of sphere or spherical portion. This radius of the circle is taken true or normal.

In construction of **Isometric Drawing or View** of object containing sphere or spherical portion all dimensions are taken true except the radius of the circle, representing isometric of sphere or spherical portion. This radius of the circle is taken after increasing it by iso-scale. Radius is increased by scale by factor $\frac{\sqrt{3}}{\sqrt{2}}$ R_{normal} is increased to R_{view} by scale as shown in Fig. 18.20(iii).

Problem 15 : In Fig. 18.21(i) F.V. of a sphere resting centrally on the top surface of a circular plate is given. Draw its (1) Isometric projection (2) Isometric drawing.

For solution see Fig. 18.21 and follow the procedure as given below :

1. Construct first the isometric of a square prism of (55 x 55 x 25h) size. Take all dimensions

from iso-scale i.e. reduced.

2. Draw two ellipses in two rhombuses by four centre method and draw two common tangents to these two ellipses to complete the isometric projection of a circular plate.
3. By drawing two centre lines parallel to two iso-axes on the top surface of the prism find the centre P.
4. From P draw a vertical line and take on it 25 mm from iso-scale i.e. (25_{ISO}) reduced value of 25 mm ($25 \times \frac{\sqrt{2}}{\sqrt{3}}$ mm), to mark the centre point C of sphere.
5. With C as the centre and 25 mm (actual or normal) radius draw a circle to get isometric projection of a sphere. Note that it will not pass through P, but P will remain inside the circle.
6. Similarly draw iso-drawing with all actual dimensions. For sphere circle take radius equal to 25 increased. To get 25 increased use scale as shown in Fig. (v).

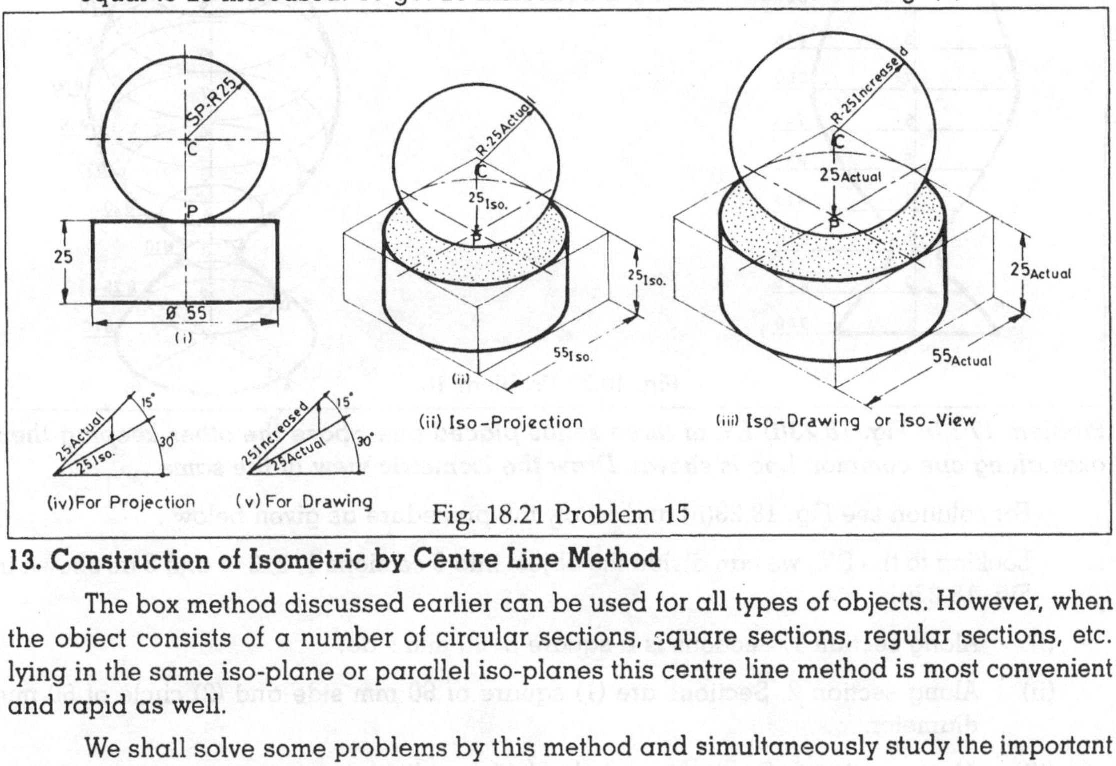

Fig. 18.21 Problem 15

13. Construction of Isometric by Centre Line Method

The box method discussed earlier can be used for all types of objects. However, when the object consists of a number of circular sections, square sections, regular sections, etc. lying in the same iso-plane or parallel iso-planes this centre line method is most convenient and rapid as well.

We shall solve some problems by this method and simultaneously study the important aspects of this method.

Problem 16 : Draw the isometric of a flower pot of 110 mm height, having circular cross-sections of 40, 25, 10, 20, 35, 45, 50, 45, 40, 30, 20 and 25 mm diameters, from bottom most plane to top most plane at an interval of 10 mm height. F.V. of the flower pot is given in Fig. No. 18.22(i)

For solution see Fig. 18.22 and follow the procedure as given below :

1. Draw vertical line 1-12 of 110 mm height and mark on it points 2, 3,...11 at equal intervals of 10 mm each.
2. Draw two centre lines parallel to two iso-axes passing through point. 1. Take (D/2) 20 mm on both the sides of the point 1 on two axes and mark points A_1, B_1, C_1 and D_1. See Fig. 18.22.

3. Through these points A_1, B_1, C_1 and D_1 draw parallel lines to iso-axes and construct rhombus. In this rhombus construct ellipse, representing isometric of a circle of 40 mm diameter. Rhombus is not shown in the figure.
4. Similarly, construct ellipses to represent isometrics of circles around points 2, 3,....12 of 25, 10,....25 mm diameters respectively.
5. Now draw two smooth curves thicker and darker tangent to all 12 ellipses already drawn. Upper most ellipse should be made thicker and darker. Visible portion of the lower most ellipse should be made thicker and darker.
6. Handles of problems 39 and 43 are drawn by this method.

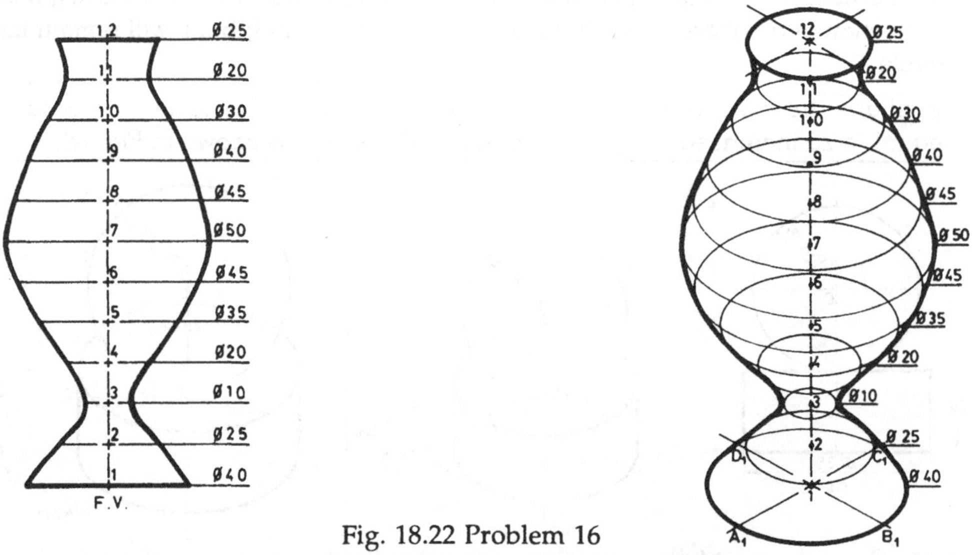

Fig. 18.22 Problem 16

Problem 17 : In Fig. 18.23(i) F.V. of three solids placed one above the other keeping their axes along one common line is shown. Draw the isometric view of the same.

For solution see Fig. 18.23(ii) and follow the procedure as given below :

1. Looking to the F.V., we can divide the object into 5 sections 1, 2, 3, 4 and 5 as shown in Fig. 18.23(i).
   (i) Along section 1, Sections is a square of 86 mm side.
   (ii) Along section 2, Sections are (1) square of 60 mm side and (2) circle of 60 mm diameter.
   (iii) Along section 3, Section is a circle of 40 mm diameter.
   (iv) At section 4, Centre of sphere is situated
   (v) Along section 5, Section is a circle of 35 mm diameter.
2. Draw vertical line and mark on it centres C_1, C_2, C_3, C_4 and C_5 at the intervals of heights 40, 50, 20 and 10 mm taken successively from bottom towards top side.
3. At section 1 i.e. around centre C_1 prepare a rhombus of 86 mm side. At section 2 i.e. around centre C_2 prepare a rhombus of 60 mm side and isometric of circle of 60 mm diameter. At section 3 prepare isometric of circle of 40 mm diameter and at section 5

prepare isometric of circle of 35 mm diameter. With centre C_4 draw a circle with $R_{increased}$ radius, (i.e. 20x11/9 mm radius) which will be touching smoothly to the ellipse of circle of 35 mm diameter. Join various sections by lines or tangent lines, depending upon object, to complete the isometric view.

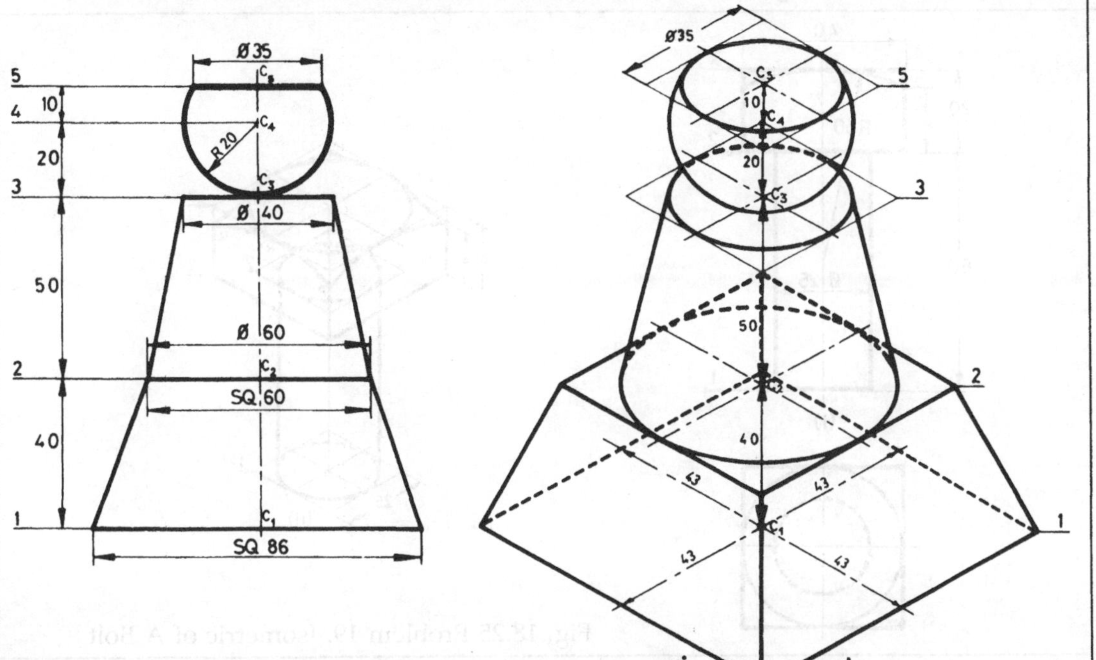

Fig. 18.23 Problem 17

Problem 18 : In Fig. 18.24(i) F.V. and T.V. of a Paper Weight are given. Draw its isometric drawing.

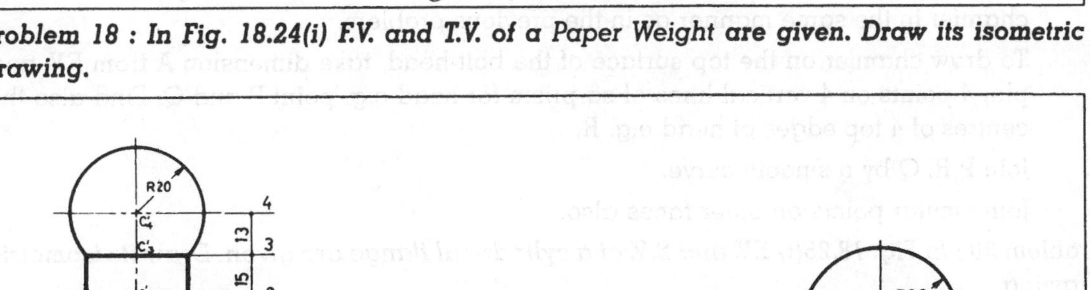

Fig. 18.24 Problem 18

Solve this problem in the same manner as problem 17. Fig. 18.24(ii) is a skeleton of construction leading to solution shown in Fig. 18.24(iii).

Problem 19 : In Fig. 18.25(i) F.V. and T.V. of a Square Headed Chamfered Bolt are given. Draw its isometric drawing.

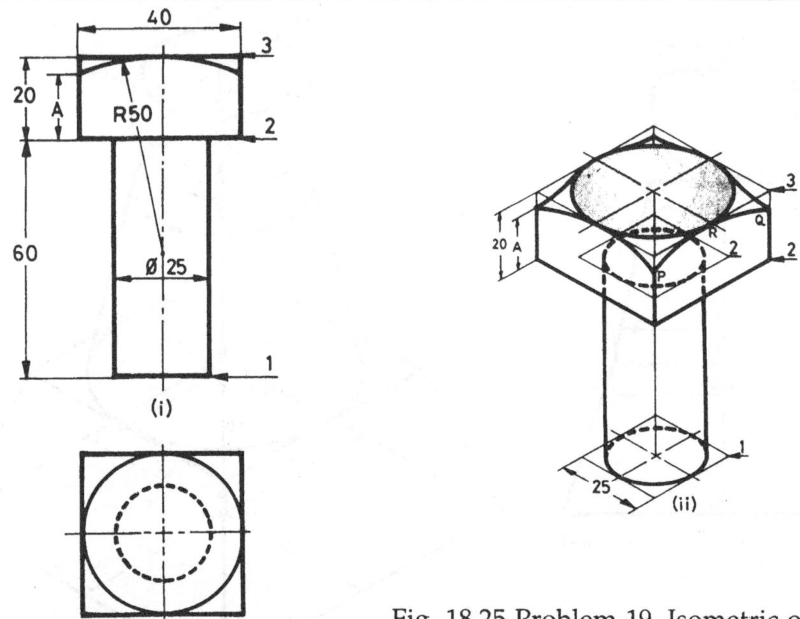

Fig. 18.25 Problem 19. Isometric of A Bolt

For solution see Fig. 18.25(ii) and follow the procedure as given below :

1. Divide the bolt in three sections and complete the isometric drawing of bolt without chamfer in the same manner as in the previous problem.
2. To draw chamfer on the top surface of the bolt-head, take dimension A from F.V. and plot 4 points on 4 vertical lines of sq-prism for head e.g. point P and Q. Find also the centres of 4 top edges of head e.g. R.
3. Join P, R, Q by a smooth curve.

   Join similar points on other faces also.

Problem 20 : In Fig. 18.26(i) F.V. and S.V. of a cylindrical flange are given. Draw its isometric drawing.

Skeleton work leading to solution is shown in Fig. 18.26 (ii)

In Fig. 18.26(ii) isometric layout of the principal centre lines is drawn. In the rhombuses shown construct ellipses by four centre method. Final solution is given in Fig. 18.26(iii).

Problem 21 : In Fig. 18.27(i) F.V. or T.V. or S.V. of a three way pipe flange is given. Draw its isometric view.

For solution see Figs. 18.27(i) and (ii) and follow the procedure as given below :

1. Take any point O and draw three lines along three iso-axes. Find points (x_1, x_2), (y_1,y_2) and (z_1, z_2) aling three different axes at distances of (40, 15) respectively and successively from O.
2. Construct isometrics of 60 diameter circles around centres X_1, Y_1 and Z_1 in planes perpendicular to corresponding axis.

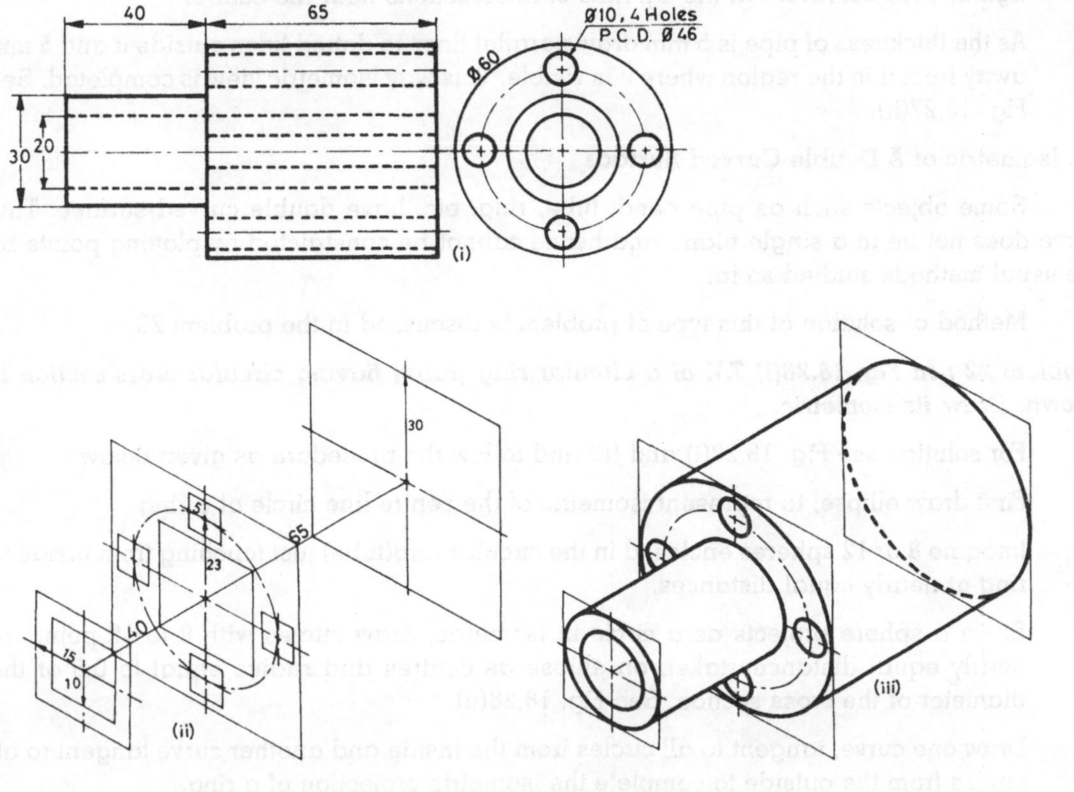

Fig. 18.26 Problem 20. Centre Line Method

3. Construct isometrics of 60 diameter and 30 diameter circles, around centres X_2, Y_2 and Z_2 in planes perpendicular to the corresponding axis.

Fig. 18.27 Problem 21. Centre Line Method

4. Draw dotted lines parallel to iso-axes tangent to corresponding ellipses shown in the figure. Give curvature at the 3 points of intersections near the centre.

5. As the thickness of pipe is 5 mm draw parallel lines to dotted lines outside it and 5 mm away from it in the region where it is visible. This way isometric view is completed. See Fig. 18.27(ii).

14. Isometric of A Double Curved Surface

Some objects such as pipe bend, tube, ring, etc. have double curved surface. This curve does not lie in a single plane and hence cannot be constructed by plotting points by the usual methods studied so far.

Method of solution of this type of problem is discussed in the problem 22.

Problem 22 : In Fig. 18.28(i) T.V. of a circular ring (tube) having circular cross-section is shown. Draw its isometric.

For solution see Fig. 18.28(i) and (ii) and follow the procedure as given below :

1. First draw ellipse, to represent isometric of the centre line circle of a ring.

2. Imagine 8 or 12 spheres enclosed in the circular ring(tube) just touching from inside to ring at nearly equal distances.

3. Since a sphere projects as a circle in isometric, draw circles with 8 to 12 points at nearly equal distances taken on ellipse as centres and radius equal to 0.5 of the diameter of the cross section. See Fig. 18.28(ii).

4. Draw one curve, tangent to all circles from the inside and another curve tangent to all circles from the outside to complete the isometric projection of a ring.

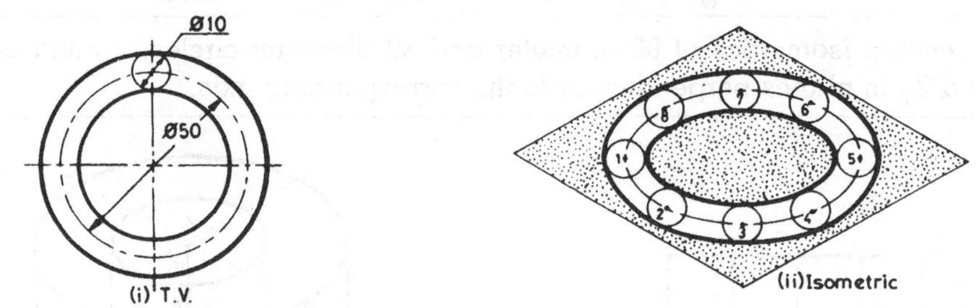

Fig. 18.28 Isometric of A Double Curved Surface

15. Construction of A Helicoid in Isometric

Problem 23 : In Fig. 18.29(i) F.V. and S.V. of a helical blade attached to a shaft on its circumference are given. Draw its isometric drawing.

For solution see Fig. 18.29(i) and (ii) and follow the procedure as given below. In isometric 1.5 times the given length is shown.

1. Draw isometric of the shaft (i.e. of cylinder)

2. Along centre line of the shaft mark points 1, 12, 11,...1, 12, 11,....7 as shown in the Fig. 18.29(ii) by transferring distances from F.V.

3. At the extreme right hand side i.e. around point 1 construct rhombus PQRS and 2

ellipses inside it to represent the isometrics of two circles of side view.

4. For dividing the two ellipses into 12 parts along circumferences mark 12 points on sides of rhombus, first by taking dimensions from S.V. along PQ, QR, RS and SP and then join them with centre point. 1. Mark points on both the ellipses. For keeping clarity of drawing only points 4, 5,....10 are marked.

5. Now from points 12, 11, 10,....1, 12,....7 of centre line draw parallel and equal length lines to 1-12, 1-11, 1-10,....1-1, 1-12,....1-7 lines obtained by construction at the extreme right. See Fig. 18.29(ii) and note how two points are marked on line 8-8.

6. Join obtained points 1, 12, 11,....1, 12,....7 by a smooth curve to get set of two curves one on the circumference of the shaft and another of the edge of helical blade.

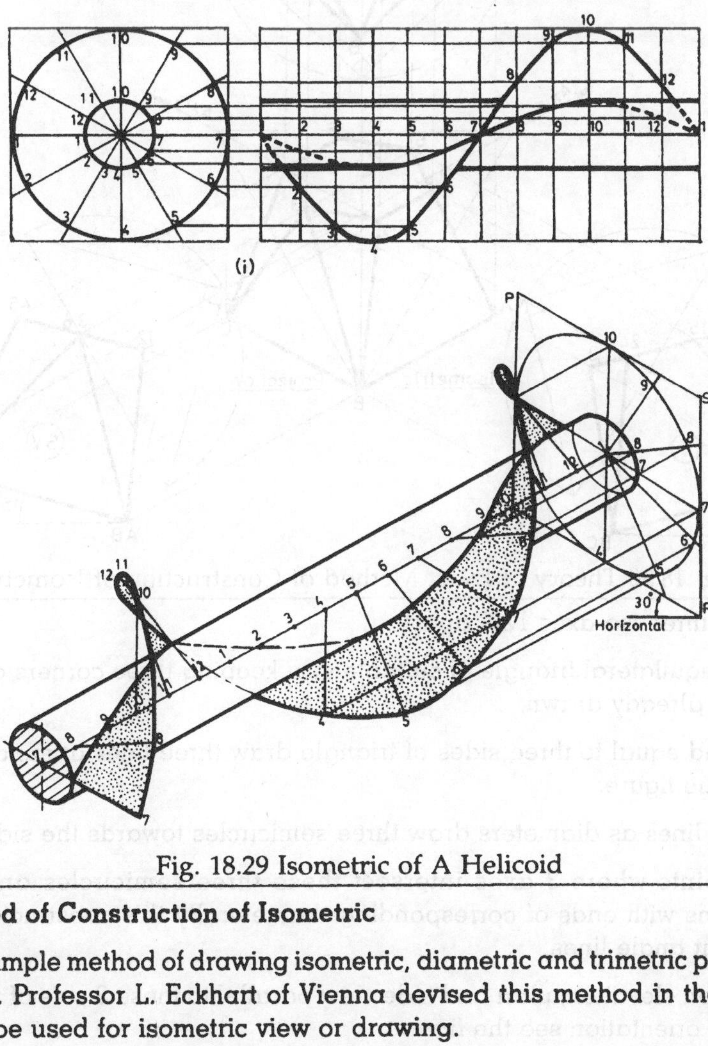

Fig. 18.29 Isometric of A Helicoid

16. Exact Method of Construction of Isometric

This is a simple method of drawing isometric, diametric and trimetric projections without the use of scale. Professor L. Eckhart of Vienna devised this method in the year 1937. This method cannot be used for isometric view or drawing.

Without going in details about the basic concepts of this method we shall study the application part of it and solve three problems.

Problem 24 : Elevation, Plan and S.V. of a lamp shade hexagonal at the top end and square

at the bottom end are given. Draw its isometric projection by the exact method.

For solution see Fig 18.30 and follow the procedure as given below :

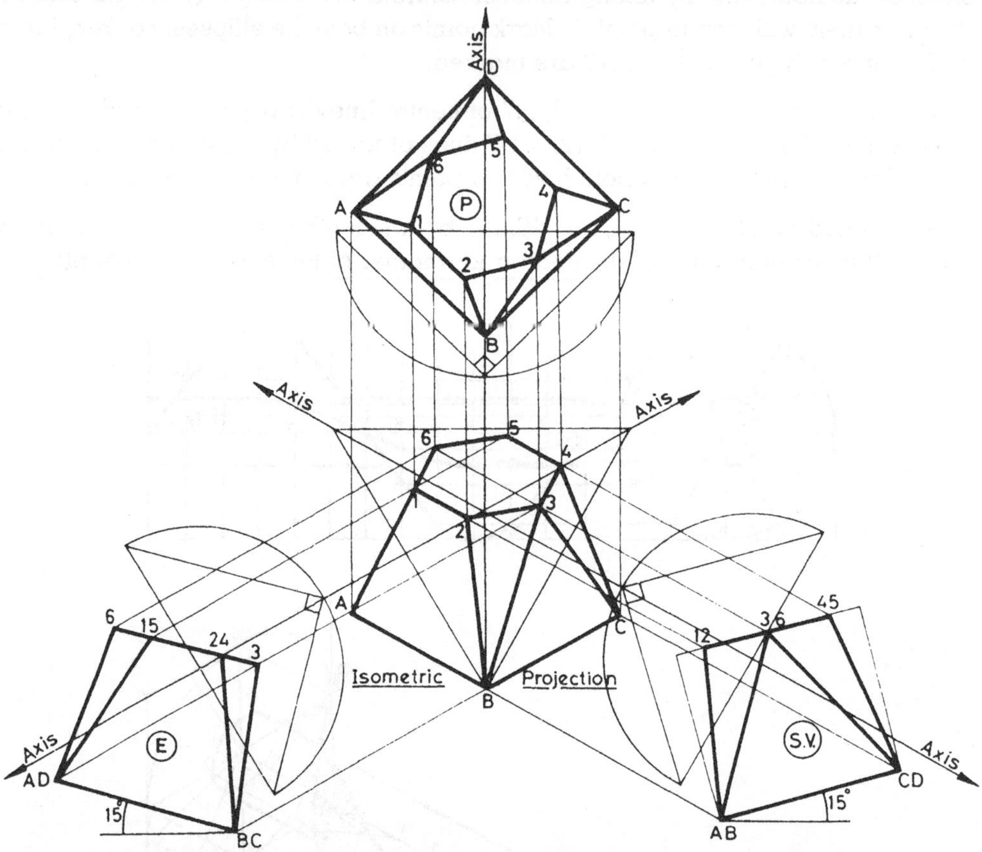

Fig. 18.30 Theory of Exact Method of Construction of Isometric

1. First draw three iso-axes 120° apart.
2. Construct equilateral triangle of suitable size keeping three corners of the triangle on three axes already drawn.
3. Parallel and equal to three sides of triangle draw three lines at suitable distance, as shown in the figure.
4. With these lines as diameters draw three semicircles towards the sides of traingle.
5. Mark 3 points where 3 axes intersect these three semicircles and join points of intersections with ends of corresponding diameter. By this construction we will get 3 sets of right angle lines.
6. Locate given elevation, plan and side view parallel to these 3 sets of right angle lines. For correct orientation see the figure.
7. Note that the base of elevation and side view will become inclined by 15° with the horizontal and plan will become equally inclined with the horizontal in the orientation.
8. From every corners of three views draw lines parallel to respective axis to get by

intersection isometric projection in the centre, see Fig. 18.30.

By the same method, if diametric projection is required then out of the three angles between axes, take two angles equal and draw iso-scales triangle instead of equilateral. Follow other procedure in the same manner.

Similarly, if trimetric projection is required then take three different angles between axes and draw a triangle of unequal sides. Follow other procedure in the same manner.

Problem 25 : Elevation and Plan of the hexagonal pyramid are given. Draw the isometric projection by the exact method.

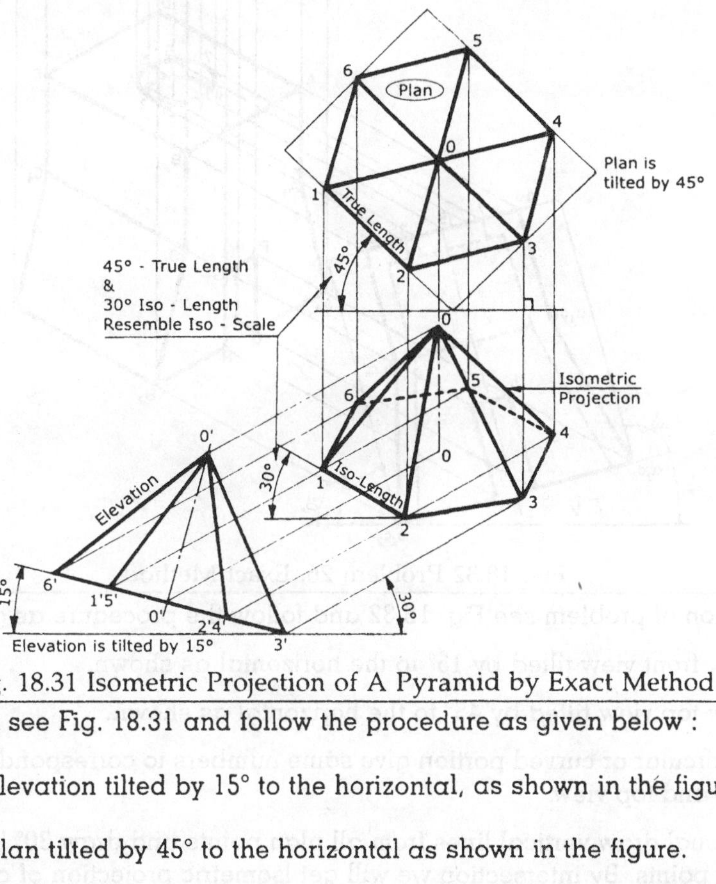

Fig. 18.31 Isometric Projection of A Pyramid by Exact Method

For solution see Fig. 18.31 and follow the procedure as given below :

1. Draw elevation tilted by 15° to the horizontal, as shown in the figure.
2. Draw plan tilted by 45° to the horizontal as shown in the figure.
3. From plan points, draw vertical lines and from elevation points draw lines at 30° to the horizontal. Intersection of the corresponding lines will generate the corresponding points in isometric projection. Join them and complete isometric projection.

Observe orientation of lines or edges and note that iso-scale is taken care of automatically.

Problem 26 : Front view and Top view of an object are given. Draw the isometric projection of an object by the exact method.

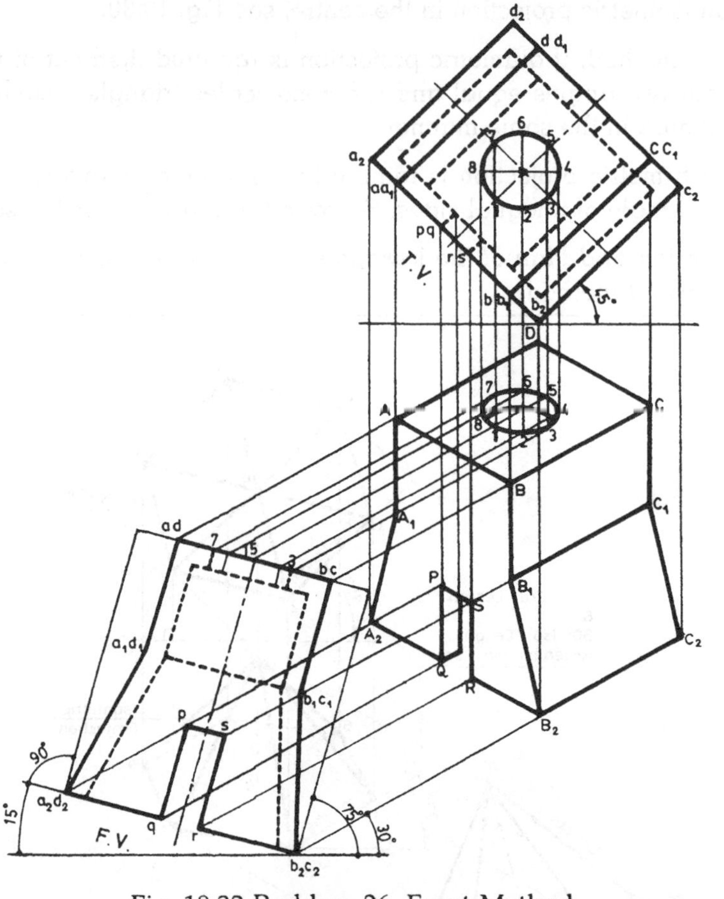

Fig. 18.32 Problem 26. Exact Method

For solution of problem see Fig. 18.32 and follow the procedure as given below :

1. Draw front view tilted by 15° to the horizontal as shown.
2. Draw top view tilted by 45° to the horizontal as shown.
3. For circular or curved portion give same numbers to corresponding points in front view and top view.
4. As usual draw vertical lines from all plan points and draw 30° lines from all front view points. By intersection we will get isometric projection of an object.

This method is extremely simple. However, imagination of the object is necessary while joining points of intersections.

17 Dimensioning in Isometric

In dimensioning isometric usual rules, followed for dimensioning orthographic projections, should be followed. Two extension lines and one dimension line used for showing one particular dimension should be in the same isometric plane. Extension lines and dimension lines are drawn parallel to the isometric axes. Dimension figures are all written sloping parallel to the extension lines and dimension lines i.e. parallel to two of the isometric axes. As per usual rules dimensions should be placed to read horizontally or from the right.

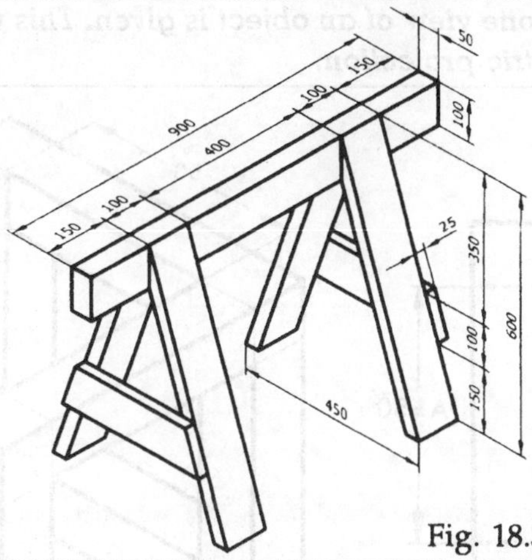

Fig. 18.33 Isometric Dimensioning

18. Practice Problems

After having a study of various methods of construction of isometric, we shall now take some typical problems for practice. Solutions of problems are so done, with some construction lines left intact, that they are self explanatory. Explanations about typical constructions are only given where found necessary. Directions of arrows represent move towards solution.

Students should cultivate the habit of imagining the three dimensional view of the object by co-relating points, lines, planes on the basis of common projectors. Try to co-relate this by putting footrule vertically and horizontally on F.V.-T.V. and on F.V.-S.V. respectively. Divide the object into many small pieces, imagine them and then arrange them in memory in their respective positions and locations to develope clear three dimensional picture in mind. Once the three dimensional picture is clear in mind, decide the method of construction and also prepare free hand sketch of it if you can. It is understood that when projection is asked reduced dimensions are taken from iso-scale except the radius for a sphere. Similarly, when view or drawing is asked dimensions are taken true or normal except the radius for a sphere. Here the radius of a sphere is increased by using iso-scale in reverse manner.

Always check every corner for visible lines out of three lines.

Problem 27 : In Fig. 18.34(i) F.V. and T.V. of a plate are given. Draw the isometric view.

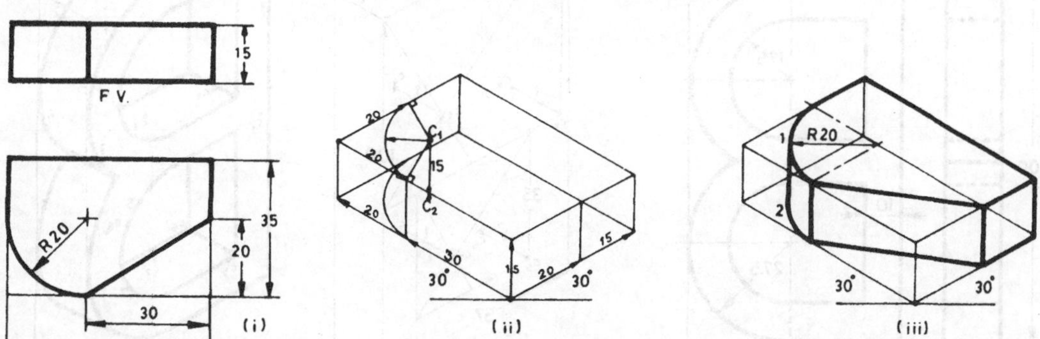

Fig. 18.34 Problem 27

For solution see Fig. 18.34 (ii) and (iii)

Problem 28 : In Fig. 18.35(i) one view of an object is given. This view represents F.V., T.V. as well as S.V. Draw the isometric projection.

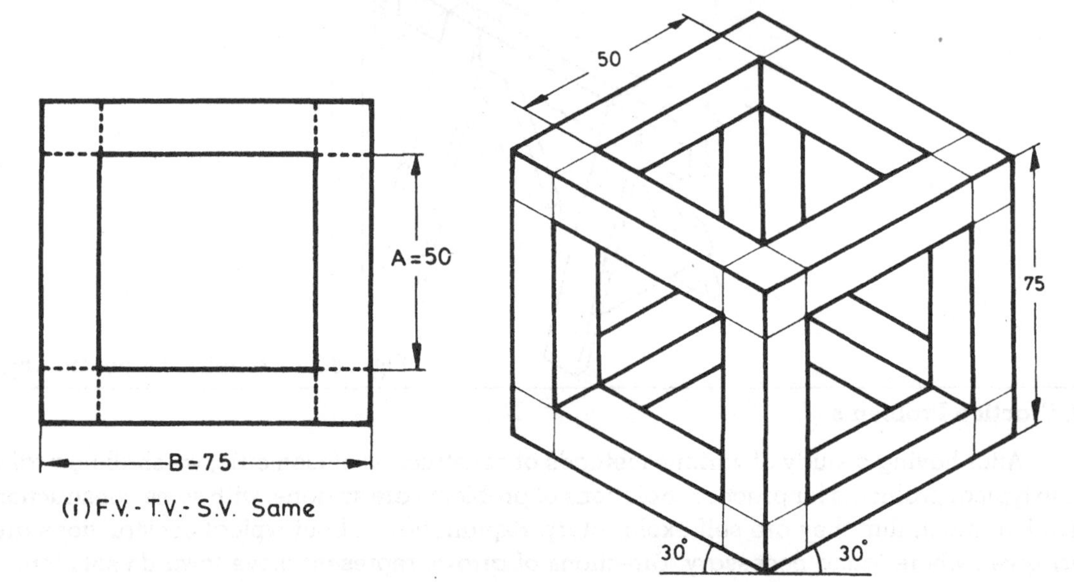

Fig. 18.35 Problem 28

For solution see Fig. 18.35 (ii). Construction lines are light.

Problem 29 : Draw the isometric drawing of the Alphabet 'B' from the given elevation and R.H.S.V. in Fig. 18.36.

For solution see Fig. 18.36(ii) and 18.36(iii).

After getting the centres in the front plane transfer them to rear plane by lines of thickness lengths along the 3rd axis.

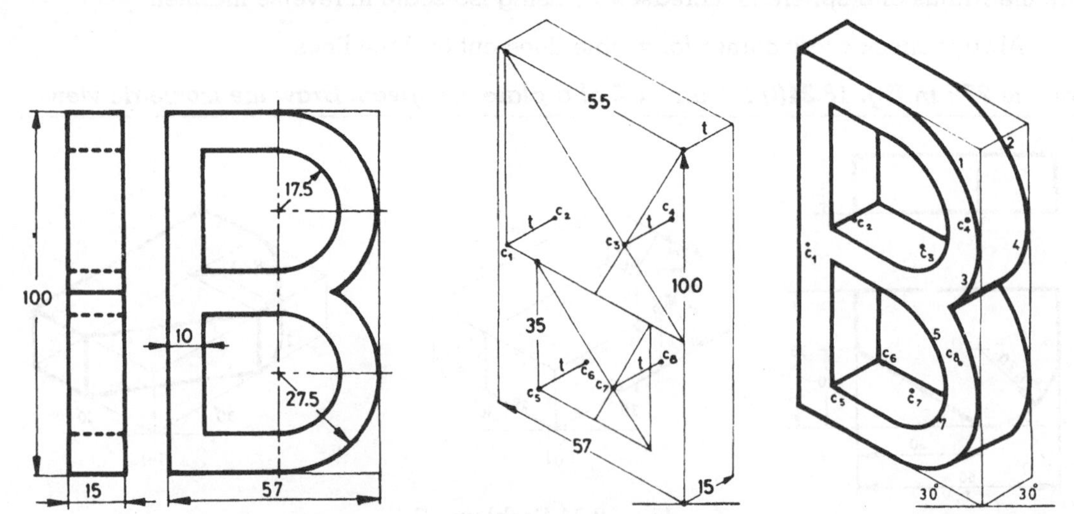

Fig. 18.36 Problem 29 Alphabet

Problem 30 : Draw the isometric of machined component from the given F.V. and T.V. in Fig. 18.37.

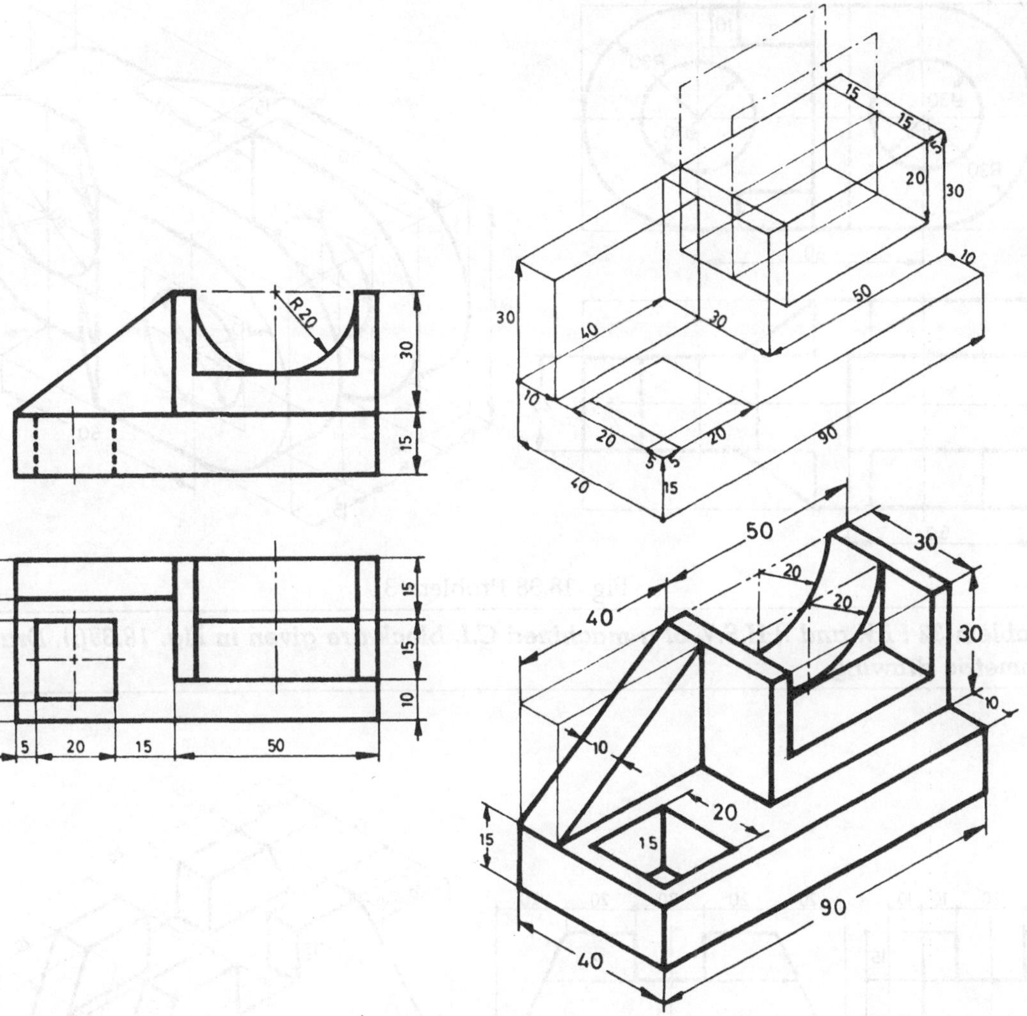

Fig. 18.37 Problem 30

For solution see skeleton of Fig. 18.37(ii) and see the final solution in Fig. 18.37(iii).

Do not forget to draw visible portion of two lines representing square for the hole at the lower surface.

Problem 31 : Draw the isometric projection of the object from the given elevation and plan in Fig. 18.38.

For solution see Fig. 18.38.

Do not forget to draw circular rear edge for the hole, inside the front edge(ellipse) for hole, when thickness is less in comparison to the size of the hole. In the two holes, edges are seen and in the third hole it is not seen.

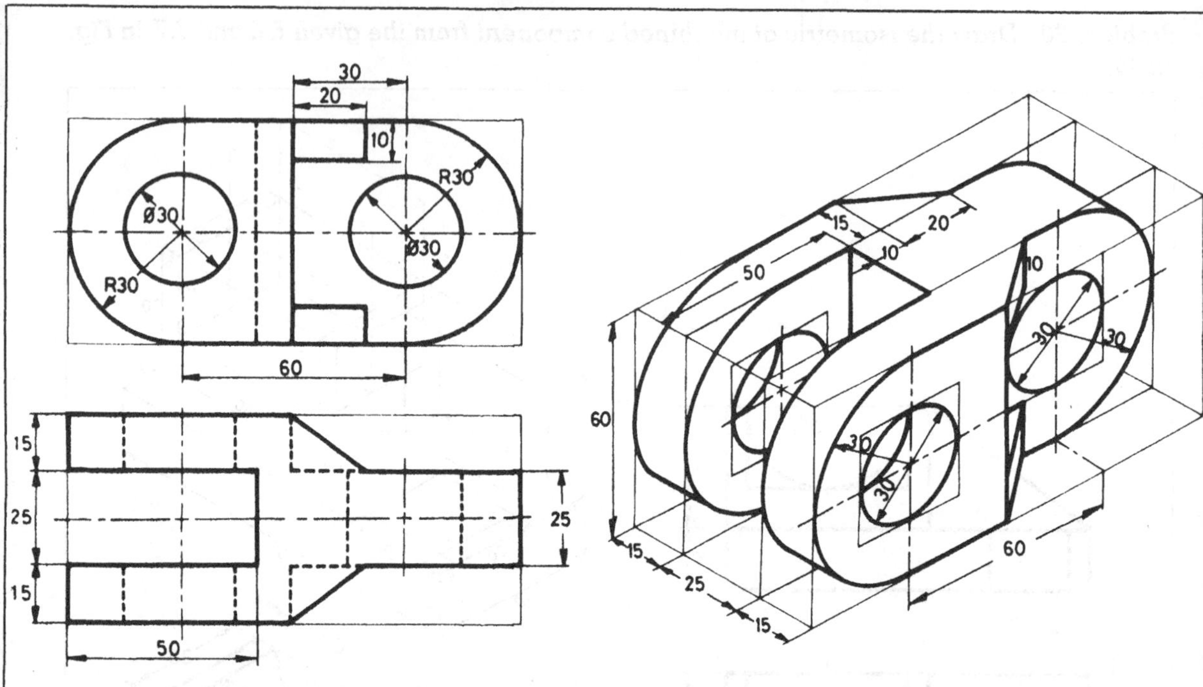

Fig. 18.38 Problem 31

Problem 32 : F.V. and R.H.S.V. of a machined C.I. block are given in Fig. 18.39(i). Draw its isometric drawing.

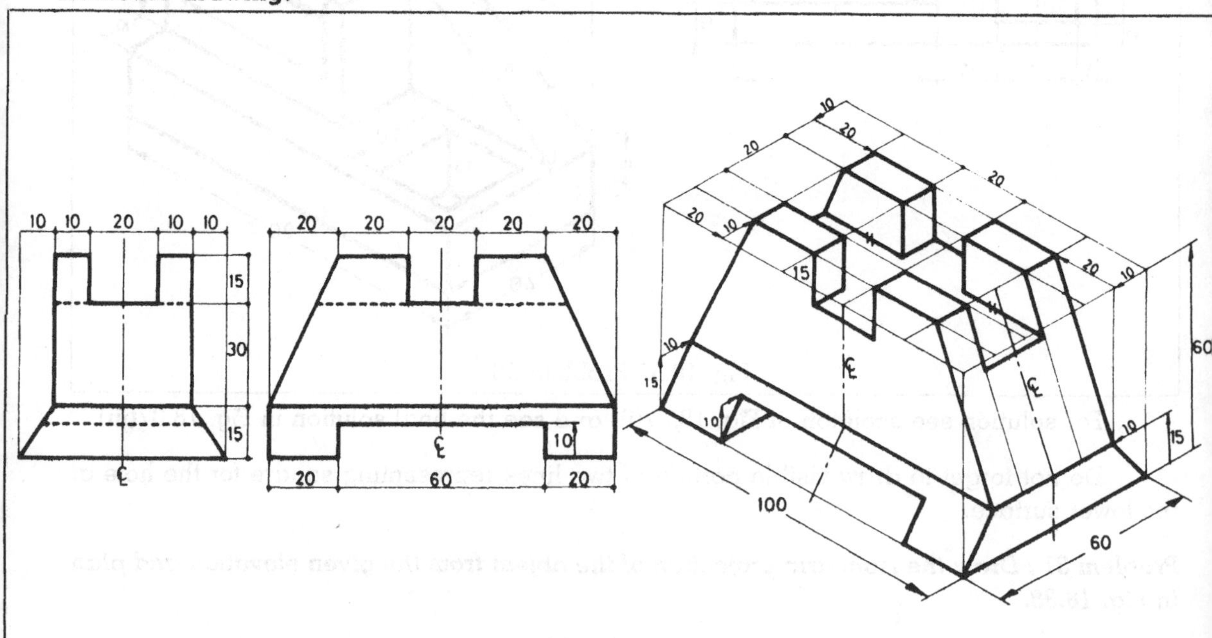

Fig. 18.39 Problem 32

For solution see Fig. 18.39(ii).

The lines for grooves on vertical surfaces are vertical and on sloping surfaces they are parallel to the centre line (C.L.). The height for groove is taken only along the vertical line.

Cutting of dimensions is shown by one sided arrow lines with dimensions on them.

Problem 33 : F.V., S.V. and T.V. of a dovetail grooved machined block are given in Fig. 18.40(i). Draw its isometric view.

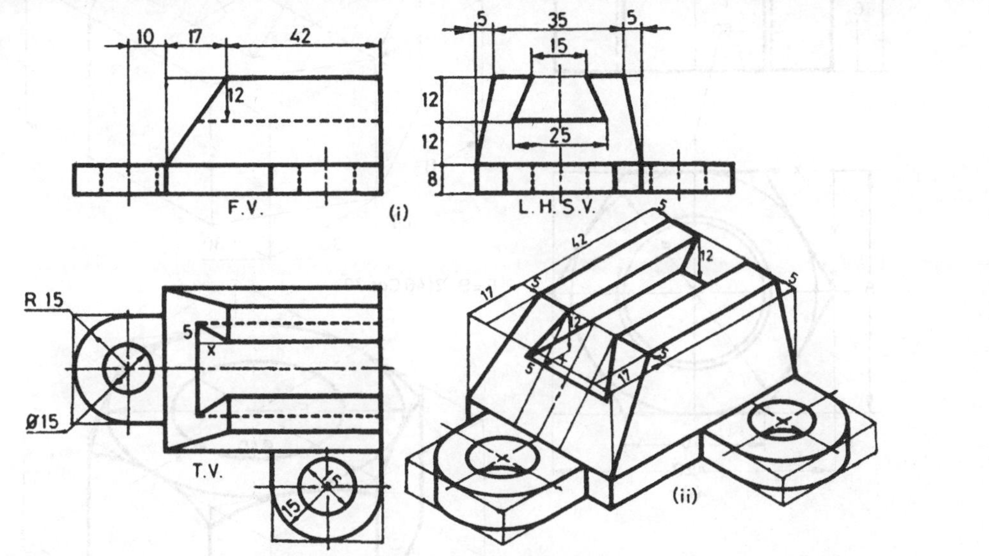

Fig. 18.40 Problem 33

For solution see Fig. 18.40(ii).

Here 3 separate boxes are constructed for 3 separate portions of the object at the appropriate location. For dovetailed groove co-ordinates 12, x and 5 are used.

Problem 34 : F.V. and T.V. of an object are give in Fig. 18.41 (i). Draw its isometric drawing.

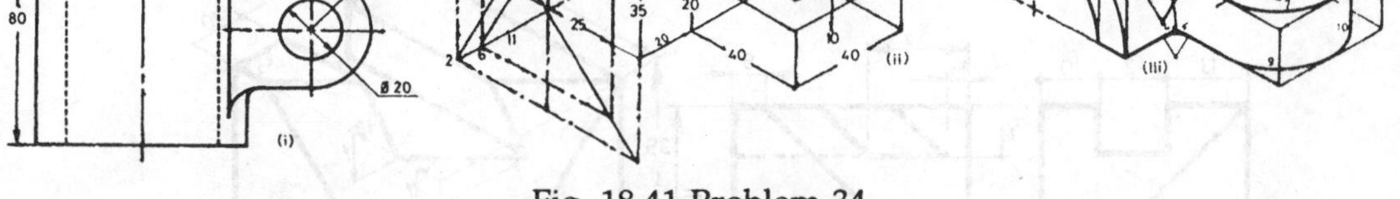

Fig. 18.41 Problem 34

For solution see Fig. 18.41(ii) and Fig 18.41(iii).

In this problem radius of the upper curve along which the upper surface is meeting the curved surface is given as 10 mm. Radius of the meeting curve at the lower suface is obtained by the construction as shown in Fig. 18.41 (iii). Fig. 18.41(ii) is a skeleton leading to solution.

Problem 35 : Draw the isometric view of a chamfered hexagonal nut for a bolt of diameter D. Avoid isometric of threads.

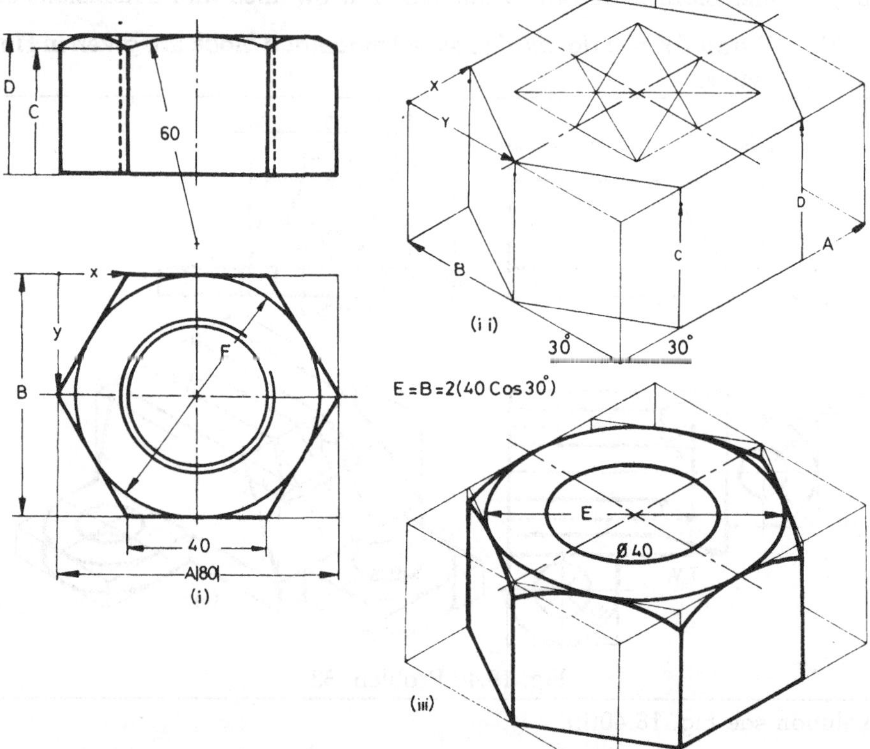

Fig. 18.42 Problem 35

For solution see Fig. 18.42(ii) and Fig. 18.42(iii)

First prepare F.V. and T.V. of a nut with standard proportions as shown in the figure. For isometric of chamfering use the dimension C and the midpoints of six upper edges of hexagonal prism. Fig. 18.42(ii) is a skeleton leading to the solution as given in Fig. 18.42(iii).

Problem 36 : In Fig. 18.43, F.V. and L.H.S.V. of a slotted truncated block are given. Draw the isometric drawing.

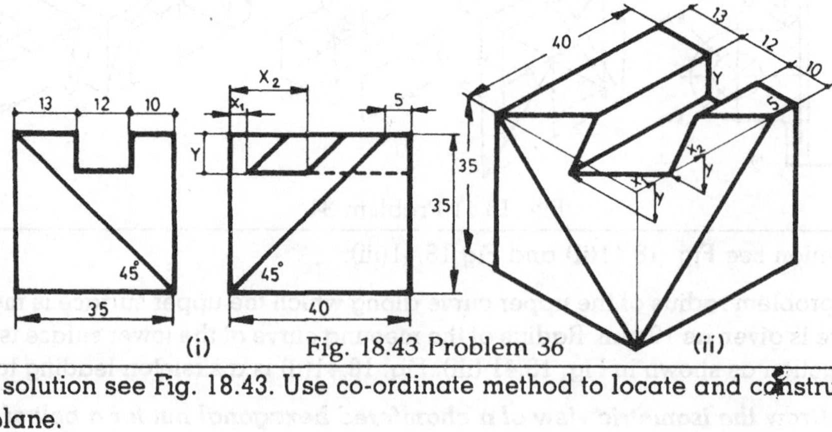

Fig. 18.43 Problem 36

For solution see Fig. 18.43. Use co-ordinate method to locate and construct slot on an-inclined plane.

Problem 37 : In Fig. 18.44, F.V. and T.V. of an object are given Draw the isometric drawing.

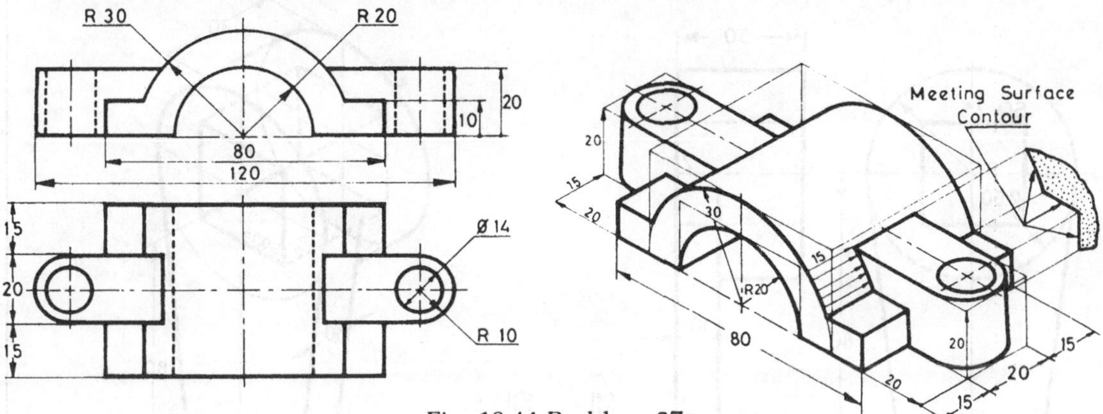

Fig. 18.44 Problem 37

For solution see Fig. 18.44.

Single sided arrows with dimensions on them represent the move towards solution. Arrows are also drawn on the outline of the object even-though theoretically it is incorrect. Try to understand meeting surface contour. Do not forget to draw visible small portion of the object.

Problem 38 : Draw the isometric view of an object from the given F.V. and T.V. of the object.

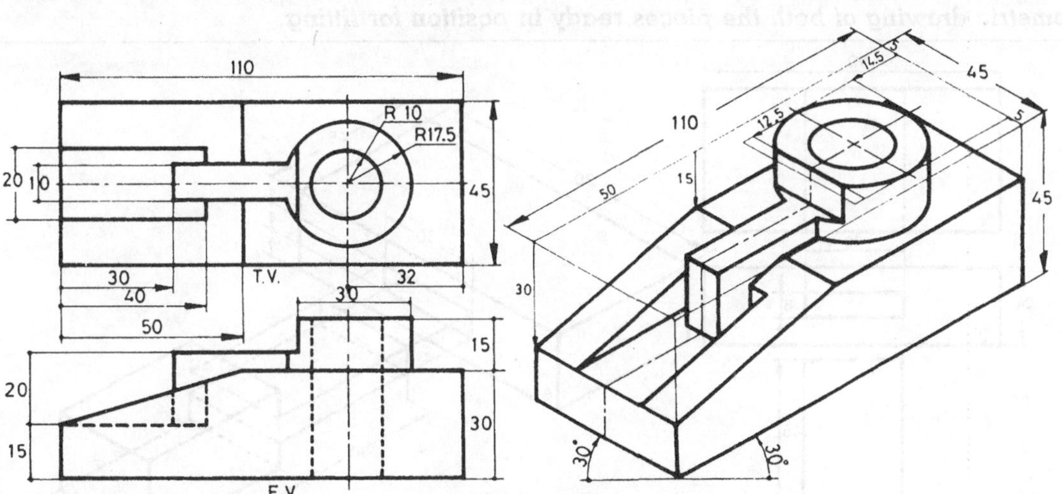

Fig. 18.45 Problem 38

For solution see Fig. 18.45.

This is an important problem from the point of view of imagining the object by co-relating details along projectors. Imagine the object piece by piece and then combine them.

Problem 39 : F.V. and L.H. side view of cranking handle are given in Fig. 18.46. Draw its isometric drawing.

For solution see Fig. 18.46.

Isometric of handle is constructed on the basis of centre line method. Dotted lines are drawn to make drawing more clear.

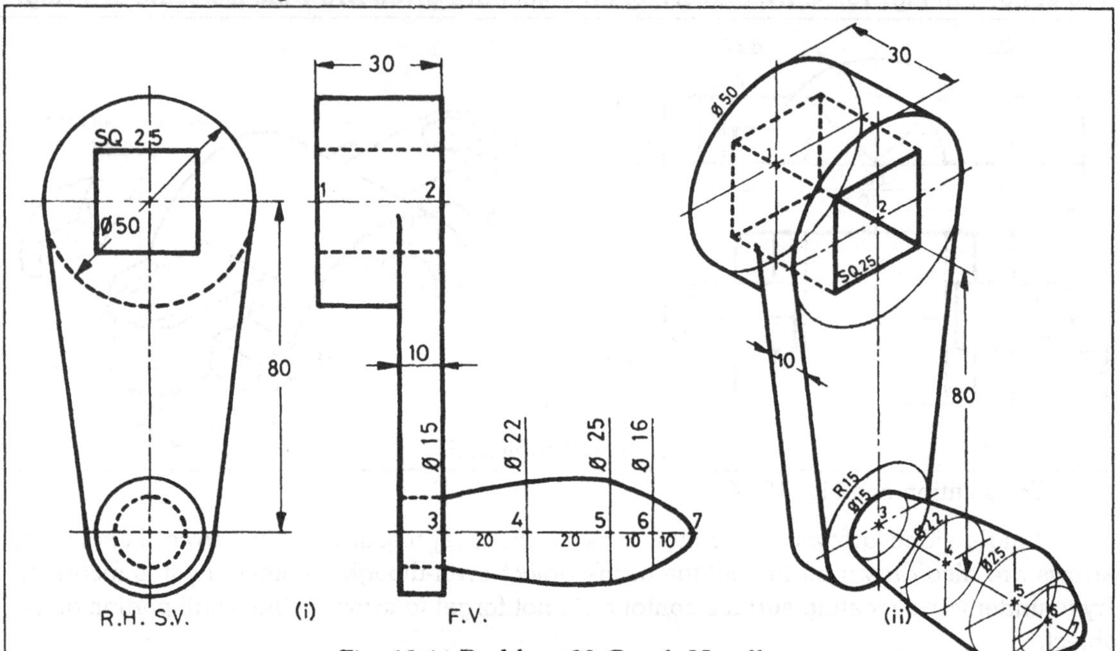

Fig. 18.46 Problem 39 Crank Handle

Problem 40 : F.V. and T.V. of two pieces of wooden joint are given in Fig. 18.47. Draw the isometric drawing of both the pieces ready in position for fitting.

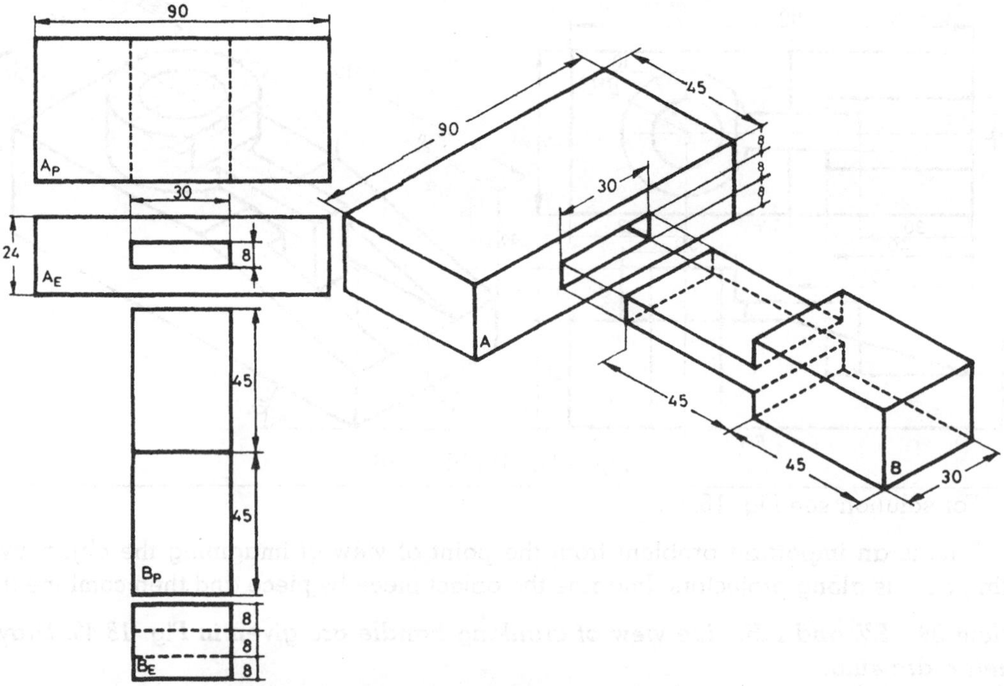

Fig. 18.47 Carpentary Joint - Problem 40

For solution see Fig. 18.47.

Problem 41 : F.V. and S.V. of a small table are given in Fig. 18.48. Draw its isometric drawing. Take suitable scale.

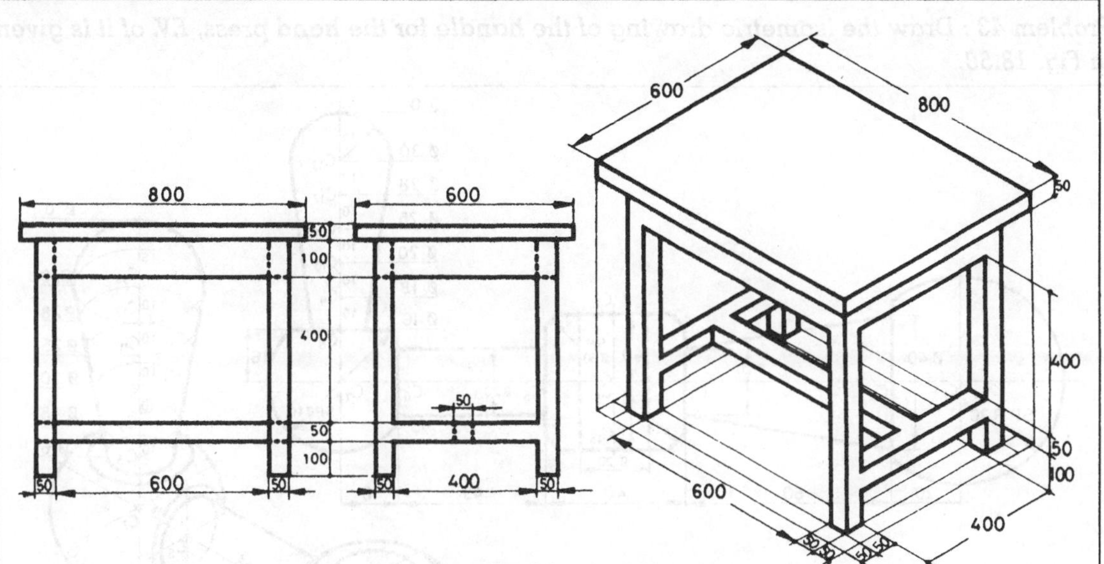

Fig. 18.48 Table. Problem 41

For solution see Fig. 18.48.

Problem 42 : F.V., T.V. and L.H.S.V. of a circular block with different cuttings are given in Fig. 18.49. Draw its isometric view.

In this problem it is desirable to draw 5 ellipses for five circular sections as shown in the figure. Dimensions are required to be taken on edges of rhombuses as shown in the figure for cutting cylinder at various places.

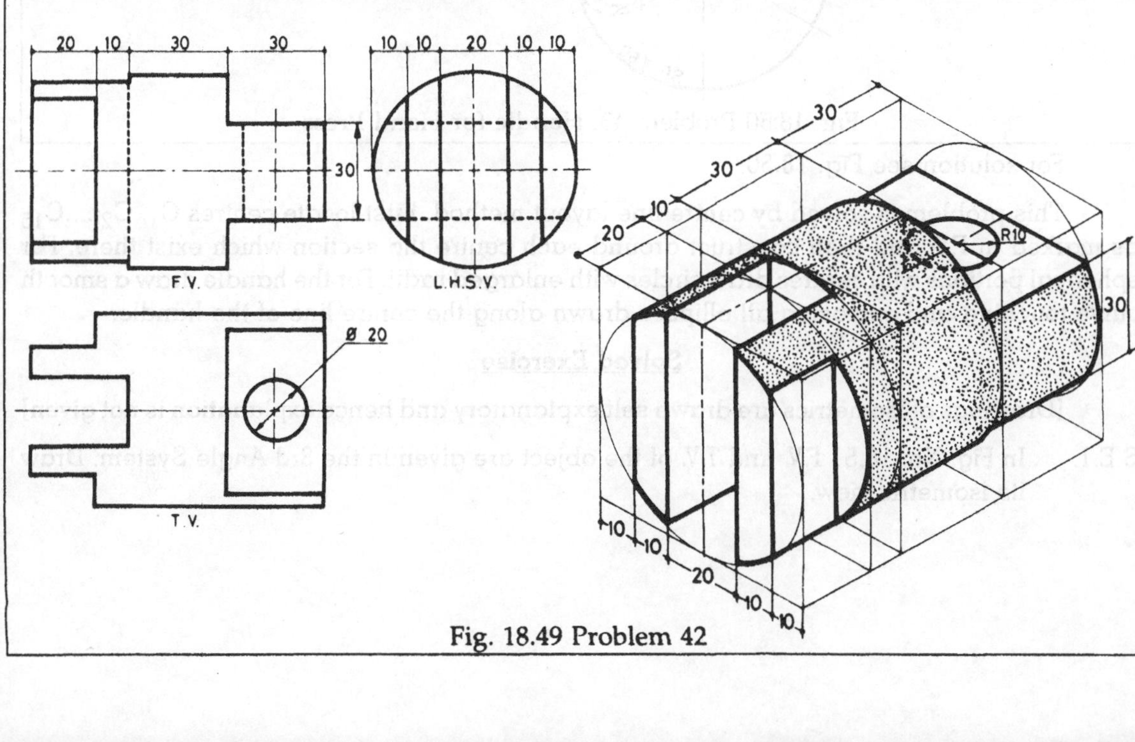

Fig. 18.49 Problem 42

For solution of problem see Fig. 18.49. To make drawing clear, curved surface of the cylinder is shaded.

Problem 43 : Draw the isometric drawing of the handle for the hand press, F.V. of it is given in Fig. 18.50.

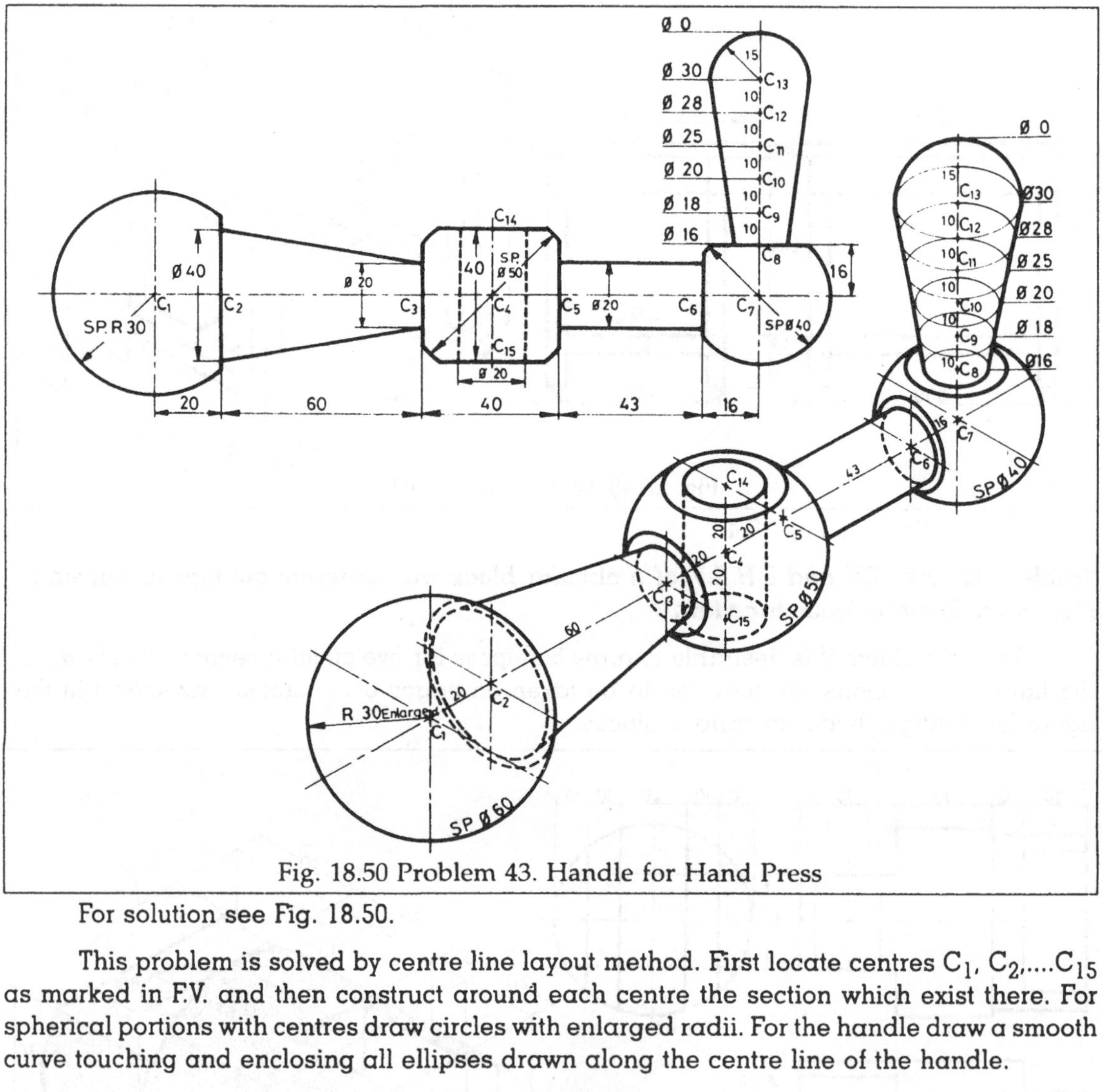

Fig. 18.50 Problem 43. Handle for Hand Press

For solution see Fig. 18.50.

This problem is solved by centre line layout method. First locate centres C_1, C_2,....C_{15} as marked in F.V. and then construct around each centre the section which exist there. For spherical portions with centres draw circles with enlarged radii. For the handle draw a smooth curve touching and enclosing all ellipses drawn along the centre line of the handle.

Solved Exercise

[Drawings of isometrics are drawn self explanatory and hence explanation is not given]

S.E.1. In Fig. No. 18.51 F.V. and T.V. of the object are given in the 3rd Angle System. Draw its isometric view.

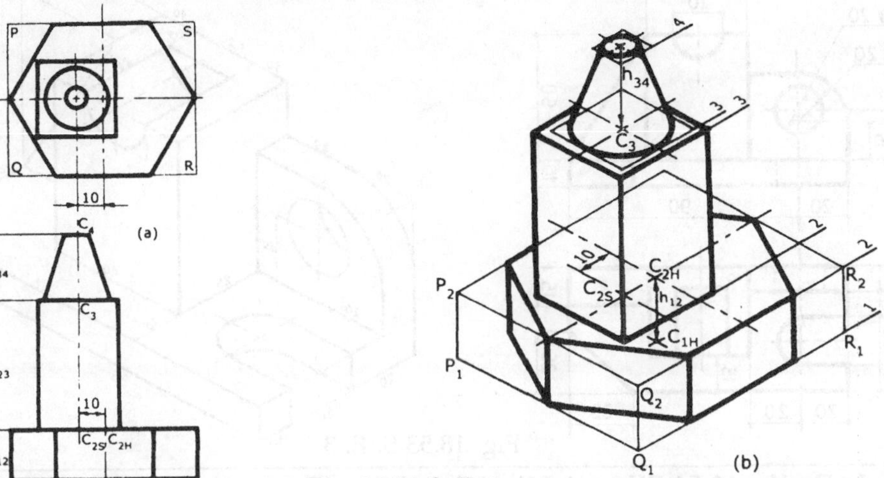

Fig. 18.51 S. E. 1

S.E.2. In Fig.No. 18.52 F.V. and L.H.S.V. of **A Hut** is given in 1st Angle System. Draw its isometric drawing.

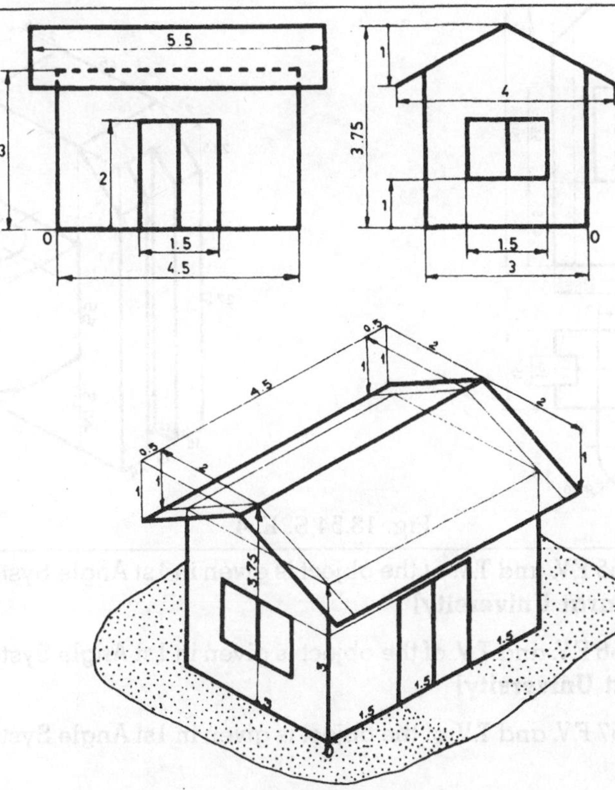

Fig. 18.52 S. E. 2

S.E.3. In Fig. No. 18.53 F.V. and T.V. of the object is given in 1st Angle System. Draw its isometric drawing [**Gujarat University**]

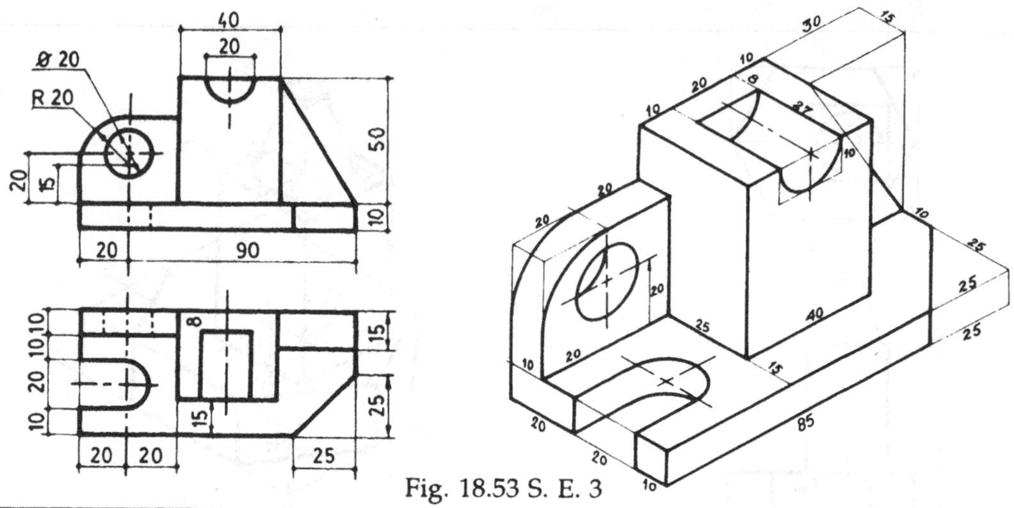

Fig. 18.53 S. E. 3

S.E.4. In Fig.No. 18.54 F.V. and T.V. of the object is given in 1st Angle System. Draw its isometric view. **[Gujarat University]**

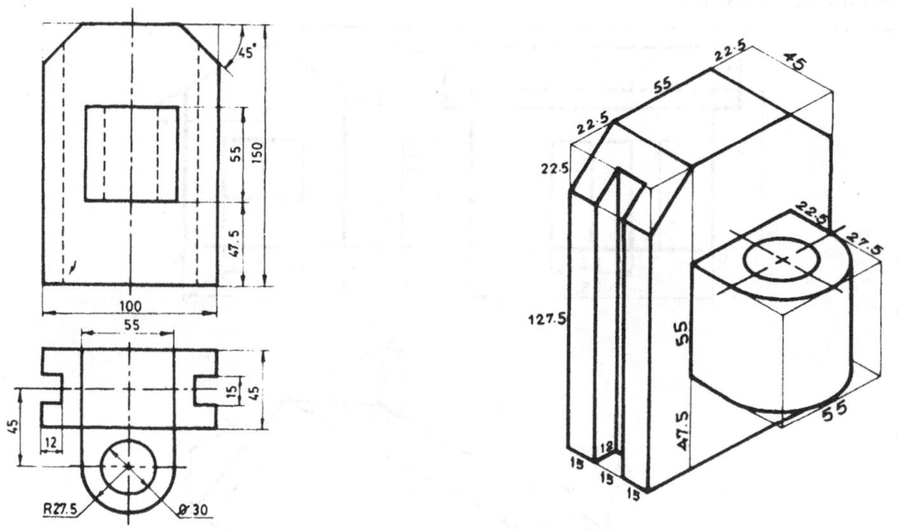

Fig. 18.54 S. E. 4

S.E.5. In Fig.No. 18.55 F.V. and T.V. of the object is given in 1st Angle System. Draw isometric drawing **[Gujarat University]**

S.E.6. In Fig.No. 18.56 F.V. and T.V. of the object is given in 1st Angle System. Draw isometric view. **[Gujarat University]**

S.E.7. In Fig.No. 18.57 F.V. and T.V. of the object is given in 1st Angle System. Draw isometric view.

S.E.8. In Fig. No. 18.58 F.V. and T.V. of the object is given in 1st Angle System. Draw isometric drawing.

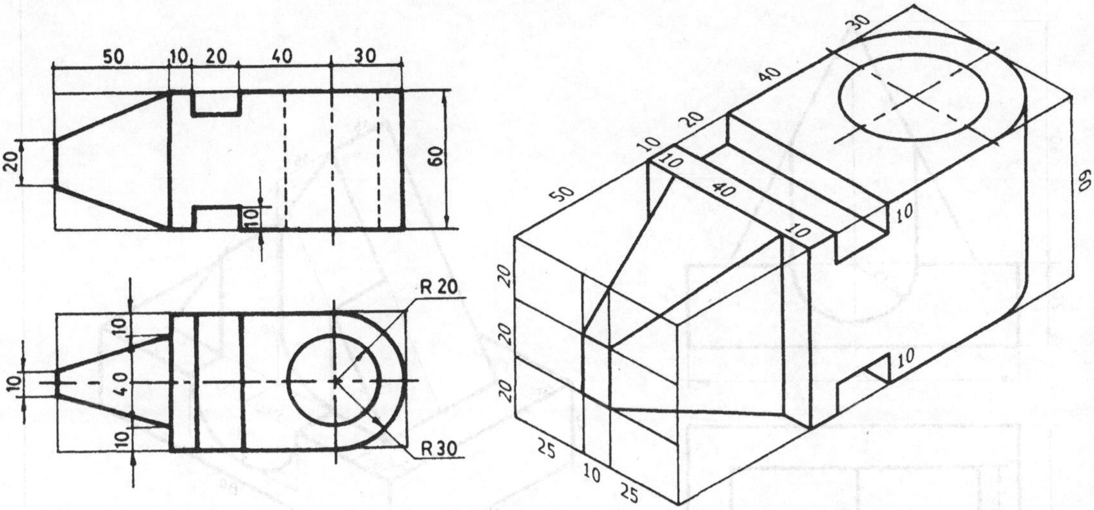

Fig. 18.55 S. E. 5

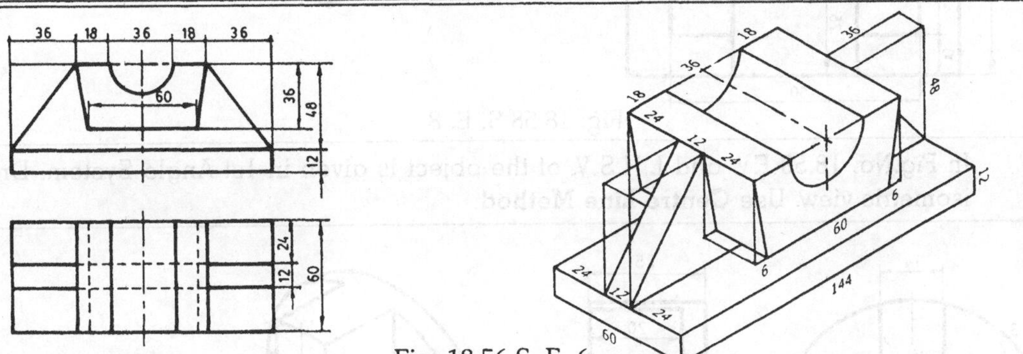

Fig. 18.56 S. E. 6

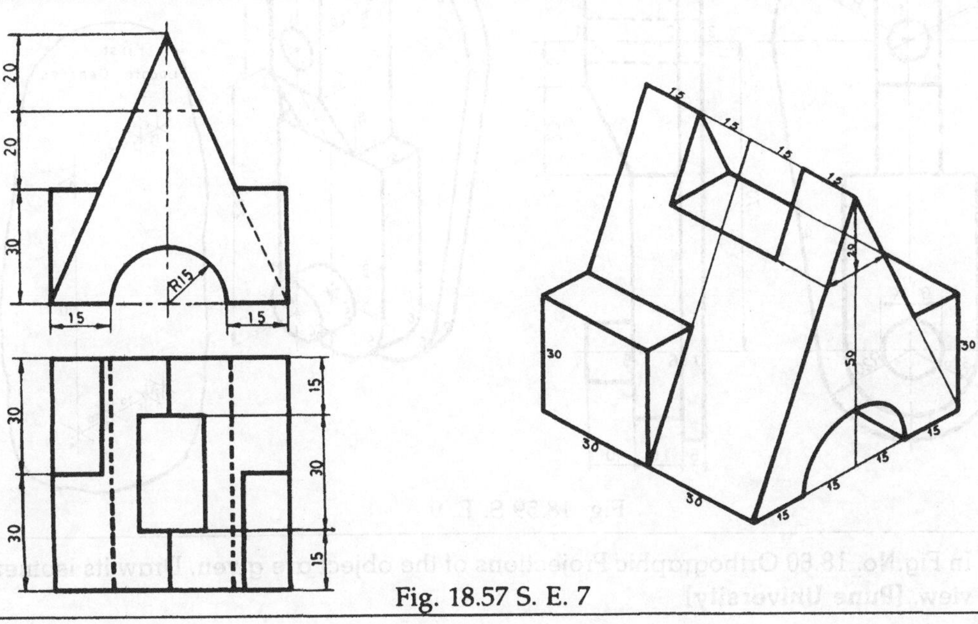

Fig. 18.57 S. E. 7

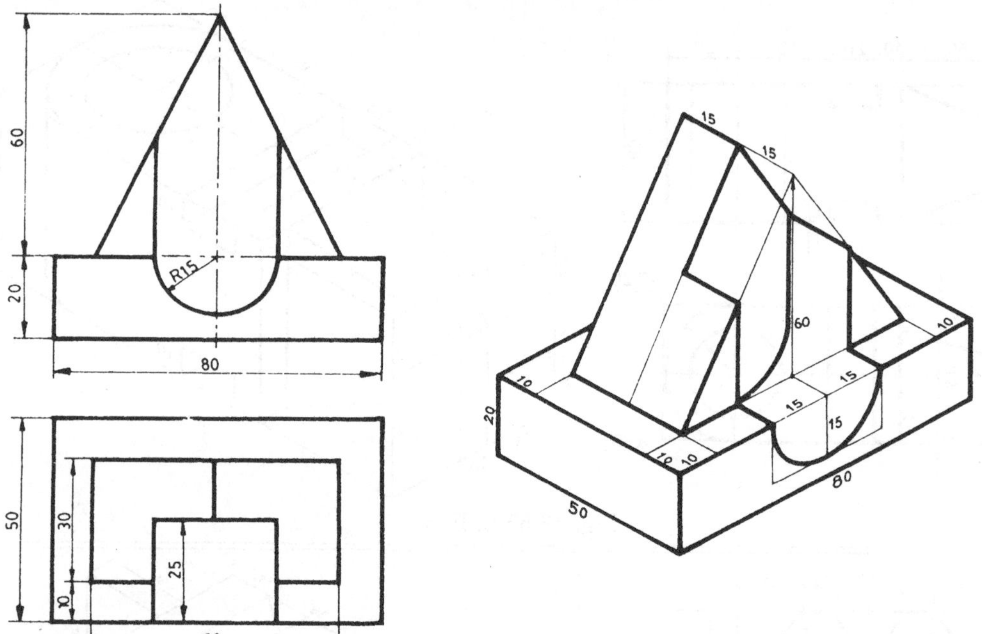

Fig. 18.58 S. E. 8

S.E.9. In Fig.No. 18.59 F.V. and L.H.S.V. of the object is given in 1st Angle System. Draw isometric view. Use **Centre Line Method**.

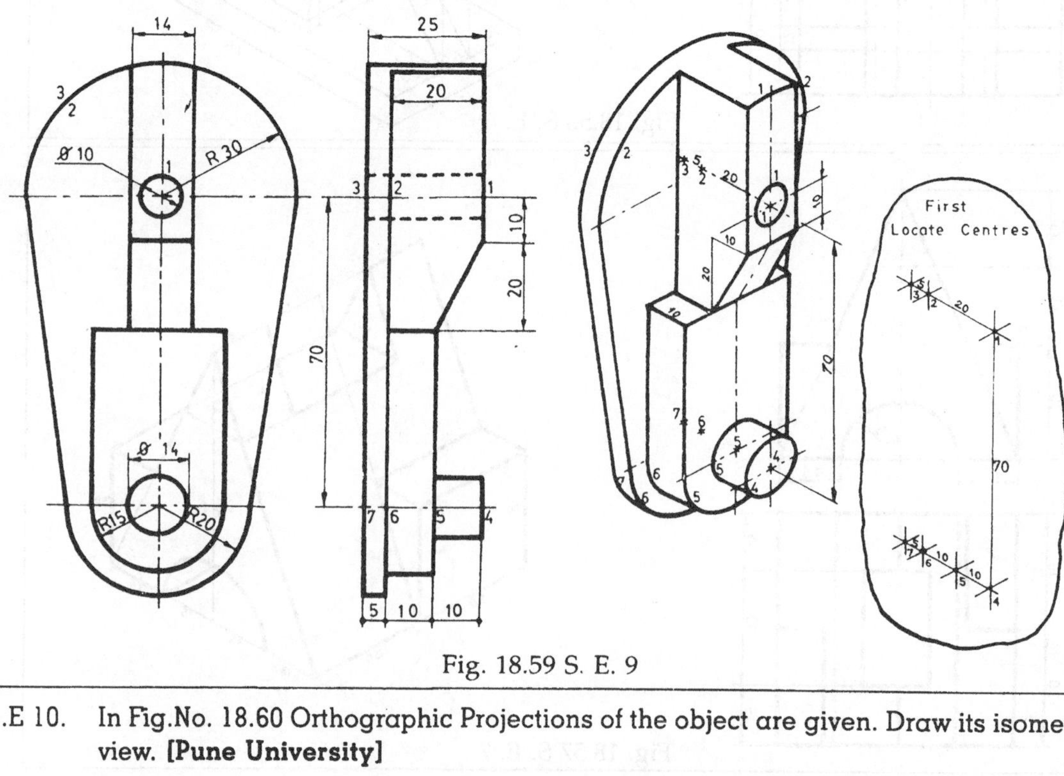

Fig. 18.59 S. E. 9

S.E 10. In Fig.No. 18.60 Orthographic Projections of the object are given. Draw its isometric view. **[Pune University]**

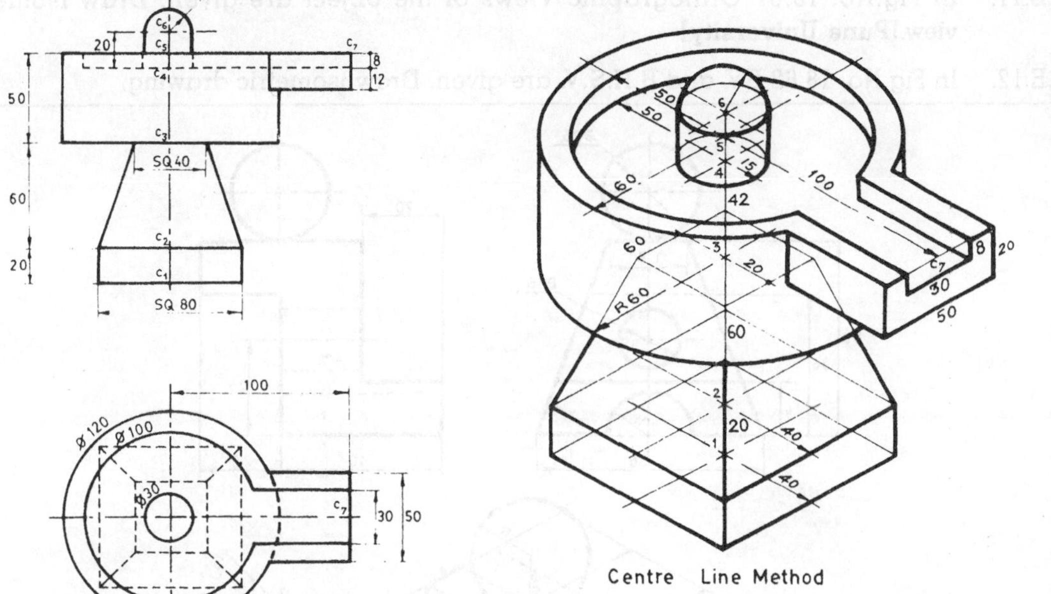

Fig. 18.60 S. E. 10

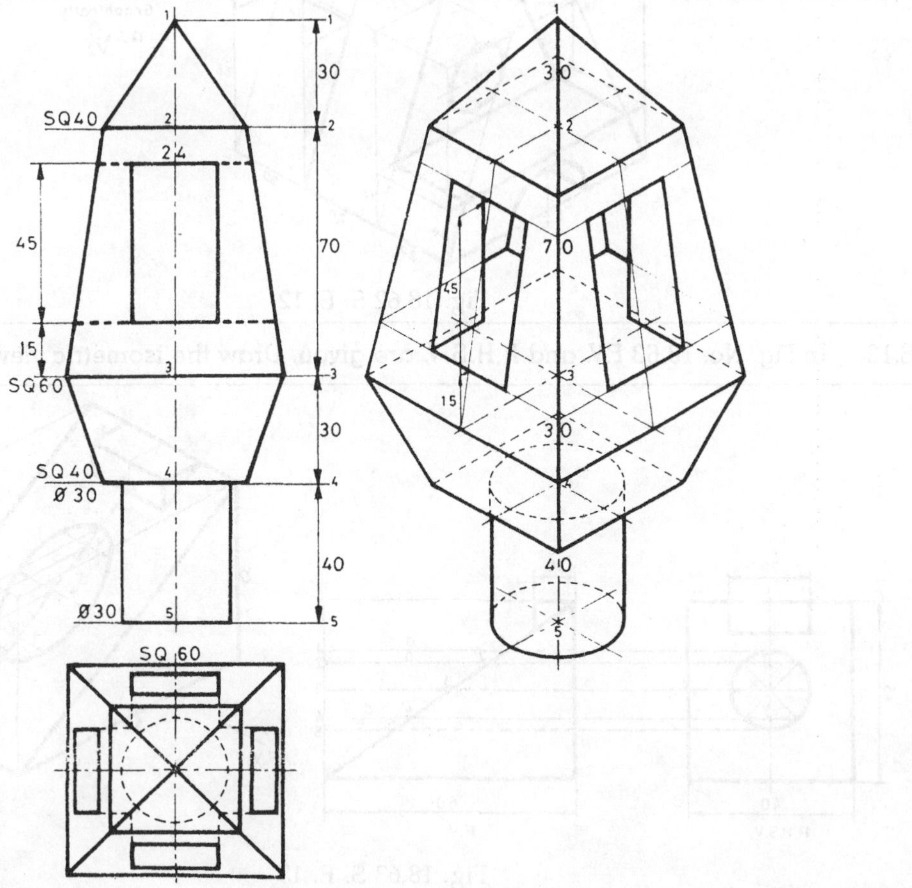

Fig. 18.61 S. E. 11

S.E.11. In Fig.No. 18.61 Orthographic Views of the object are given. Draw isometric view.**[Pune University]**

S.E.12. In Fig.No. 18.62 F.V. and R.H.S.V. are given. Draw isometric drawing.

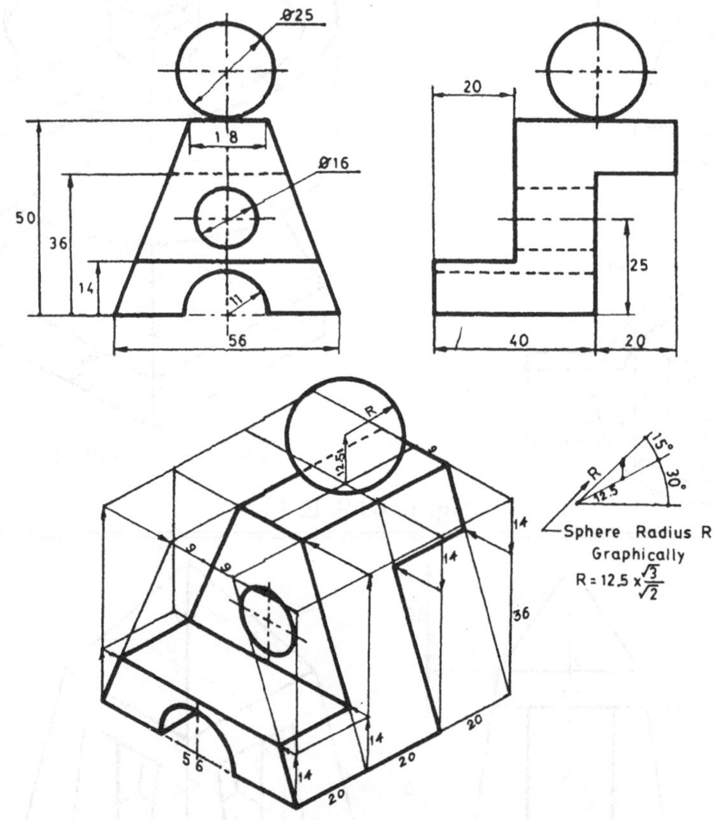

Fig. 18.62 S. E. 12

S.E.13. In Fig. No. 18.63 F.V. and R.H.S.V. are given. Draw the isometric view of the object.

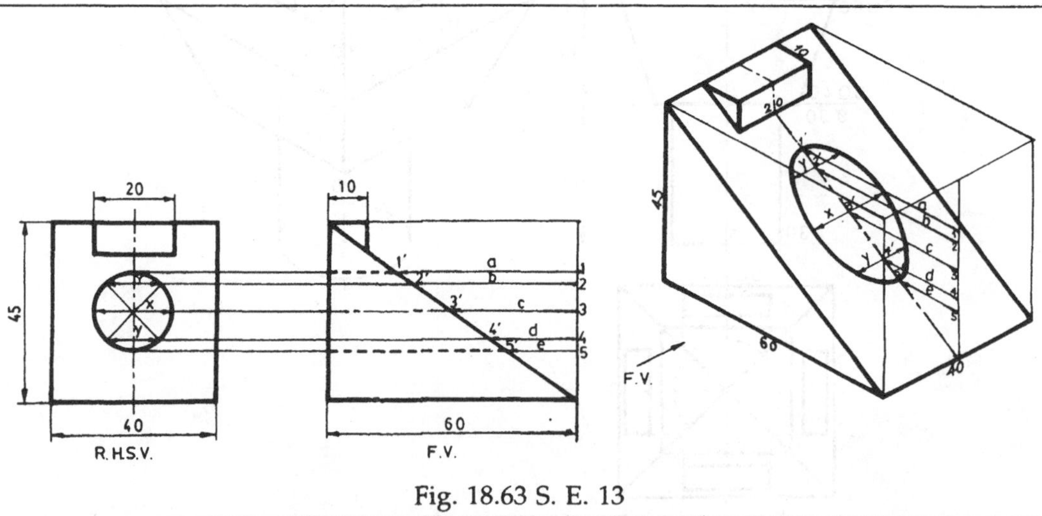

Fig. 18.63 S. E. 13

Few practice problems with solutions are given on next 4 full pages.

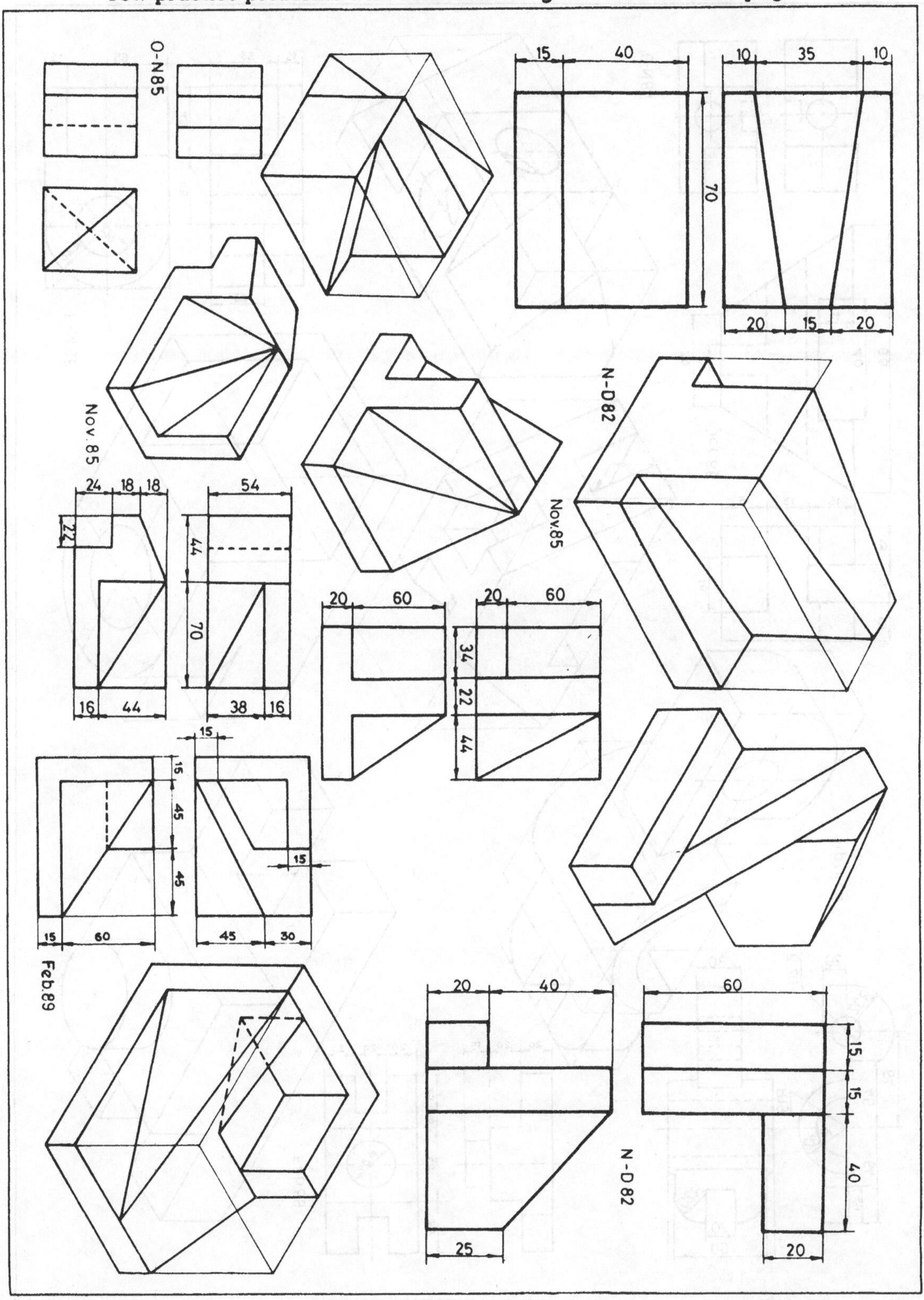

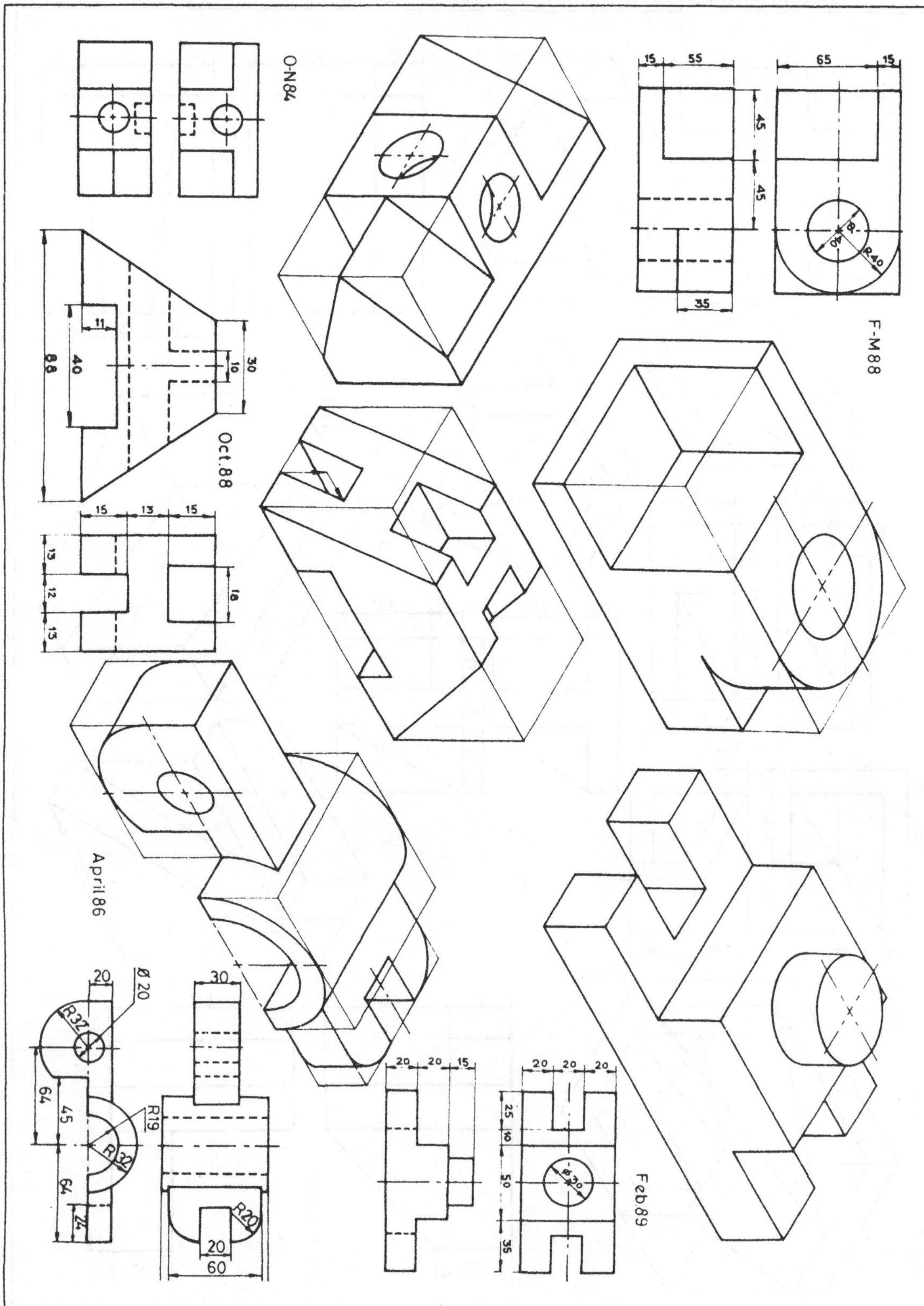
O-N84
Oct.88
April86
F-M88
Feb.89

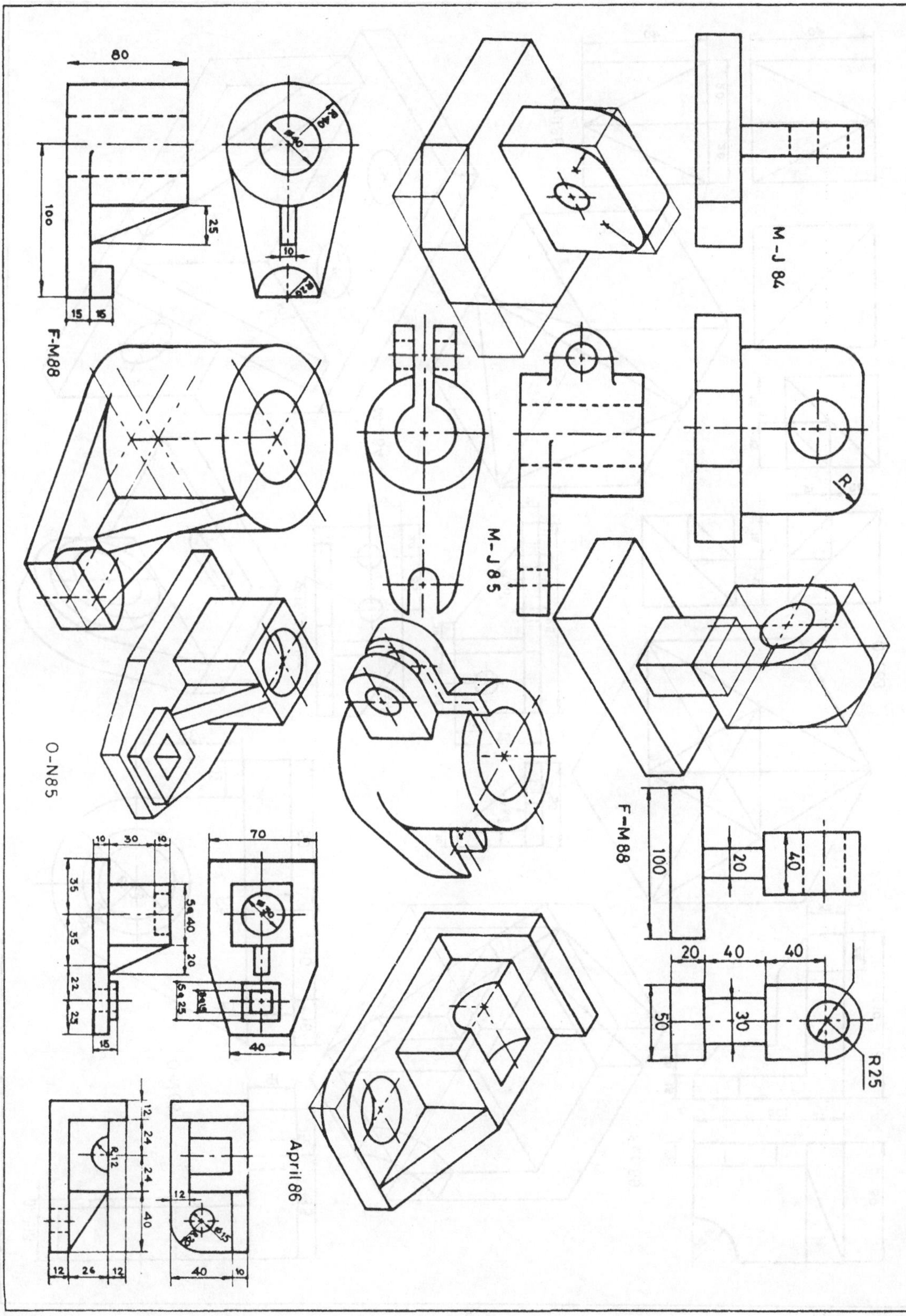
F-M88
M-J 84
M-J 85
O-N85
F-M88
April 86

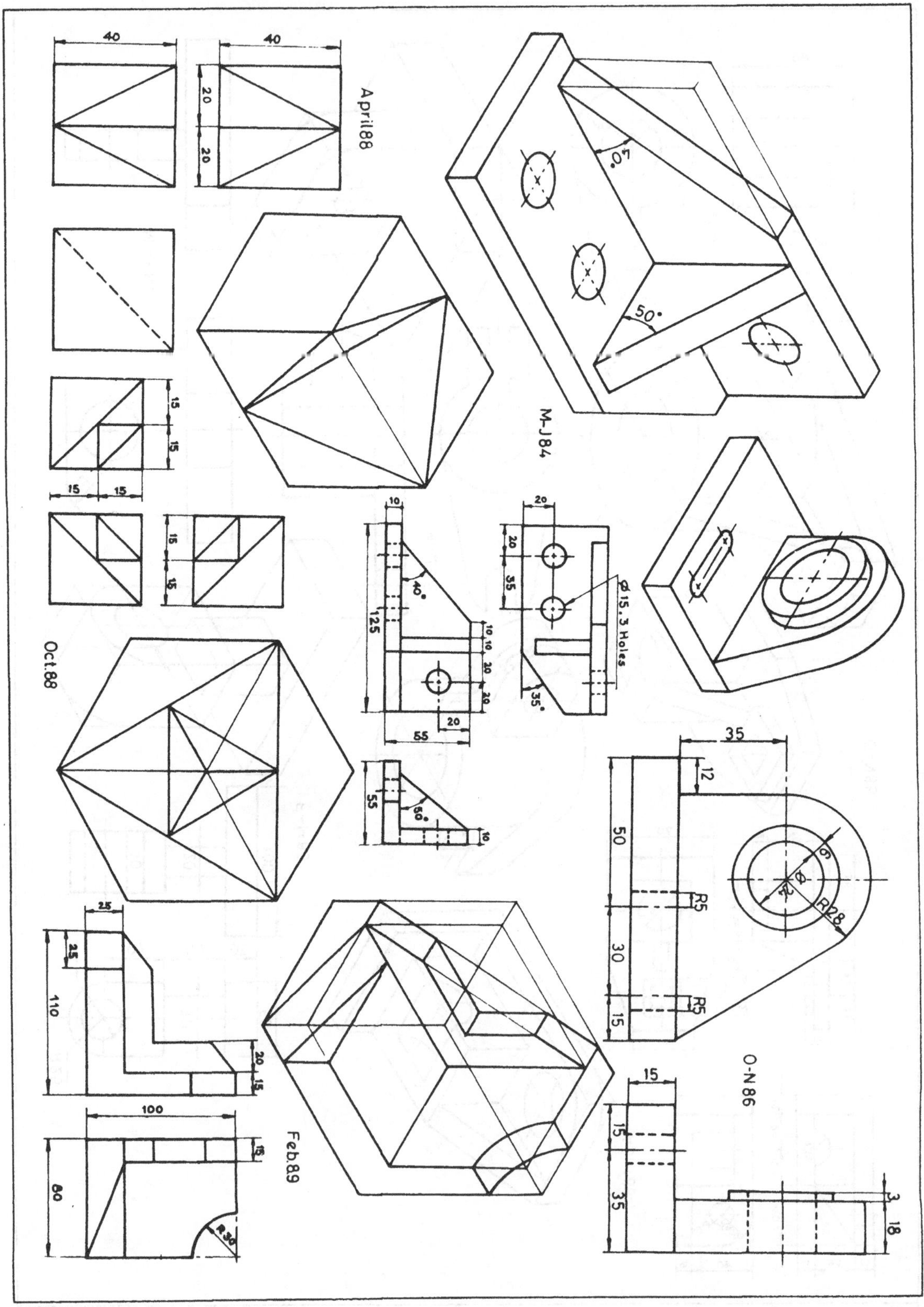
April88
M-J84
Oct.88
⌀15.3 Holes
O-N86
Feb.89

Ex. 1 to 51. Projections of various objects 1 to 51 are given in Fig. 18.64(a) and Fig. 18.64(b). Draw the isometric view/projection of each object. Assume each division of graph of 10 mm length.

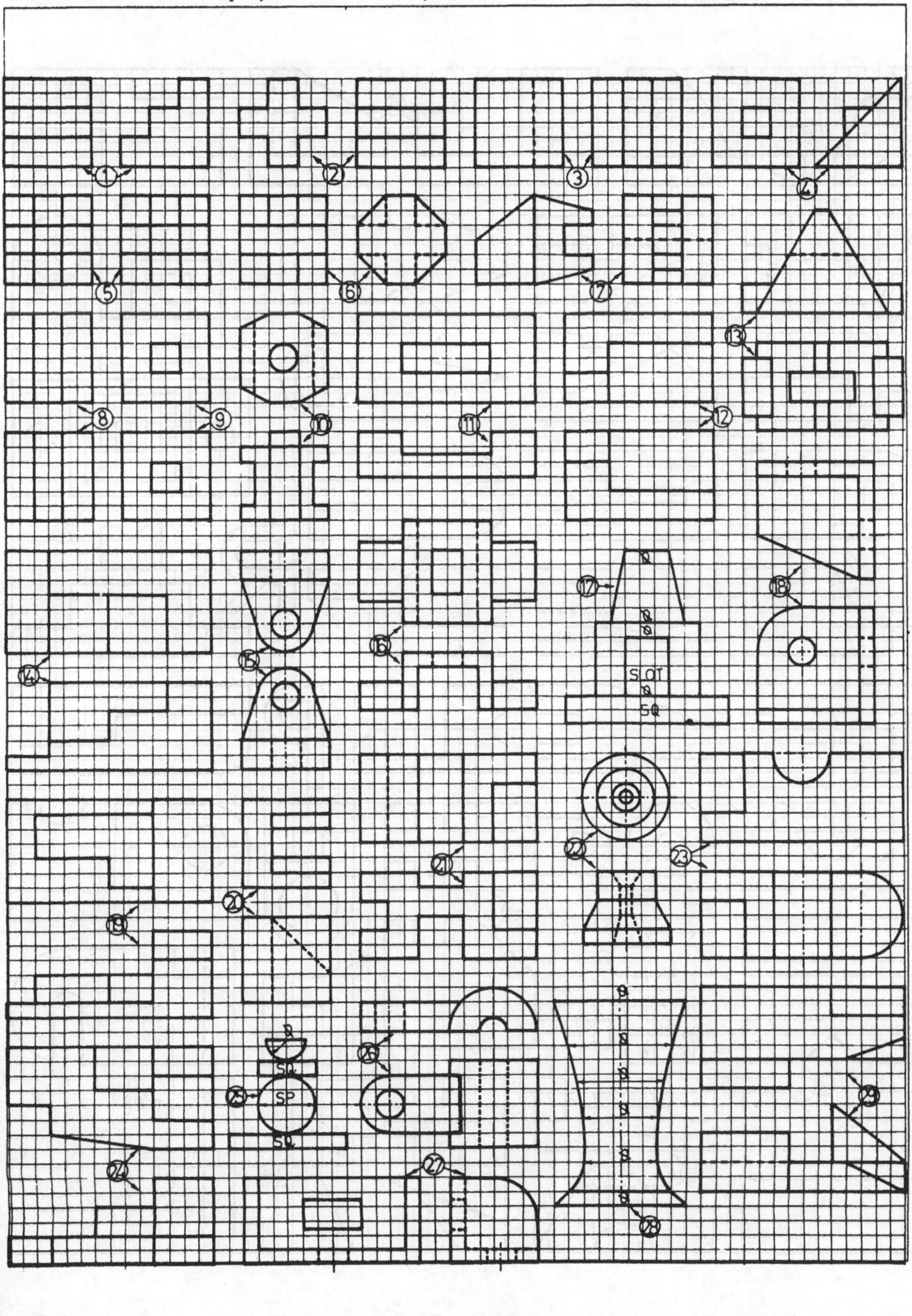

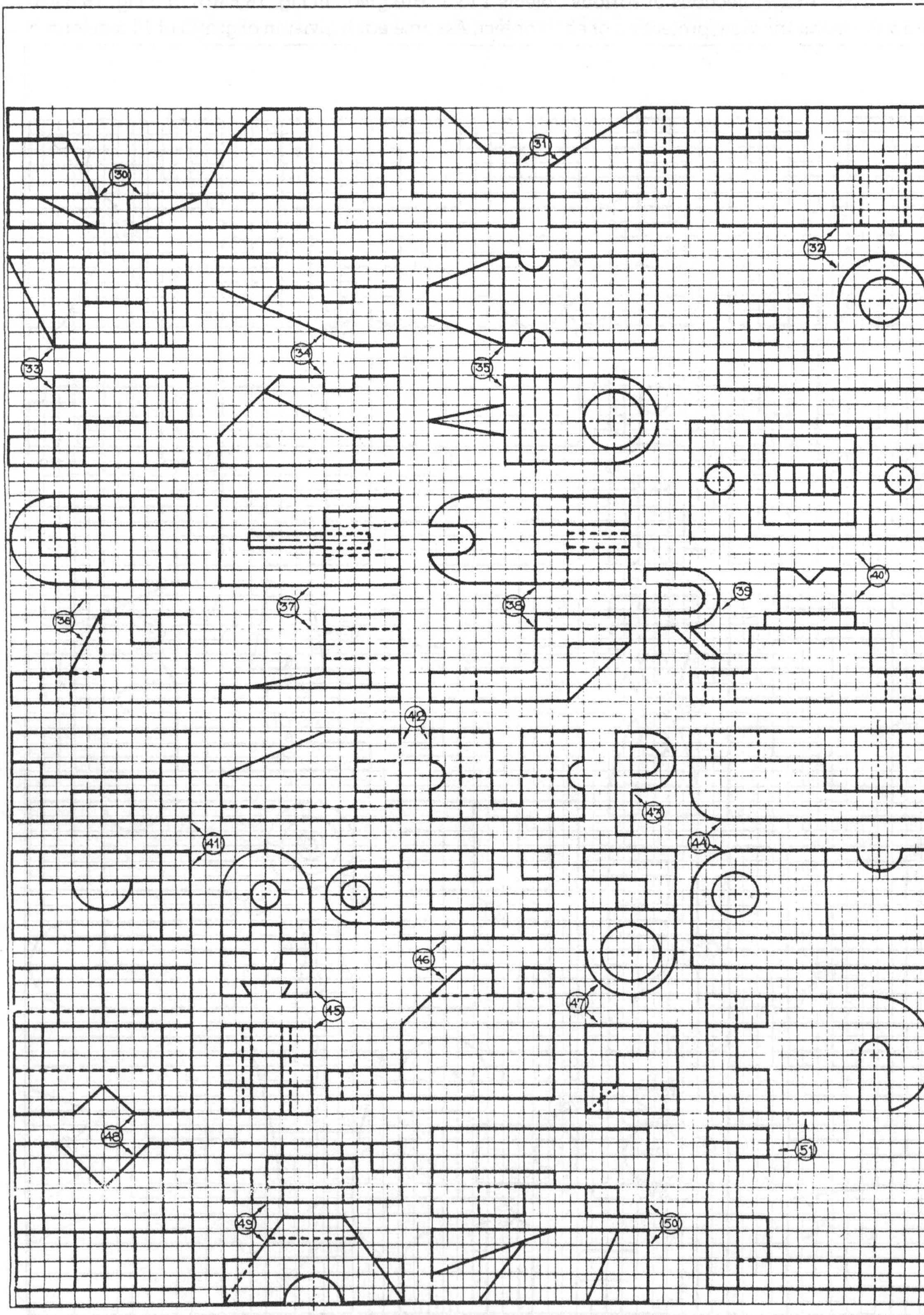
30
31
32
33
34
35
36
37
38
39
40
41
42
43
44
45
46
47
48
49
50
51

Problem : Figure shows F. V. and S. V. of an object. Draw its **Isometric Projection.** [14]

[B.T.E.M.S. N/D 1999]

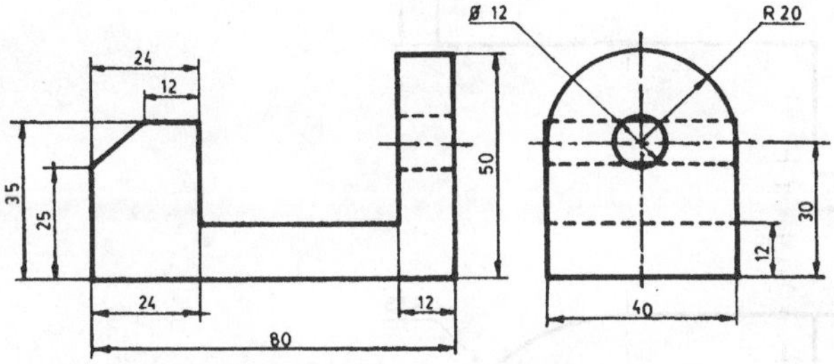

(a) Given Data

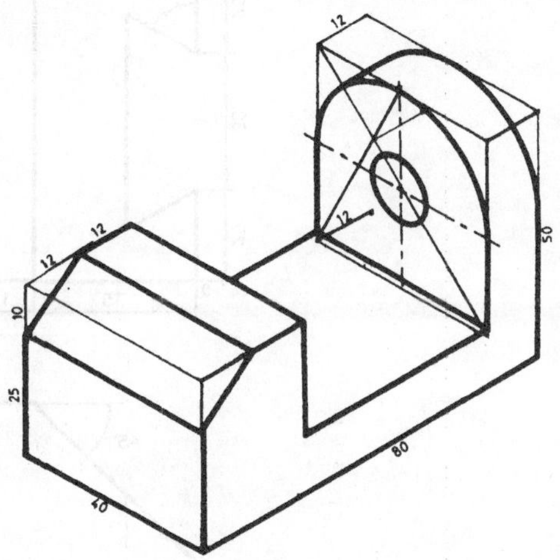

(b) Isometric

Problem : Fig. shows two orthographic views of an object by first angle method. Draw **An Isometric View** with Natural Scale. [14]

[B.T.E.M.S. N/D 1996]

[B.T.E.M.S. A/M 1998]

[Pune University Dec. 1992]

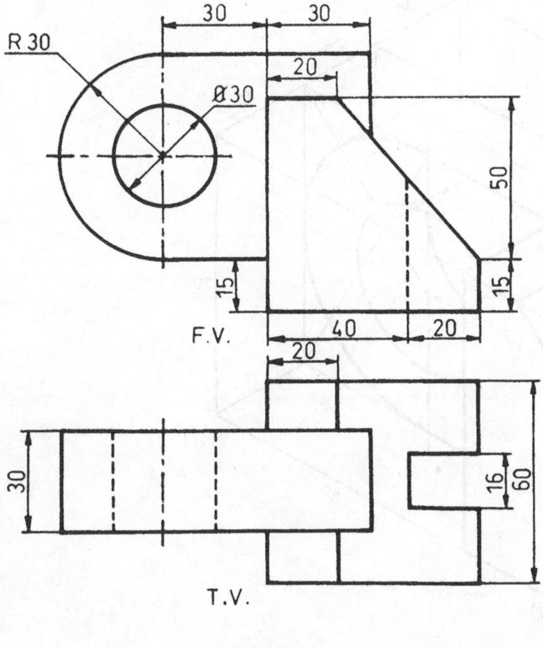

(a) Given Data

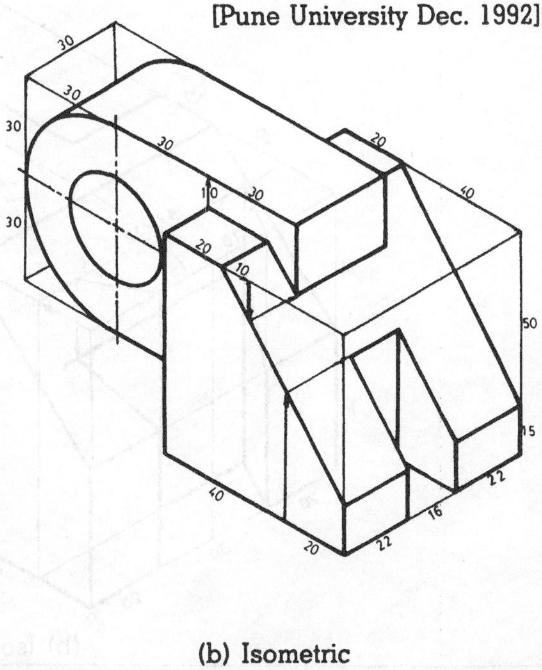

(b) Isometric

Problem : Two orthographic views are given in Figure by third angle method. Draw **An Isometric View**, taking O as origin. Use Natural Scale. [14]

[B.T.E.M.S. N/D 1998]

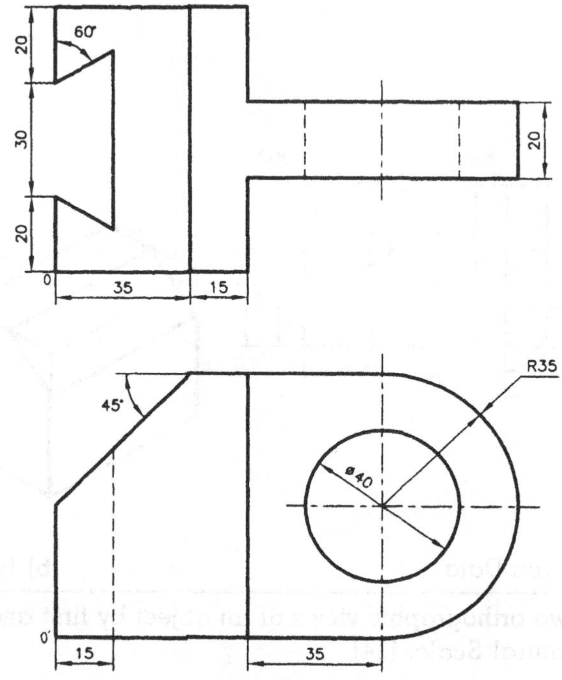

(a) Given Data

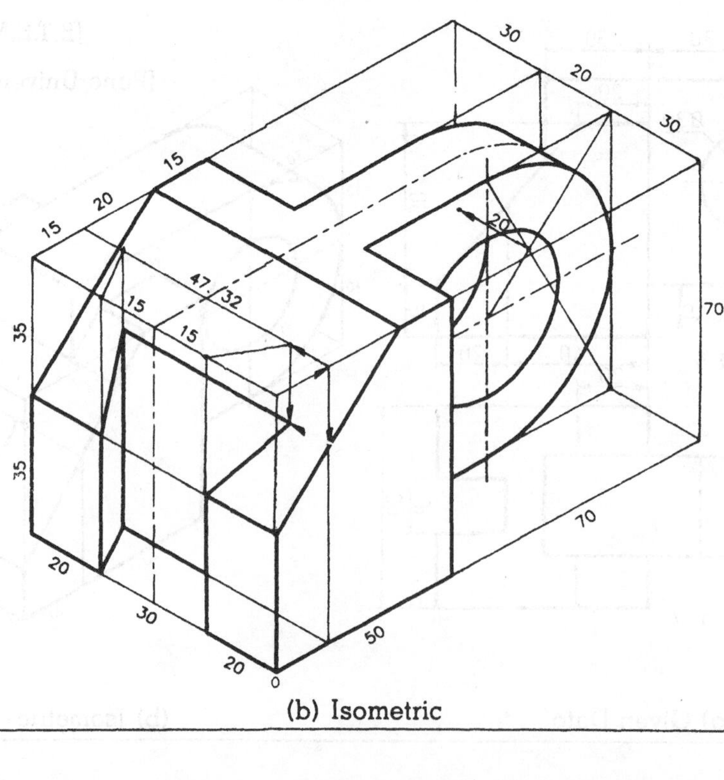

(b) Isometric

Problem : Orthographic views of an object are shown in Figure. Using Natural Scale, draw its **Isometric View.**

[Pune University]

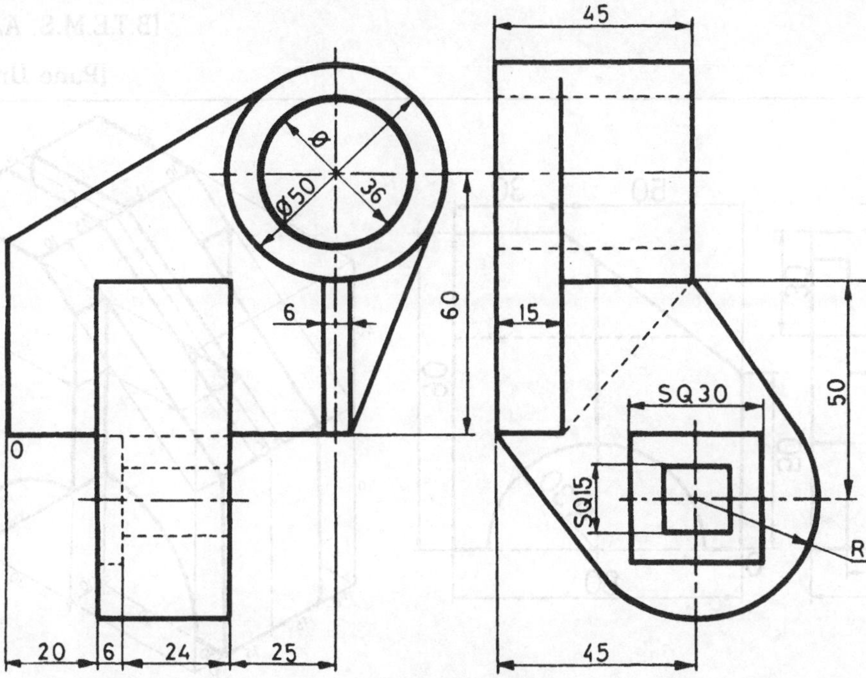

(a) Given Data

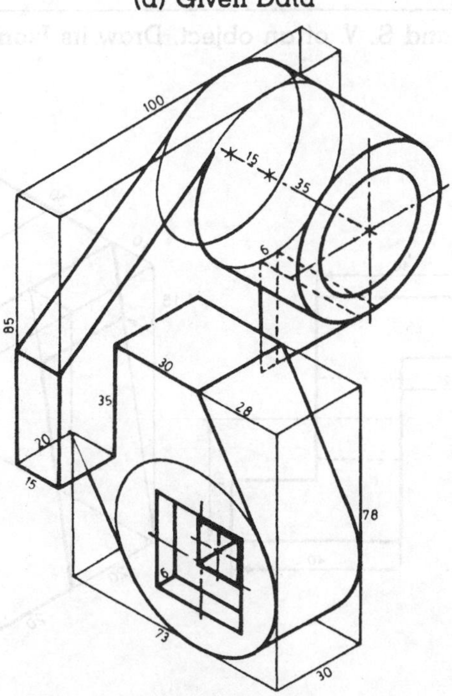

(b) Isometric

Problem : Two orthographic views of an object are given in Fig. by third angle method. Draw **An Isometric View**, taking O as origin. Use Natural Scale. [14]

[B.T.E.M.S. A/M 1998]

[B.T.E.M.S. A/M 2000]

[Pune University]

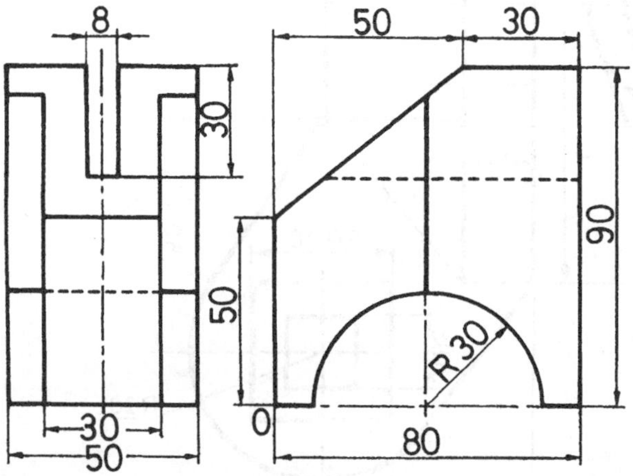

(a) Given Data

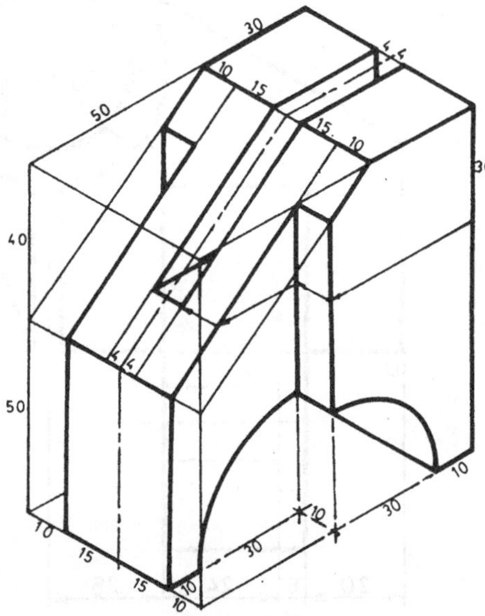

(b) Isometric

Problem : Fig. shows F. V. and S. V. of an object. Draw its **Isometric Projection** about an origin O. [14]

[B.T.E.M.S. A/M 1997]

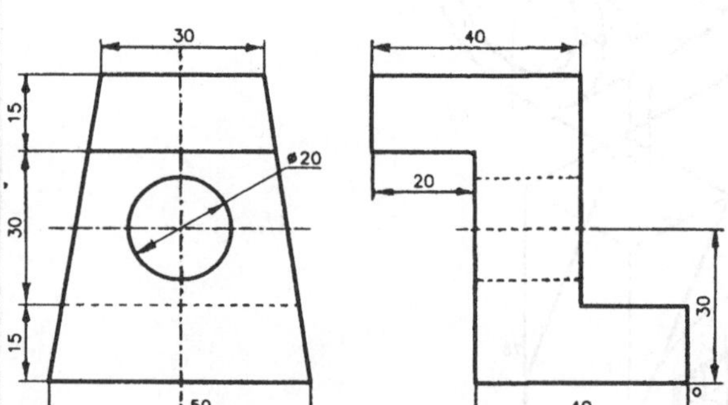

(a) Given Data

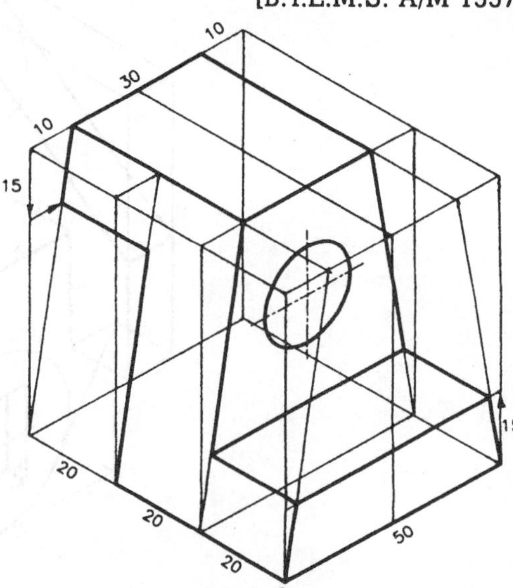

(b) Isometric

Problem : Refer figure and draw **The Isometric View** of the given object using Natural Scale.

[Mumbai University, Dec. 2000]

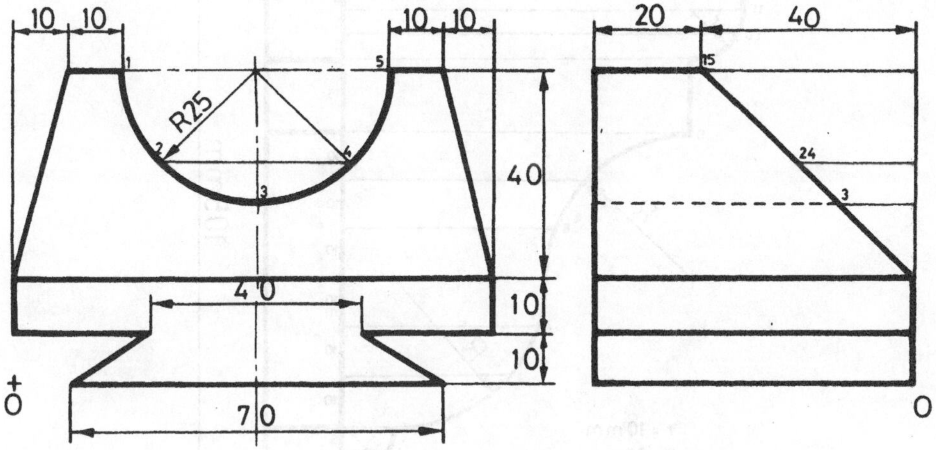

(a) Given Data

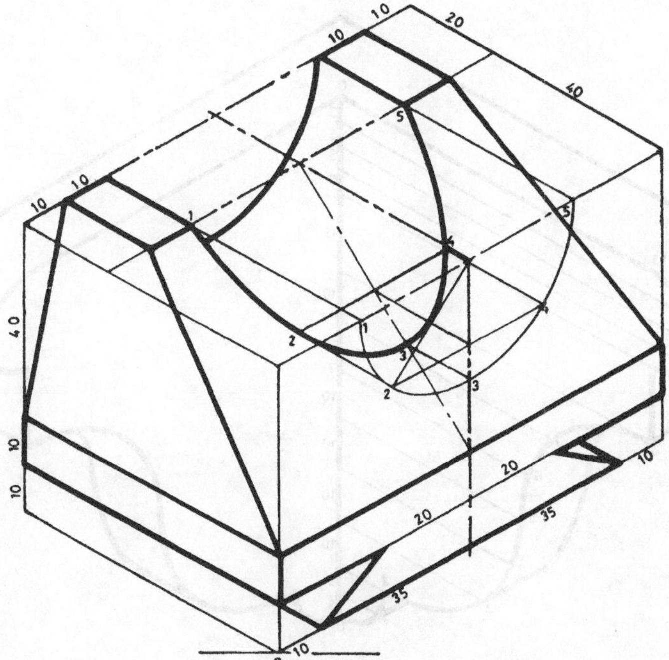

(b) Isometric

Problem : From the given view, draw **An Isometric View.** Co-ordinate method is used in drawing Isometric.

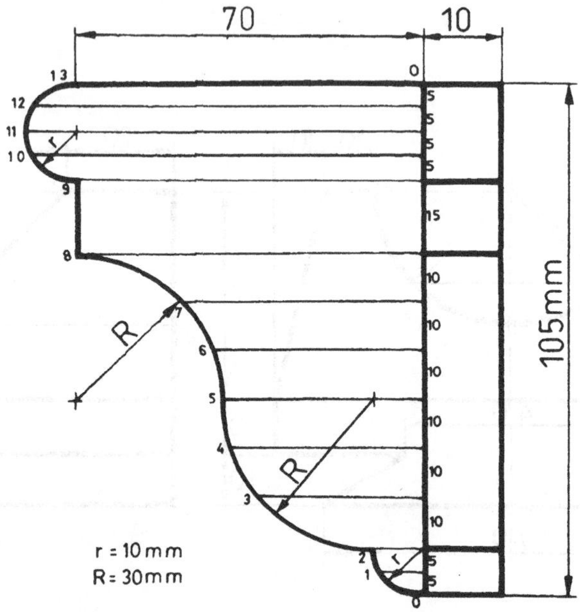

(a) Given Data

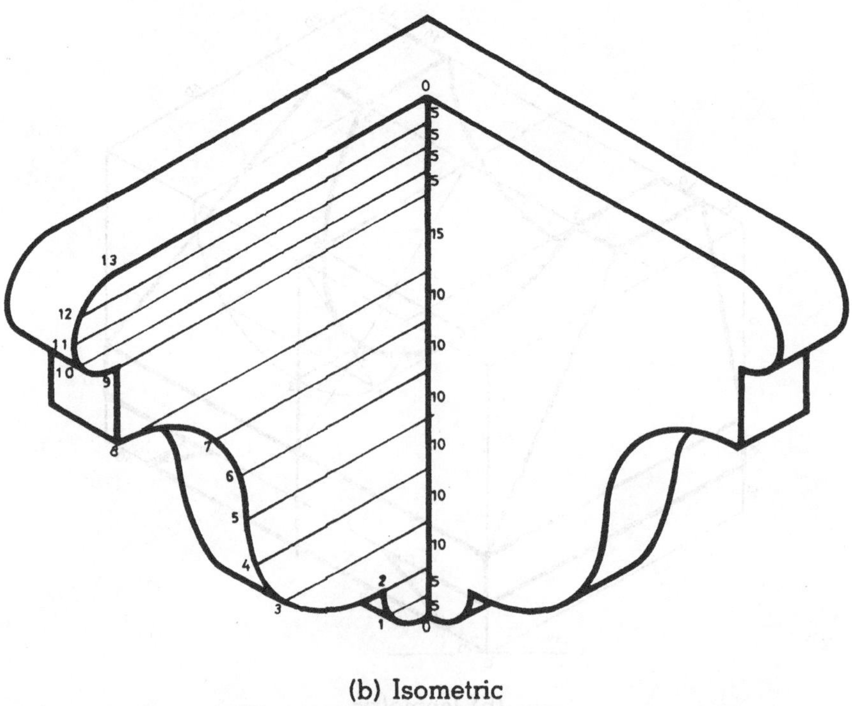

(b) Isometric

Problem : Fig shows F. V. and L. H. S. V. of an object. Draw **The Isometric View** of the object Using Natural Scale.

[Mumbai University, May 1997]

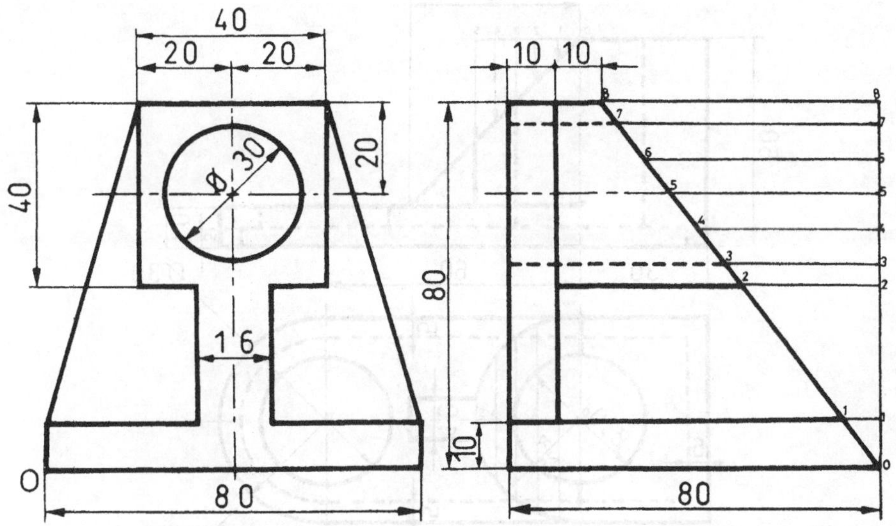

(a) Given Data

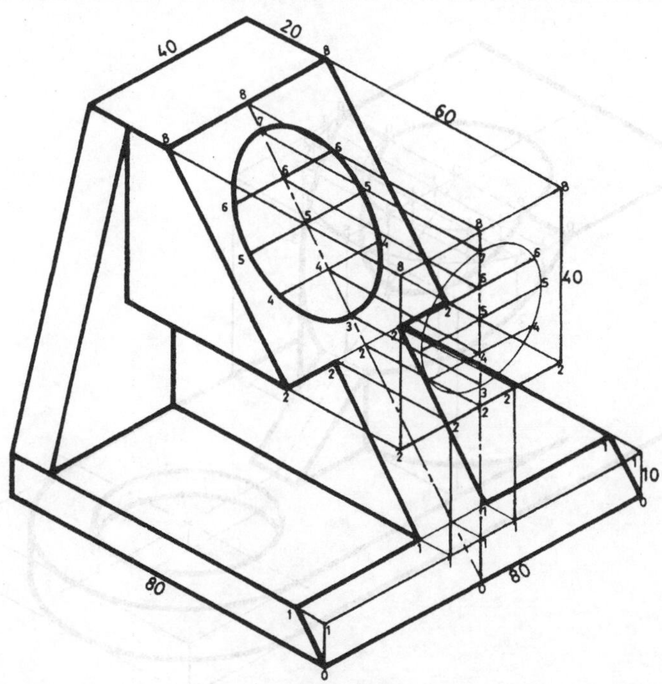

(b) Isometric

Problem : Fig. shows F. V. and T. V. of an object. Draw its **Isometric View** of the object using Natural Scale.

[Mumbai University Nov./Dec. 1996]

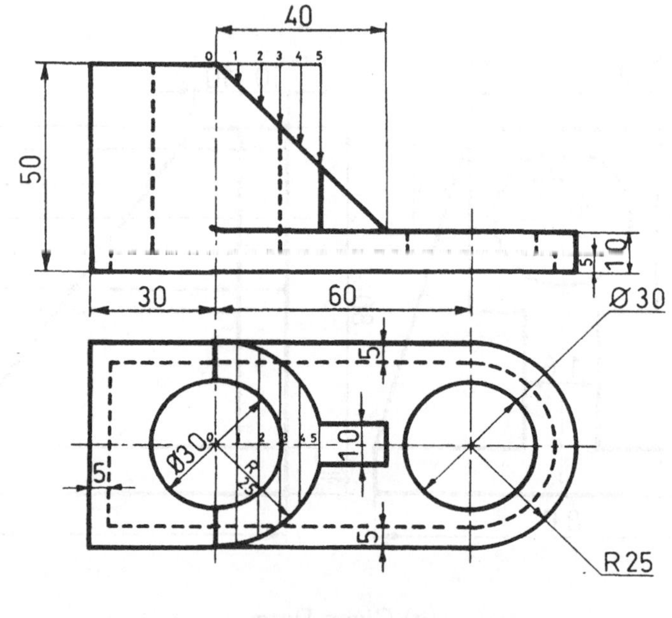

(a) Given Data

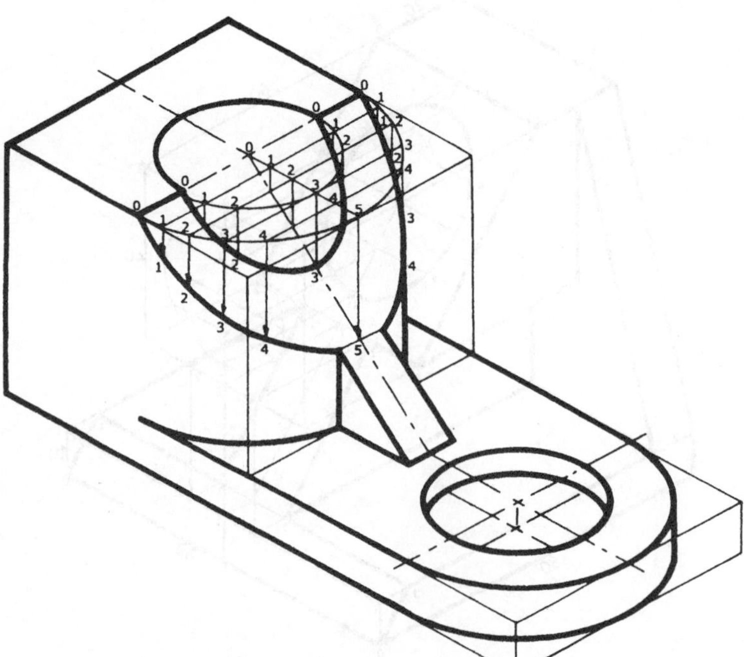

(b) Isometric

CHAPTER 19
DETAIL AND ASSEMBLY DRAWINGS

1. Assembly Drawings :

Assembly drawing of a mechanism or a machine shows all parts in their relative working position. There are many types of assembly drawings.

(i) Working assembly drawings,

(ii) Sub-assembly drawings,

(iii) Design assembly drawings, etc.

Working assembly drawing is made for simple assembly. In this drawing each part is fully dimensioned and hence detail drawing of the same is not required.

Sub-assembly is an assembly of one of the units of the complete machine e.g. Assembly of a gear box of an automobile. It is also known as unit assembly.

In assembly drawing each part is identified by part number. Generally hidden lines are not drawn in assembly drawings to increase the clarity and readability of the drawings. In assembly drawings overall dimensions are given. Detail dimensions of all parts are not shown. If needed assembly drawing contains specific instructions to assembler. Generally, one view is drawn sectional to illustrate the various parts and their respective positions in assembly.

2. Detail Drawings :

Detail drawing must contain complete information regarding manufacture of part. Detail dimensions of all parts along with surface roughness symbols should be given in detail drawing. Choice and number of views should be selected so as to describe the component completely. Tolerance, if any, must be mentioned on matching parts. Detail drawing must also contain part number and scale of drawing. In this chapter six (6) drawings are given in different manner for preparing detail and assembly drawings.

Problem 1:

Detail drawings of all parts along with pictorial view of an assembly of a '*Lathe Tool Post*' are given in Fig.

Draw (1) Half Sectional F. V. of an Assembly.

(2) T. V. of an Assembly.

(3) S. V. of an Assembly.

Detail Drawing of a '*Lathe Tool Post*'

(1)	Main Body	-	Medium Carbon Steel	- 1
(2)	Square Block	-	Cold rolled steel	- 1
(3)	Ring	-	Medium carbon steel	- 1
(4)	Set Screw	-	Tool steel	- 1
(5)	Wedge	-	Medium carbon steel	- 1

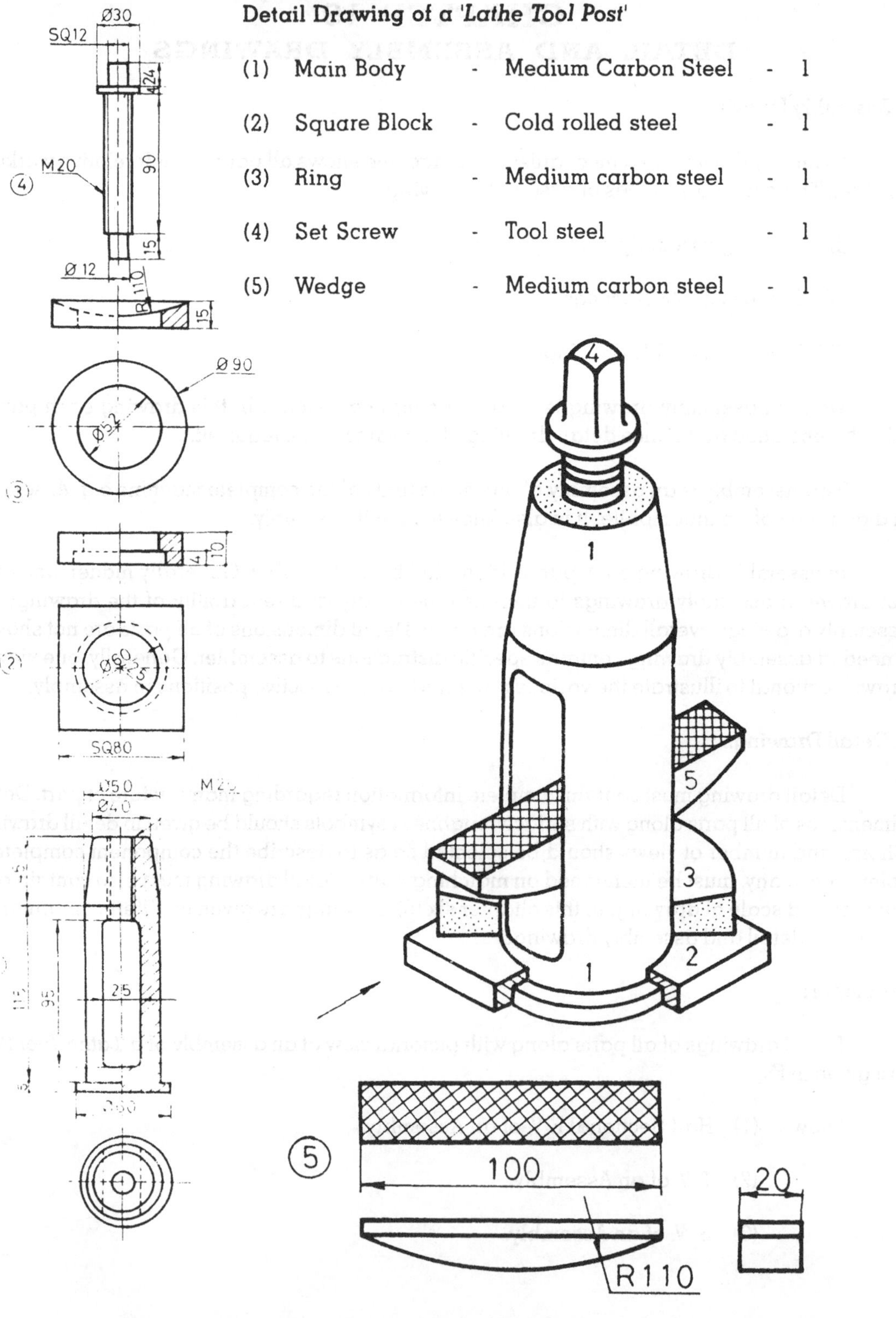

Problem 2 :

Detail drawings of all parts of a '***Screw Jack***' are given in Fig.

Draw (1) Half Sectional F. V. of an Assembly.

(2) T. V.

Details of A Screw Jack

(1)	Main Body	-	C. I.	1
(2)	Nut	-	Gun Metal	1
(3)	Screw	-	M. S.	1
(4)	Cup	-	Cast Steel	1
(5)	Washer	-	M. S.	1
(6)	CSK Screw	-	M. S.	1
(7)	Tommy Bar or Handle	-	M. S.	1

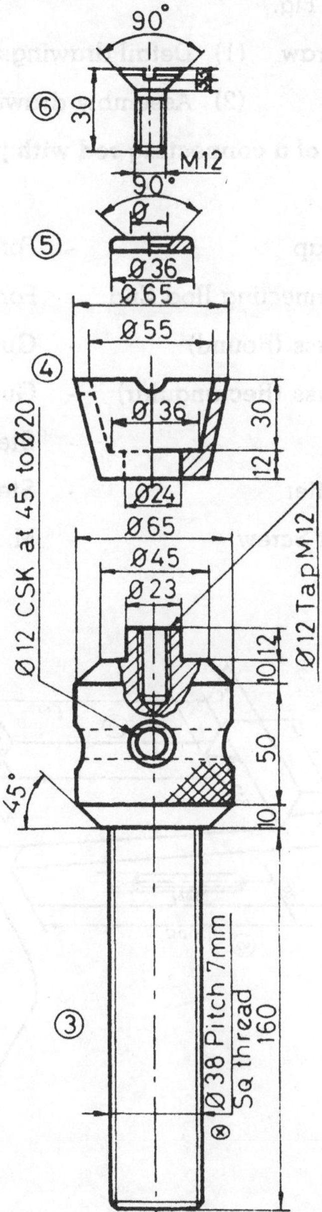

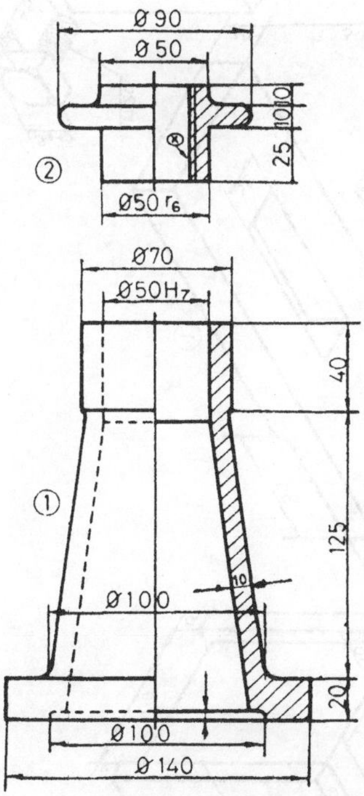

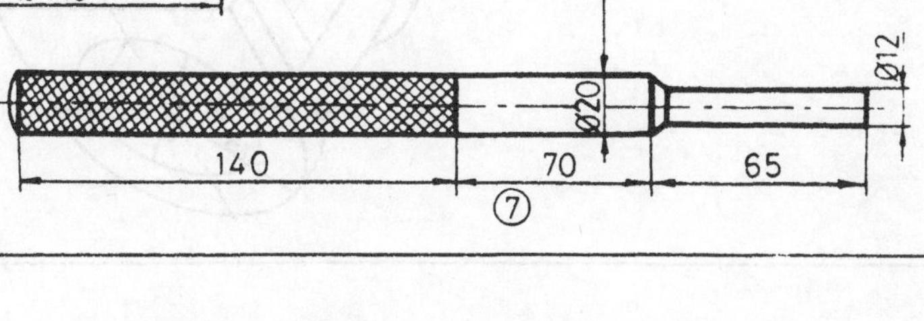

Problem 3 :

Pictorial views of all parts of a '***Big End of A Connecting Rod, with jib and cotter***' are given in Fig.

Draw (1) Detail drawings of all parts.

(2) Assembly drawings. (2 views)

Big end of a connecting rod with jib and cotter Part List.

(1)	Strap	Forged Steel	-	1
(2)	Connecting Rod End	Forged Steel	-	1
(3)	Brass (Round)	Gun Metal	-	1
(4)	Brass (Rectangular)	Gun Metal	-	1
(5)	Jib	Steel	-	1
(6)	Cotter	Steel	-	1
(7)	Set Screw	M. S.	-	1

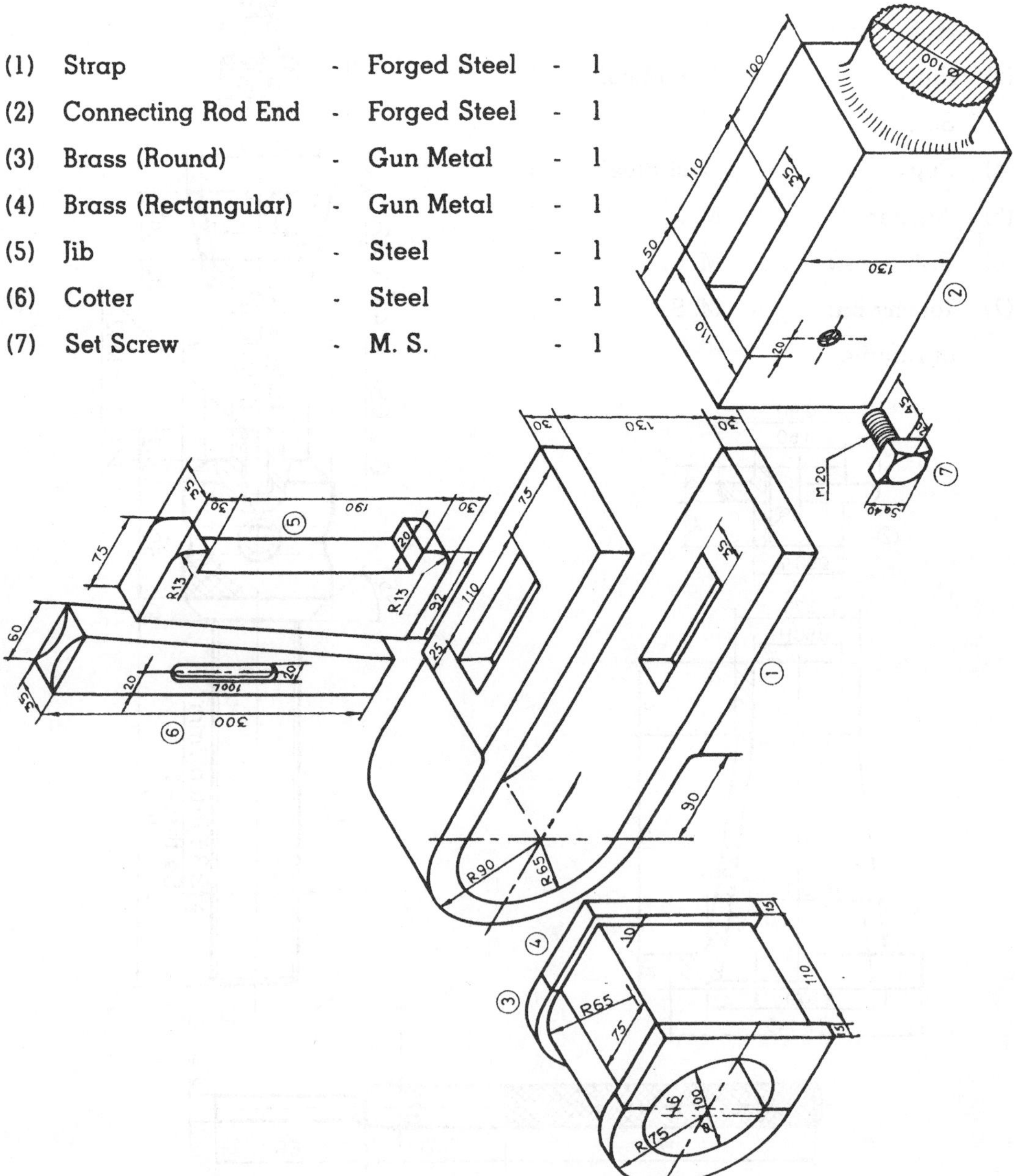

Problem 4 :

Pictorial views of all parts of a '*Friction Coupling*' are given in Fig.

Draw (1) Detail drawings.

(2) Assembly drawings. (2 views)

Friction Coupling Part List

(1)	Disc - 1	-	1	Cast Iron
(2)	Disc - 2	-	1	Cast Iron
(3)	Shaft for Disc - 1	-	1	Mild Steel
(4)	Shaft for Disc - 2	-	1	Mild Steel
(5)	Gib Headed Key	-	1	Mild Steel
(6)	Parallel Key	-	1	Mild Steel

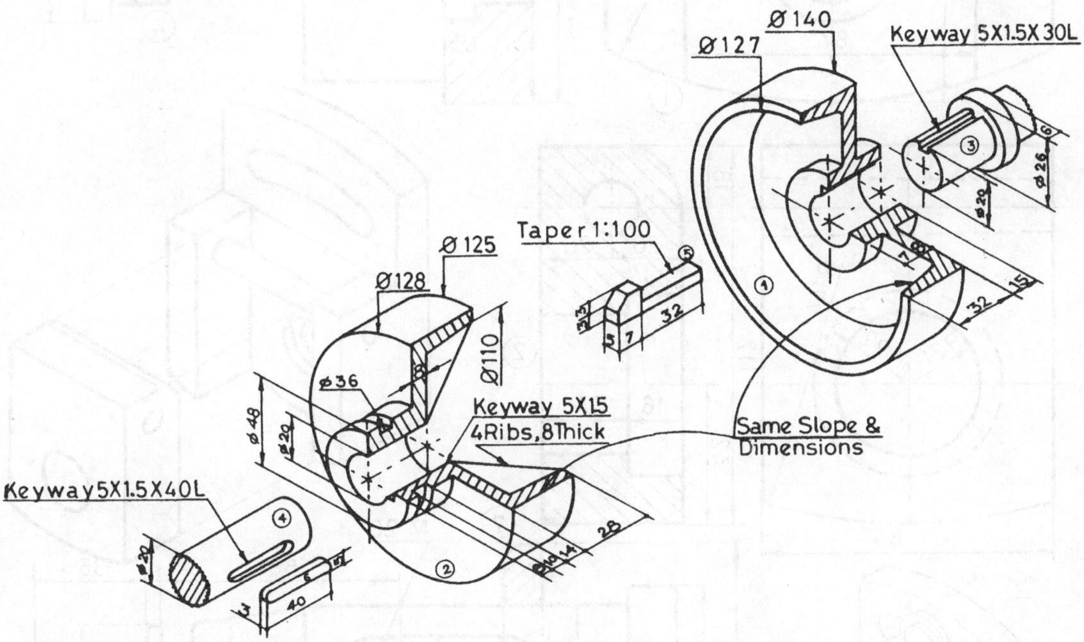

Problem 5 :

Detail drawings of all parts along with pictorial views of sub-assembly of a '**Tool Head of A Shaping Machine**' are given in Fig.

Draw (1) F. V.

(2) S. V.

and (3) T. V. of an Assembly.

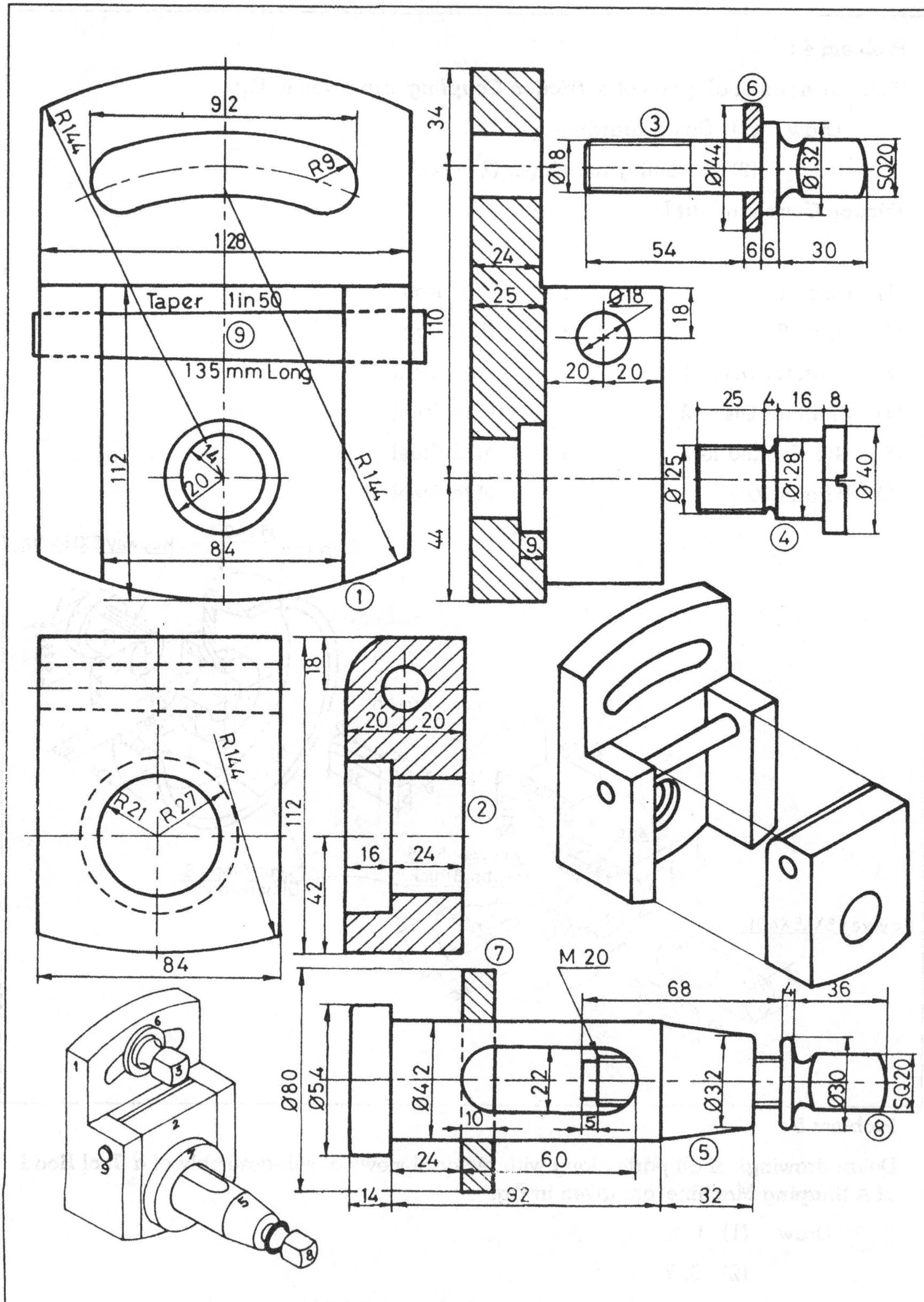

92
R144
R9
128
34
Ø18
③
⑥
Ø44
Ø32
SQ20
54
6 6
30
24
25
Ø18
18
Taper 1in 50
⑨
110
135 mm Long
20 20
25 4 16 8
112
14
20
R 144
Ø 125
Ø 28
Ø 40
④
44
9
84
①
18
20 20
R144
R21 R27
112
②
16 24
42
84
⑦
M 20
68
4 36
Ø80
Ø54
Ø42
22
Ø32
Ø30
SQ20
10
5
⑤
⑧
24
60
14
92
32

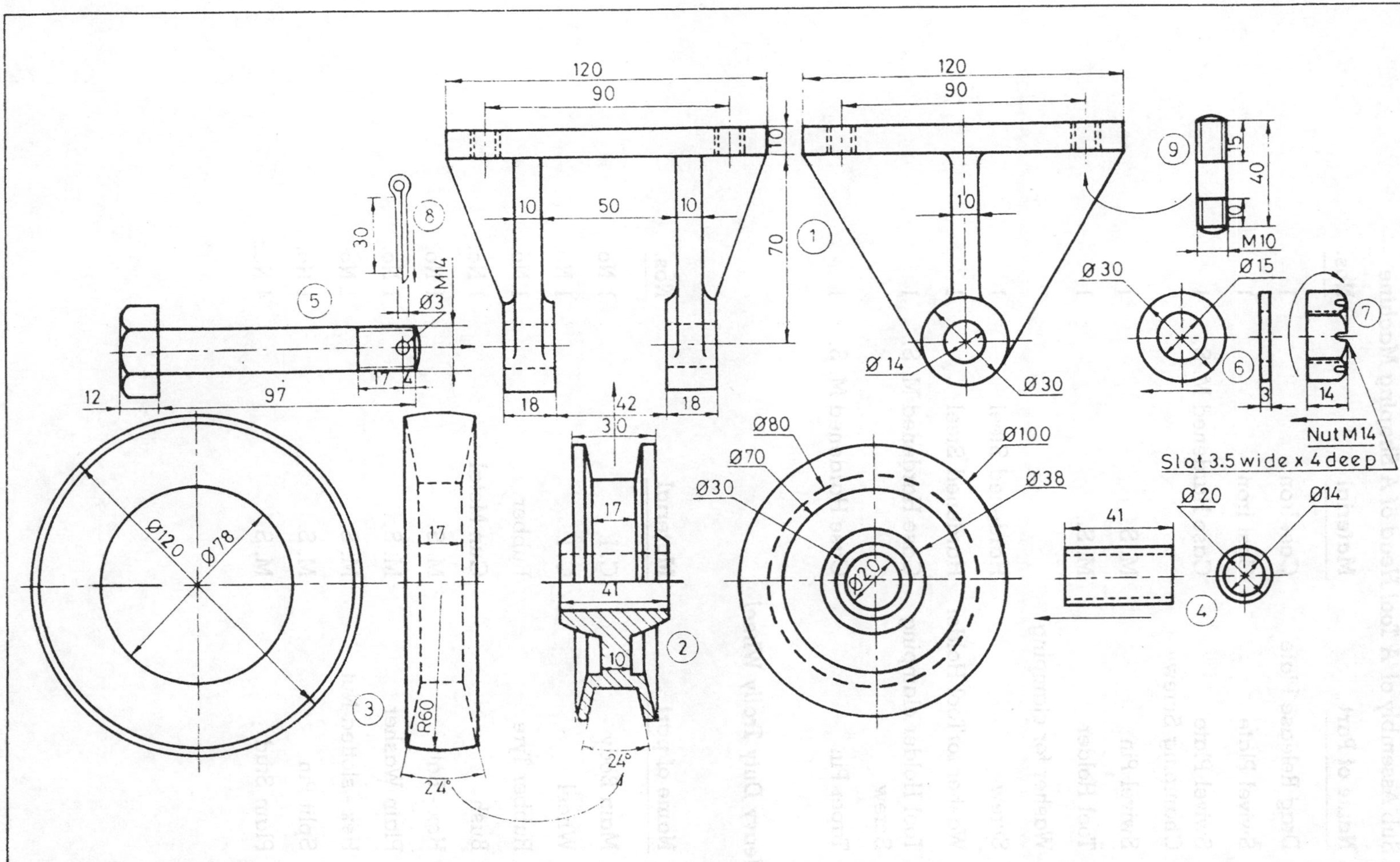

Problem 6 :

Detail drawings of all parts of a '***Heavy Duty Trolley Wheel***' are given in Fig.

Draw (1) F. V. and S. V. of an Assembly.

Details of A Sub Assembly of '***A Tool Head of A Shaping Machine***'

Sr. No.	Name of Part	Material	Nos.
(1)	Drag Release Plate	Cast Iron	1
(2)	Swivel Plate	Cast Iron	1
(3)	Swivel Plate Champing Screw	Case Hardened M. S.	1
(4)	Swivel Pin	M. S.	1
(5)	Tool Holder	M. S.	1
(6)	Washer for clamping Screw	Hardened Steel	1
(7)	Washer forTool Holder	Hardened Steel	1
(8)	Tool Holder clamping Screw	Case Hardened M. S.	1
(9)	Taper Pin	Case Hardened M. S.	1

Details of ***Heavy Duty Trolly Wheel***

Sr. No.	Name of part	Material	Nos.
(1)	Main Body	C. I.	1 No.
(2)	Wheel	C. I.	1 No.
(3)	Rubber Tyre	Rubber	1 No.
(4)	Bush	Gun Metal	1 No.
(5)	Hex - Bolt	M. S.	1 No.
(6)	Plain Washer	M. S.	1 No.
(7)	Hex - slotted Nut	M. S.	1 No.
(8)	Split Pin	M. S.	1 No.
(9)	Plain Stud	M. S.	4 Nos.

CHAPTER 20

DIMENSIONING

1. General :

Dimensioning is necessary in order to convey a clear and definite size-description of an object. A description of shape of object without dimensioning is meaningless. Manufacturer of component or object must know the exact length, width and height of a piece, diameters of shafts and holes and their locations, the amount of tolerance for machining and many other such details relating to manufacture of a particular component. Dimensioning supplies such information.

Perfect dimensioning is that which helps to ensure fabrication of piece exactly as intended by the designer. In perfect dimensioning, a worker is not required to do

(1) Calculations

(2) Assumption about missing dimensions

(3) Scaling of drawing, etc.

2. Dimensioning Elements : See Fig. 20.1

Following are the elements of dimensioning :

(i) Dimension Lines - (A) : The dimension lines are lighter lines than outlines of an object. They are drawn either full with dimension figures placed above lines or broken at or near the centre to allow for the insertion of the dimension figures. A dimension line is shown by letter (A) in Fig. 20.1

(ii) Extension Lines or Projection Lines - (B) : Extension lines are lighter lines used to indicate the extremities of the distance measured by the dimension line. According to new practice there should not be any gap between the extension lines and the outlines of the object. Extension lines should extend about 3mm beyond the dimension lines. These lines are drawn at right angles to the features being dimensioned. Centre lines may be extended to serve as extension lines. Extension lines are shown by letter (B) in Fig. 20.1

(iii) Arrowheads - (C) : See Figs. 20.1, 20.2 and 20.3

The dimension line is terminated by two arrowheads that touch the extension lines. The length of the arrowheads may vary from 3mm for small drawings to 5 mm for large drawings. The width/depth of the base of the arrowhead should be approximately 1/3 of the length of arrowheads. An arrowhead is drawn freehand, with two concave strokes made in the direction of the point or may be made triangular in shape. (See Fig. 20.2) Arrowheads should be drawn within the limits of the dimensioned feature. When enough space is not available, they should be placed outside. Adjacent arrowheads may be replaced by a dot. A dot may also replace a single arrowhead to indicate the common origin of successive dimensions. (See Fig. 20.3) Some of the wrong practices are shown in Fig. 20.3A.

(iv) **Dimension Figures** - (D) : The dimension figure may be placed in the centre above dimension line or in the break of the dimension line. All dimensions are mostly given in mm. The dimension figure is shown by letter (D) in Fig. 20.1

(V) **Leaders or Pointer Lines** - (E) : See Fig. 20.1

The leader is a thin straight line that extends from a dimension figure or note towards the form to which it applies. One end of the leader ends in an arrowhead that touches the outline of the object and other end of it terminates with shoulder of about 3 mm length. Leaders are never drawn horizontal or vertical to avoid confusion with lines of the object but they are drawn at 30°, 45° or 60°. A leader is shown in Fig. 20.1 by letter (E).

(vi) ***Notes*** - (F) : Instruction and basic informations relating to the manufacture of a part, such as "5 mm drilled & tapped", "6 mm drill and ream" or "chamfer 3 x 45° are placed adjacent to the features to which they apply and be connected to them by leaders. In Fig. 20.1 Note " Ø 20 CSK AT 90° to Ø 30" is shown.

(vii) ***Surface Roughness Symbols*** - (G) : See Fig. 20.4.

The basic symbol consists of two legs of unequal length inclined at approximately 60° to the line representing the surface under consideration as shown in Fig. 20.4.1

If the removal of material by machining is required, a bar is added to the basic symbol, as shown in Fig. 20.4.2.

If the removal of material is not permitted a circle is added to the basic symbol as shown in Fig. 20.4.3

When special surface characteristics have to be indicated a line is added to the longer leg of any of the above symbols as shown in Fig. 20.4.4

The value or values defining the principal criterion of roughness are added to the symbols given in Fig. 20.4.1 to 20.4.3 as shown in Fig. 20.4.5 to 20.4.7. Surface roughness specified as in

(a) Fig. 3.4.5, may be obtained by any production method

(b) Fig. 3.4.6 shall be obtained by removal of material by machining

(c) Fig. 3.4.7 shall be obtained without removal of material

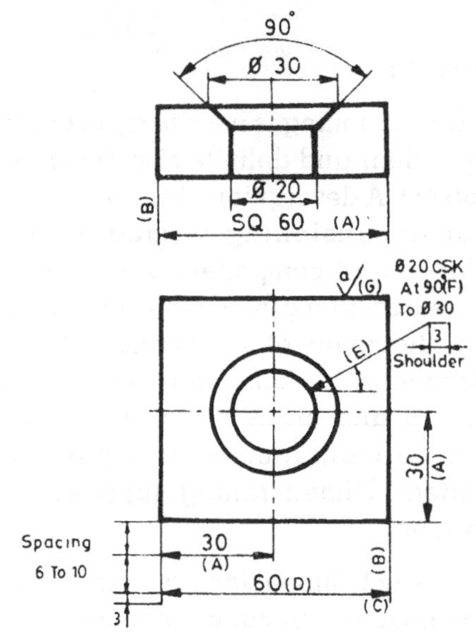

Fig. 20.1

Fig. 20.2

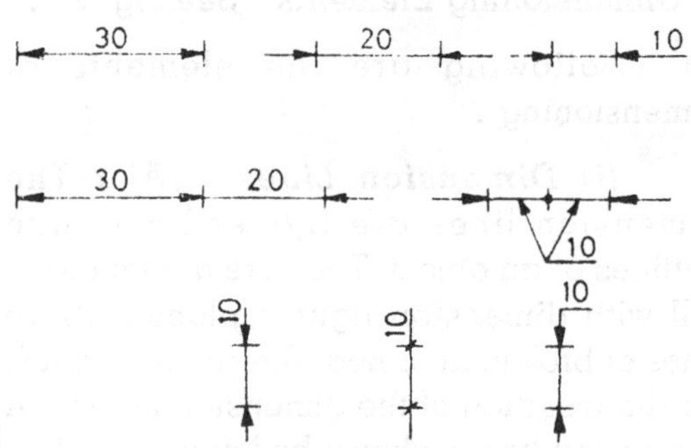

Fig. 20.3

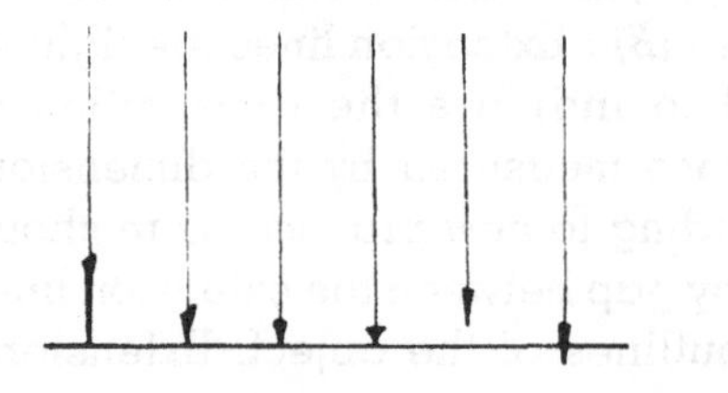

Fig. 20.3 A

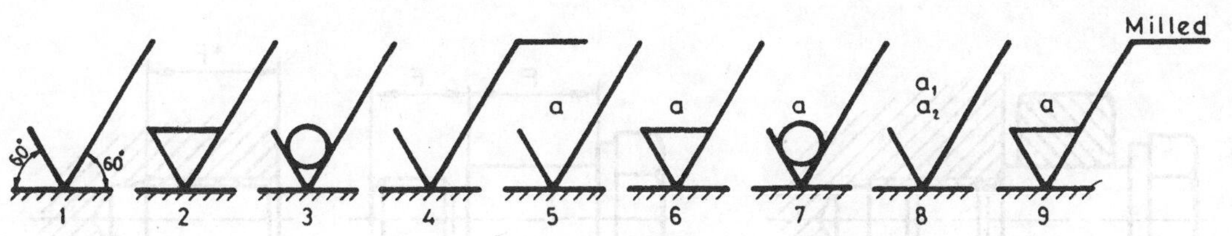

Fig. 20.4

When only one value is specified it represents the maximum permissible value of surface roughness. If it is necessary to impose maximum and minimum limits of the principal criterion of surface roughness, both values should be shown, as in Fig. 20.4.8 with the maximum limit a_1 above the minimum limit a_2.

If it is required that the final surface roughness be produced by one particular production method, this method should be indicated in plain language on an extension of the longer leg of the symbol given in Fig. 20.4.4 as shown in Fig. 20.4.9

3. Size and Location Dimensions : See Fig. 20.5

There are two kinds of dimensions -

(i) Size (S) dimensions, which describe the size of a feature

(ii) Location (L) dimensions, which specify the location of a feature.

In Fig. 20.5 size dimensions are shown by letter (S) in bracket and location dimensions are shown by letter (L) in bracket.

Height, length, width, thickness, diameter of pin or hole, etc. are known as size dimensions and distance between centre lines of holes or between centre line of hole and one surface are known as location dimensions.

4. Functional and Datum Dimensions: See Fig. 20.6

A functional dimension is a dimension which is essential for the function of a part.

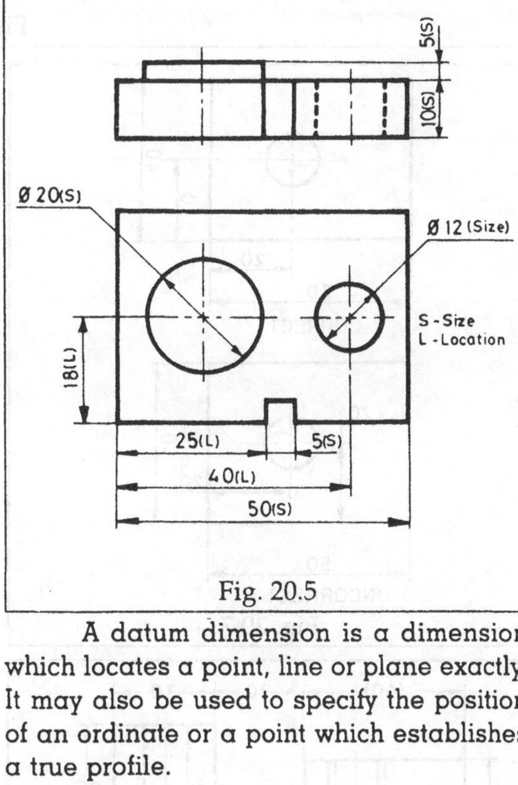

Fig. 20.5

A datum dimension is a dimension which locates a point, line or plane exactly. It may also be used to specify the position of an ordinate or a point which establishes a true profile.

5. Placing of Dimensions :

As far as possible dimensions, notes, etc. should be placed outside the outlines of the view. However, sometimes it becomes necessary to place the dimension on or near the feature for clarity of presentation. If the feature is well away from the outline, it is impossible to place the dimension outside the outline of the view and hence it may be placed inside (See Fig. 20.7 and Fig. 20.8)

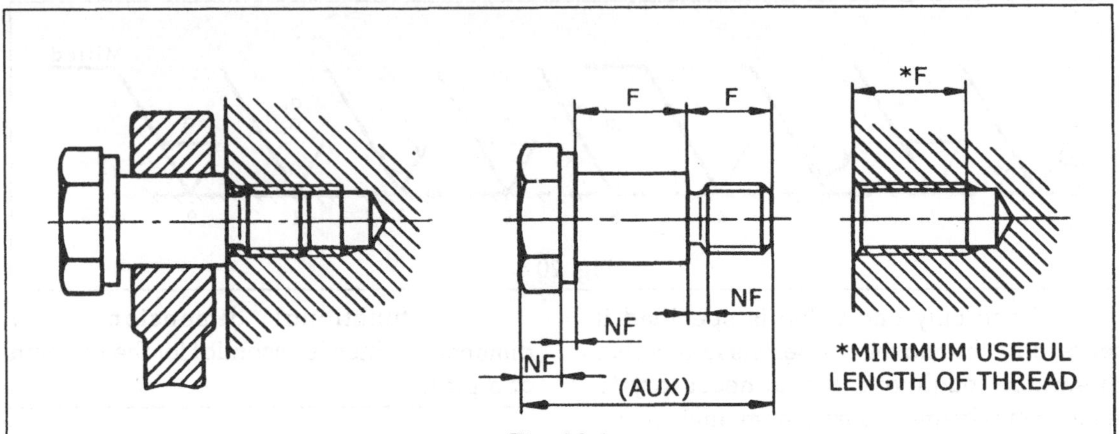

Fig. 20.6

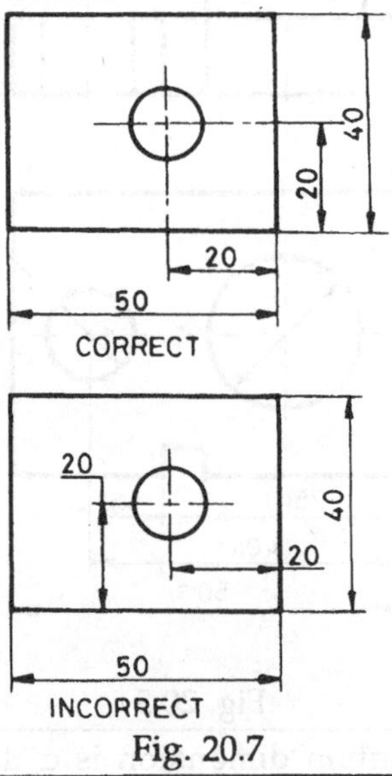

Fig. 20.7

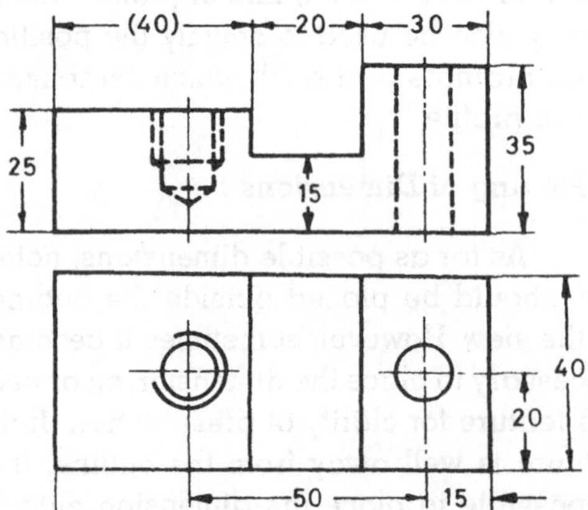

Fig. 20.8

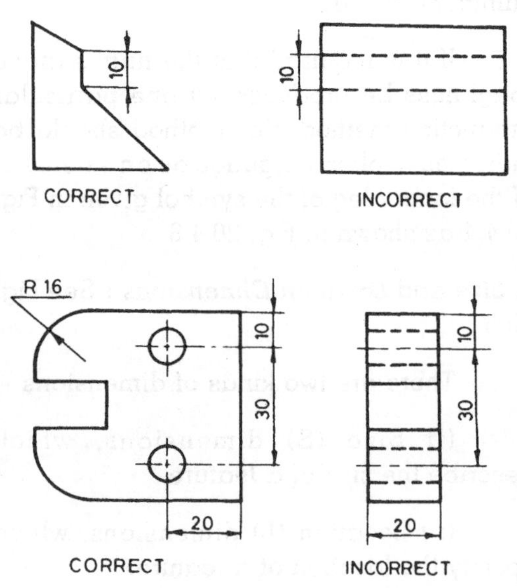

Fig. 20.9

Dimensions should be placed on the view which shows the relevant features most clearly. (See Fig. 20.9) Dimensions, placed between two views, should apply to one view only. Dimensions should not be placed too close to each other or to the parts being dimensioned.

Dimension lines should be placed 6 mm away from the outlines of the object. Smaller dimension should be placed nearer to the outline and larger dimension should

be placed away from the outline. Gap between two dimension lines should be nearly 6 mm. This method minimizes the possibility of crossed extension lines. (See Figs. 20.10 and 20.11)

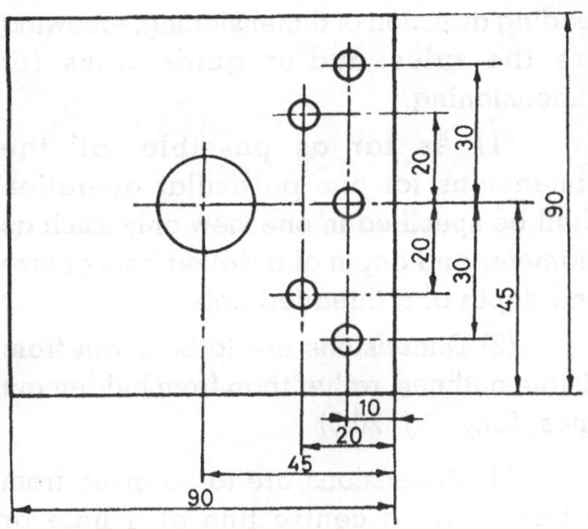

Fig. 20.10

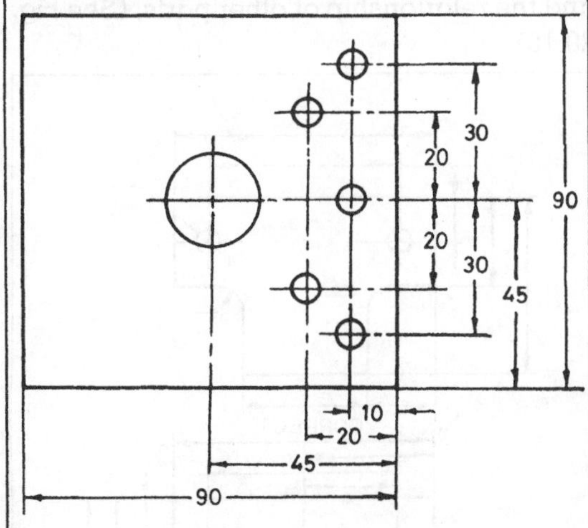

Fig. 20.11

Continuous dimensions are preferred to staggered dimensions. Due to lack of space, figures are staggered. (See Fig. 20.12) Mostly dimension figures (numbers) are kept in the centre of the dimension lines. However, when number of parallel dimension lines are many they may be staggered. Crowding of dimensions in small space should be avoided by placing dimensions in different views.

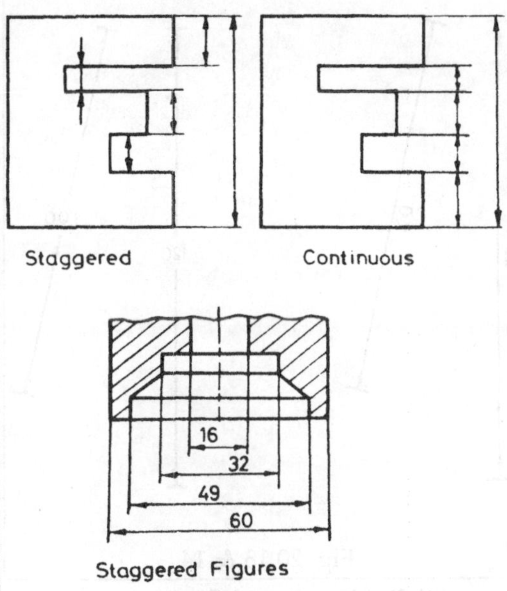

Fig. 20.12

Dimensioning to hidden lines should be avoided. In fact that dimensions should be placed where that feature is seen. (see Fig. 20.9)

6. *Reading Direction of Dimensions :*

The two recommended directions of placing dimensions are :

(i) Aligned system, See Figs. 20.13 and 20.10.

(ii) Unidirectional system, see Figs. 20.14 and 20.11.

(i) Aligned System :

In this system all dimension figures are so placed that they may be read from the bottom or from the right hand side of the drawing sheet and all dimension figures should be placed at right angles to and above the dimension lines.

In aligned system of dimensioning as far as possible placing of dimensions in the shaded area shown in the Fig. 20.15 should be avoided. Because dimensions in this area violate the principle of readability from bottom or from right hand side.

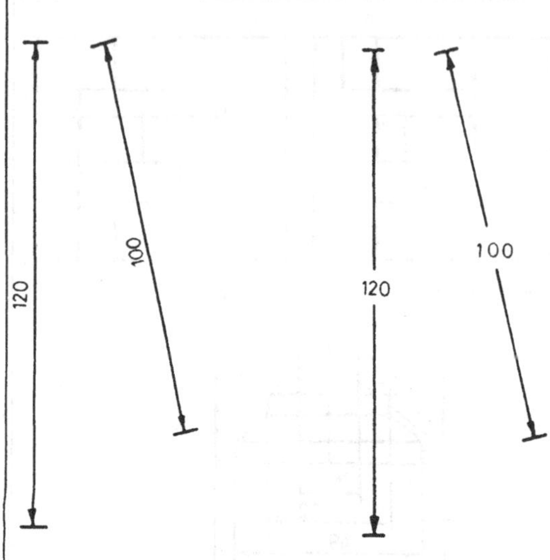

Fig. 20.13 & 14

(ii) Unidirectional System :

In this system all dimensions are so placed that they may be read from the bottom edge of the drawing sheet. In this system there is no restriction about the direction of dimension lines. This system is advantageous on larger drawings, where it is inconvenient to read dimensions from the right hand side. It is extensively used in preparing drawings of aircraft and automobile industries.

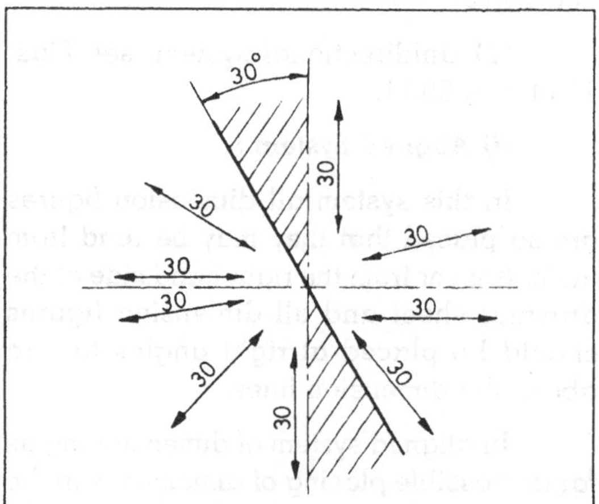

Fig. 20.15

7. General Rules, Methods and Guidelines for Dimensioning :

Dimensioning is an art. One must study these rules and guidelines to master the art of dimensioning. Over and above what we have studied in dimensioning elements, placing of dimensions and reading direction of dimensioning, following are the rules and/or guide lines for dimensioning.

(1) As far as possible, all the dimensions for one particular operation shall be specified in one view only, such as diameter and depth of a drilled hole or size and depth of a threaded hole.

(2) Dimensions are to be given from visible outlines, rather than from hidden out lines. (See Fig. 20.9)

(3) Dimensions are to be given from a base line, a centre line of a hole or cylindrical part, an important hole or a finished surface which may be readily established, based on design requirements and the relationship of other parts, (See Fig. 20.16)

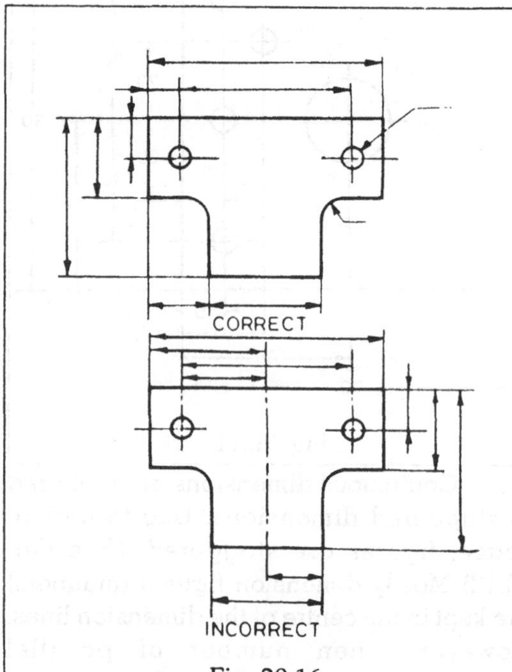

Fig. 20.16

(4) Dimensioning to a centre line should be avoided except when the centre line passes through the centre of a hole, or a cylindrical part, as shown in Fig. 20.16

(5) Dimensions for different operation on a part, for example drilling and bending should be given separately, as shown in Fig. 20.17

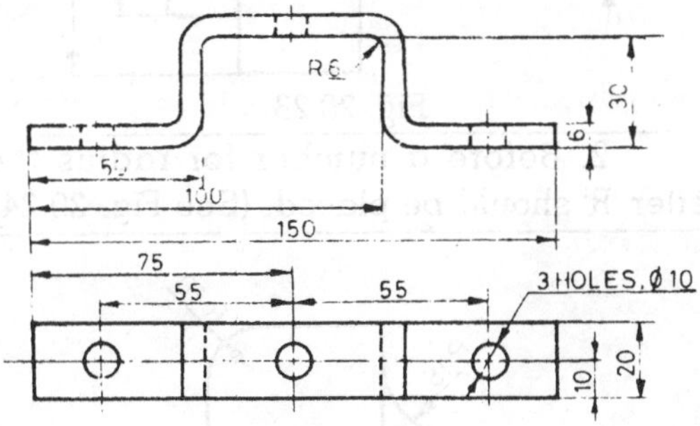

Fig. 20.17

(6) An axis or a contour line should never be used as a dimension line but may be used as an extension or projection line. (See Fig. 20.18)

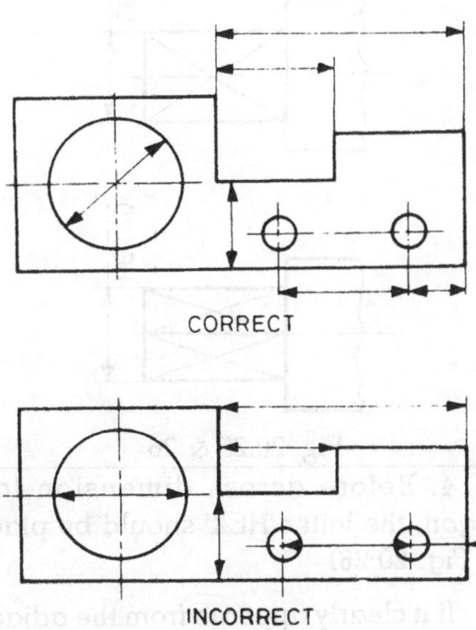

Fig. 20.18

(7) Intersection of extension lines should be avoided as far as possible. If however, the intersection of two extension lines is unavoidable, the lines should not be broken. However, dimension lines should never cross.

(8) Overall dimensions should be placed outside the intermediate dimensions, where an overall dimension is shown, one of the intermediate (less important) dimensions is redundant and should not be dimensioned. (See Fig. 20.19)

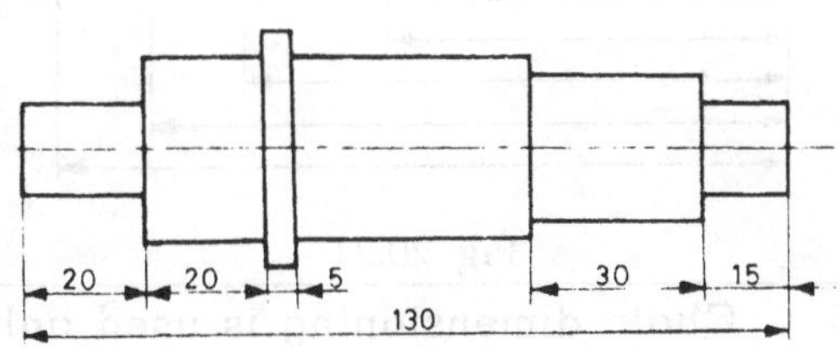

Fig. 20.19

(9) Dimensions should be quoted once only. It should not be repeated in other views.

(10) As far as possible, all dimensions in one particular drawing be expressed in one unit only. The recommended unit is millimeter. There is no necessity to write mm with the number.

(11) The decimal point in a dimension should be bold and should be in line with the bottom line of the figures. When the dimension is less than unity, a zero should precede the decimal point, such as 0.45.

(12) Depending upon requirements three arrangements of dimensions are practised.

(i) Chain dimensioning, see Fig. 20.20.

(ii) Parallel dimensioning, see Fig. 20.21.

(iii) Combined dimensioning.

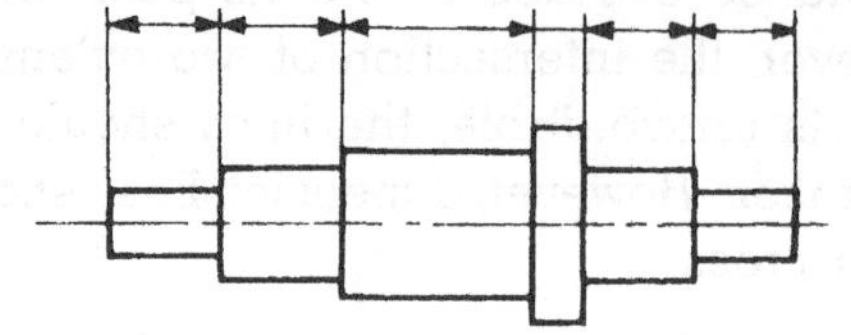

Fig. 20.20

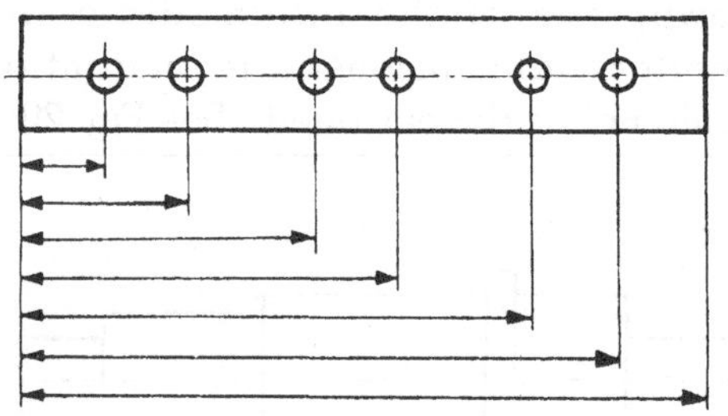

Fig. 20.21

Chain dimensioning is used only where the possibility of accumulation of tolerances does not endanger the functional requirements of the part. Parallel dimensioning is used where number of dimensions of the same direction have a common datum feature.

Combine dimensioning is nothing but combination of chain and parallel dimensioning.

(13) Numerals and letters should be large enough to ensure easy reading.

(14) Dimensions of parts which are not drawn to scale should be underlined. (See Fig. 20.22) However, when the whole drawing is not to scale, the abbreviation 'NTS' may be used.

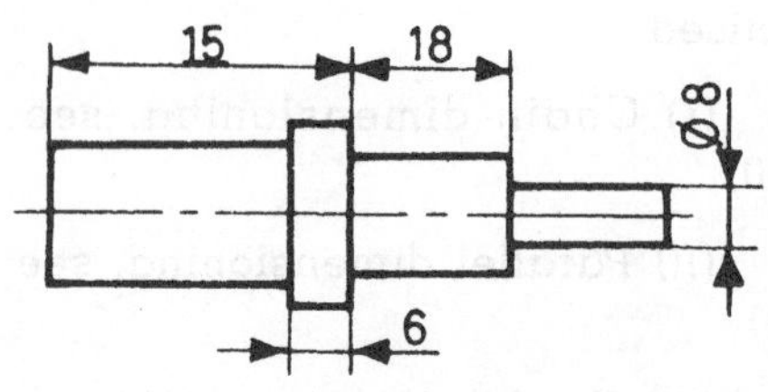

Fig. 20.22

(15) The following letters and symbols should be placed along with number to identify shape of a feature :

1. Before a number for diameter, the symbol 'Ø' should be placed. (See Fig. 20.23)

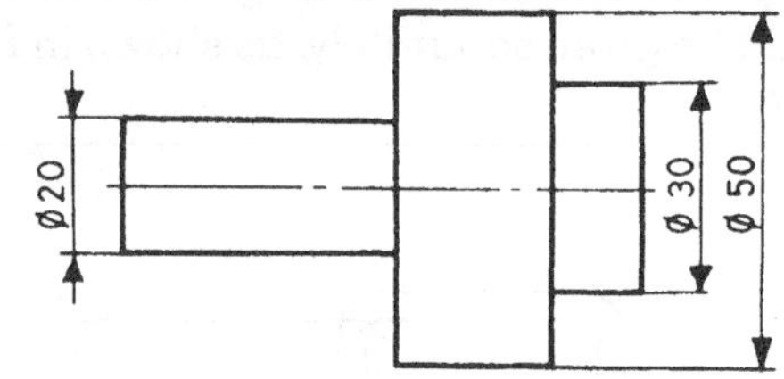

Fig. 20.23

2. Before a number for radius the letter 'R' should be placed. (See Fig. 20.24)

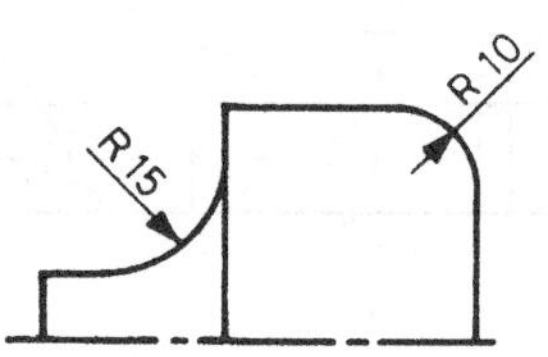

Fig. 20.24

3. Before a number for square, the letters 'SQ' should be placed. (See Fig. 20.25)

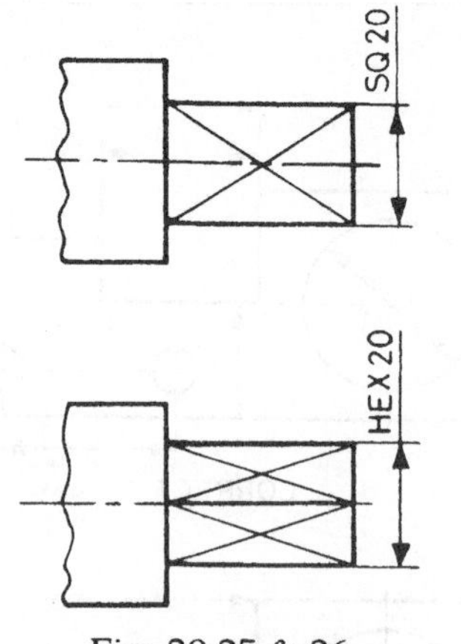

Fig. 20.25 & 26

4. Before across dimension for a hexagon, the letter 'HEX' should be placed. (See Fig. 20.26)

If it clearly appears from the adjacent view that the dimension applies to a square or an hexagonal section, the letters 'SQ' or 'HEX' may be omitted. The radius or diameter dimension of a spherical surface should be preceded by the word 'SPHERE'. (See Figs. 20.27 and 20.28)

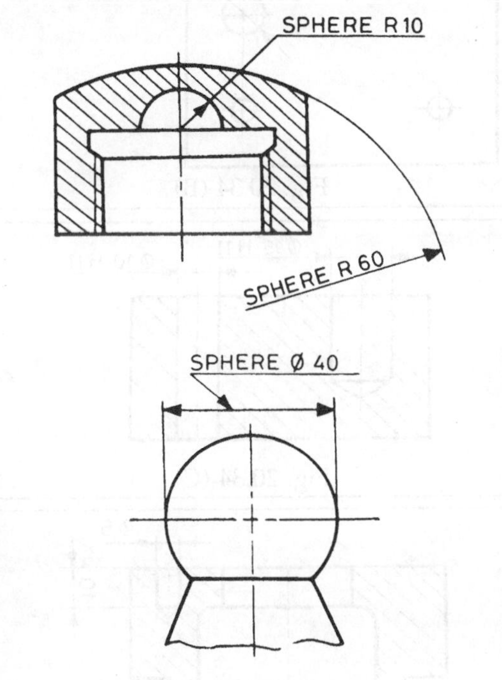

Fig. 20.27 & 28

(16) Where the space is limited, method shown in Fig. 20.29 is adopted.

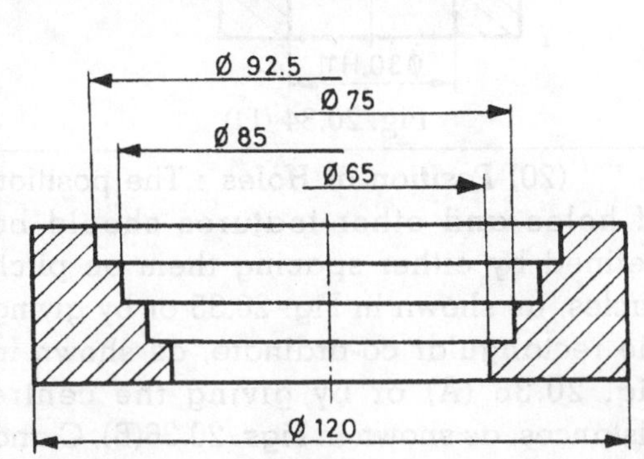

Fig. 20.29

(17) Circles should be dimensioned by any one method shown in Fig. 20.30. Selection of method depends upon size of a circle.

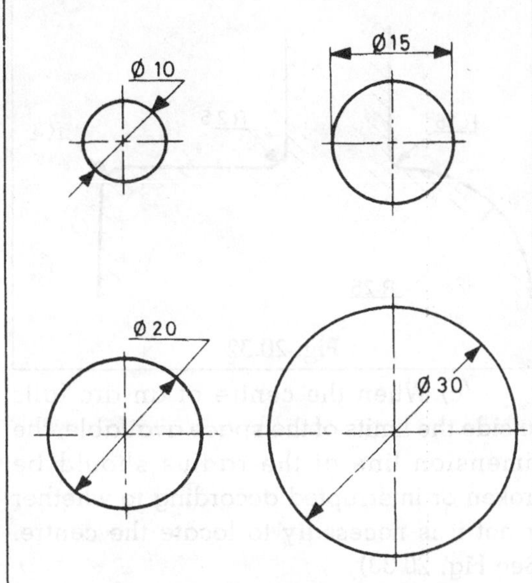

Fig. 20.30

(18) *Radii* :

(a) As far as possible the dimension line of a quarter circle radius should pass through the centre of arc. Where the centre is required for its location, it should be marked by a clear dot. While dimensioning small radii, the arrow should be reversed, as shown in Fig. 20.31.

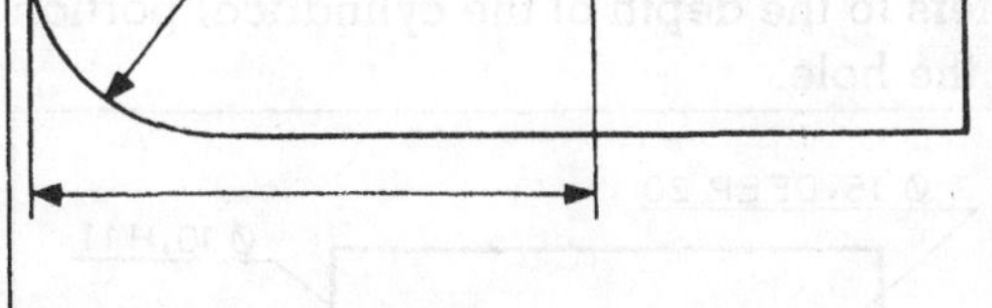

Fig. 20.31

(b) Radii of arcs which need not have their centres located should be dimensioned as shown in Fig. 20.32 and Fig. 20.33.

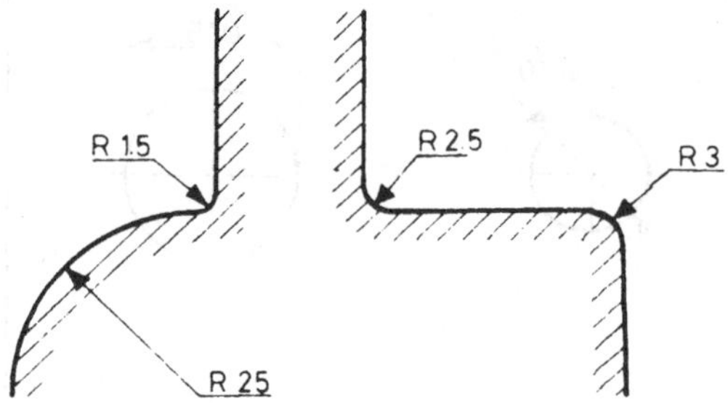

Fig. 20.32

(c) When the centre of an arc falls outside the limits of the space available, the dimension line of the radius should be broken or interrupted according to whether or not it is necessary to locate the centre. (See Fig. 20.33)

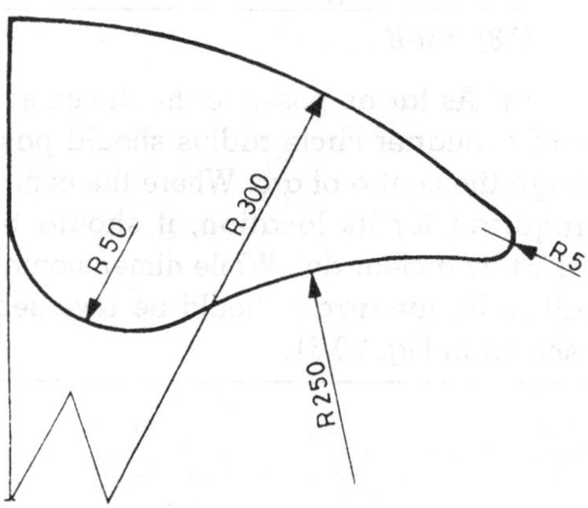

Fig. 20.33

(19) ***Hole Sizes*** : Figures 20.34 A, B, C and D are some of the typical examples for dimensioning hole sizes. Whenever depth of a hole is specified in the note form, it refers to the depth of the cylindrical portion of the hole.

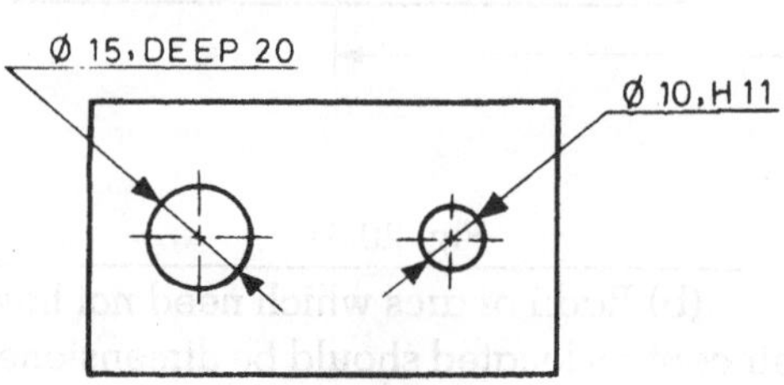

Fig. 20.34 (A)

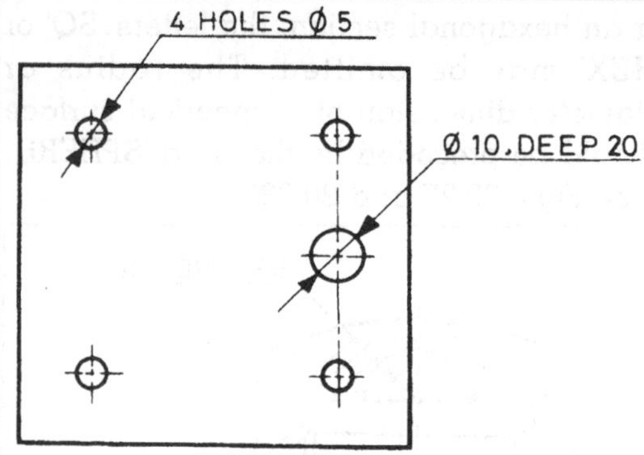

Fig. 20.34 (B)

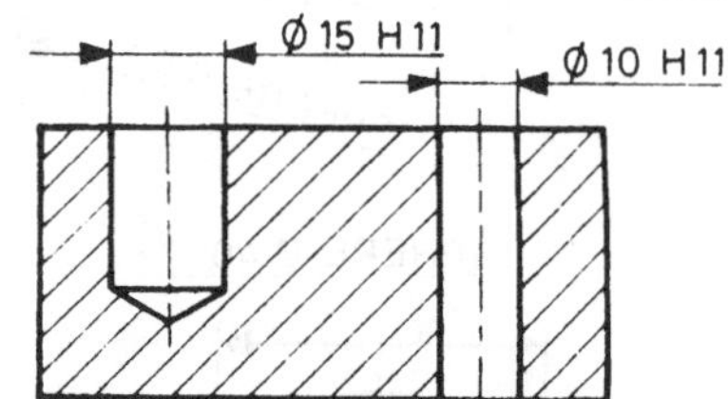

Fig. 20.34 (C)

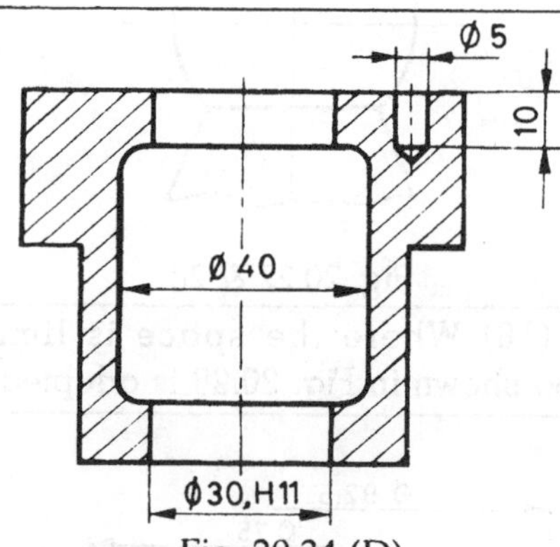

Fig. 20.34 (D)

(20) ***Position of Holes*** : The position of holes and other features should be defined by either spacing them on pitch circles, as shown in Fig. 20.35 or by giving the rectangular co-ordinate, as shown in Fig. 20.36 (A) or by giving the centre distances, as shown in Figs. 20.36(B), C and D. Where holes symmetrical to edges are to be dimensioned, the convention shown in Fig. 20.36 (E) is to be followed. Holes are dimensioned by using coordinates, even though they are lying on a pitch circle, see Fig. 20.36 (A).

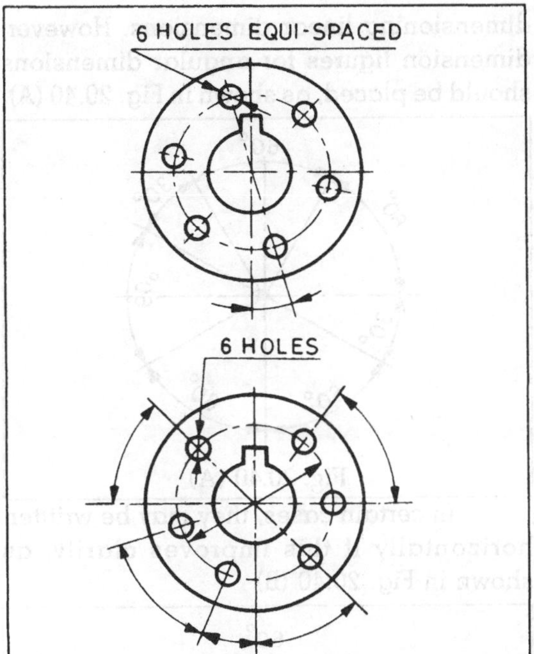

Fig. 20.35

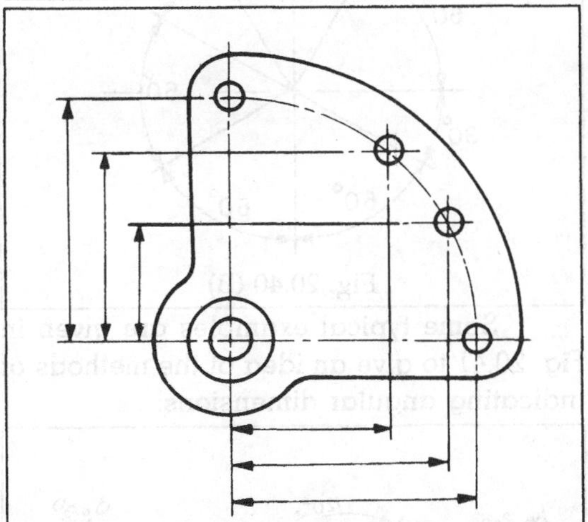

Fig. 20.36 (A)

Holes are dimensioned by using co-ordinates from the centre of a hole. [See Fig. 20.36 (B)]

Holes are positioned relative to one edge only. [See Fig. 20.36 (C)].

Symmetry of holes defined. [See Fig. 20.36 (D)]

Symmetry of hole with edges of plate is implied. [See Fig. 20.36 (E)]

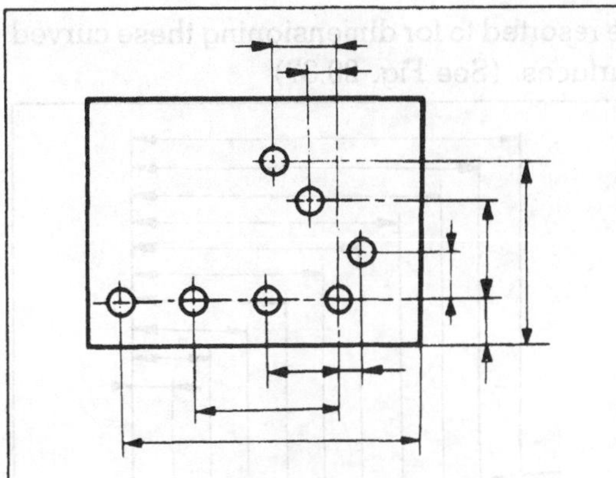

Fig. 20.36 (B)

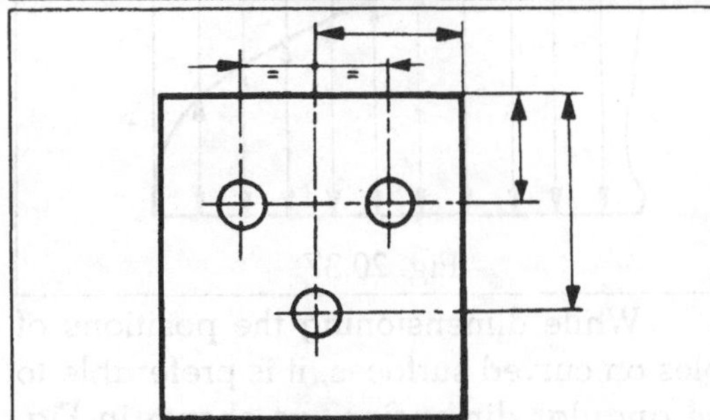

Fig. 20.36 (C)

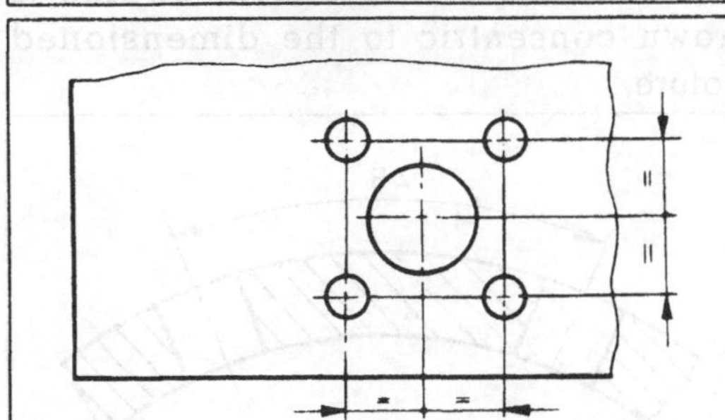

Fig. 20.36 (D)

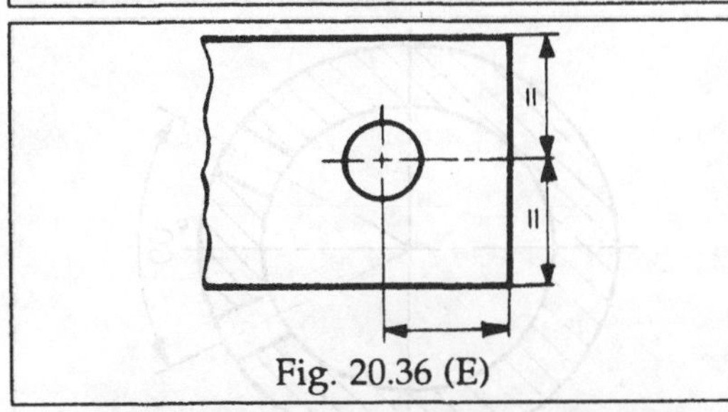

Fig. 20.36 (E)

(21) ***Dimensioning Curved Surfaces*** :

A curved line or surface composed of circular arcs, be preferably dimensioned by using radii. However, if this method is impracticable the use of co-ordinates may be resorted to for dimensioning these curved surfaces. (See Fig. 20.37)

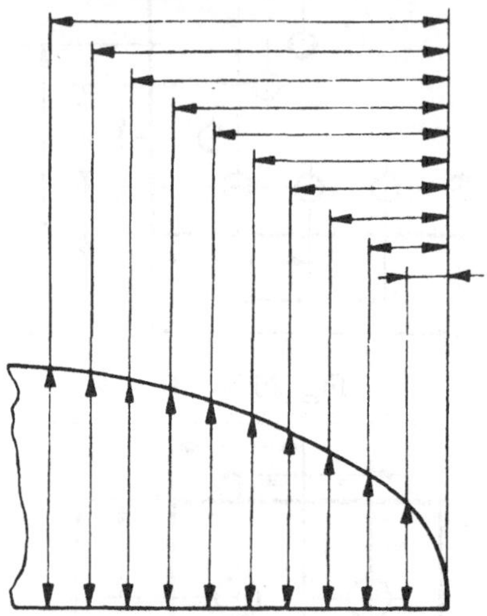

Fig. 20.37

While dimensioning the positions of holes on curved surfaces, it is preferable to use angular dimension, as shown in Fig. 20.38 and 20.39. The dimension line is drawn concentric to the dimensioned feature.

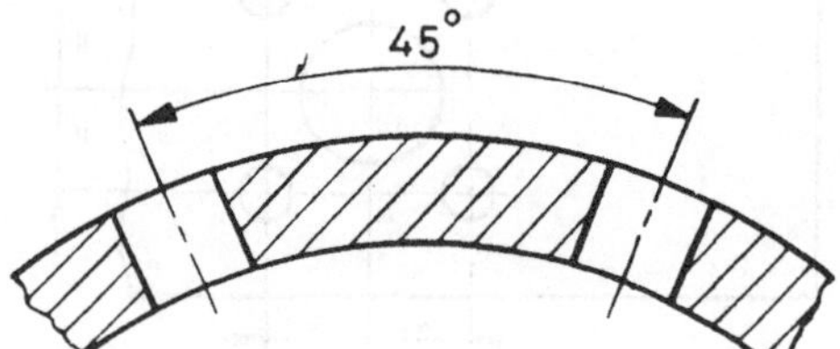

Fig. 20.38

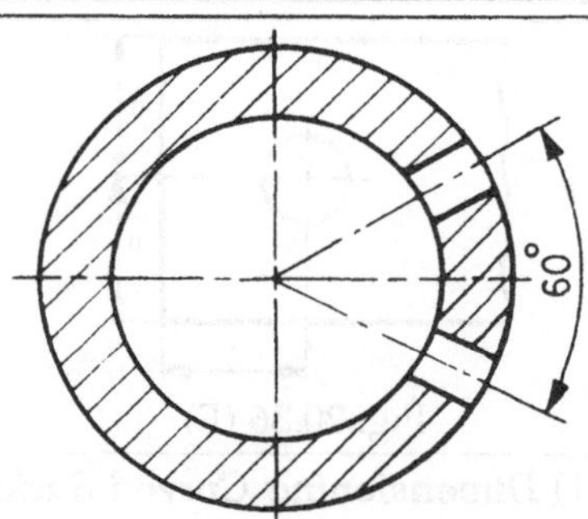

Fig. 20.39

(22) ***Angular Dimensioning*** : Angular dimensions and their deviations are dimensioned in the same manner as that of dimensioning linear dimensions. However, dimension figures for angular dimensions should be placed, as shown in Fig. 20.40 (A).

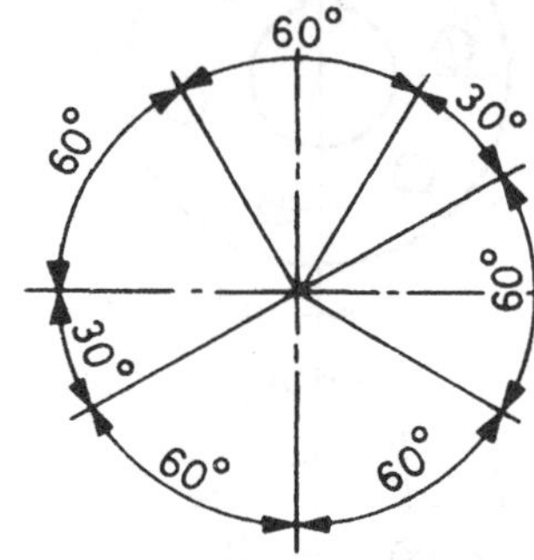

Fig. 20.40 (A)

In certain cases, they may be writtten horizontally if this improves clarity, as shown in Fig. 20.40 (B)

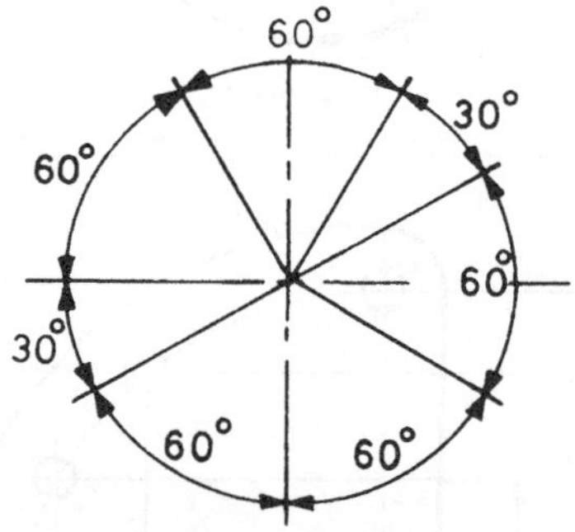

Fig. 20.40 (B)

Some typical examples are given in Fig. 20.41 to give an idea of the methods of indicating angular dimensions.

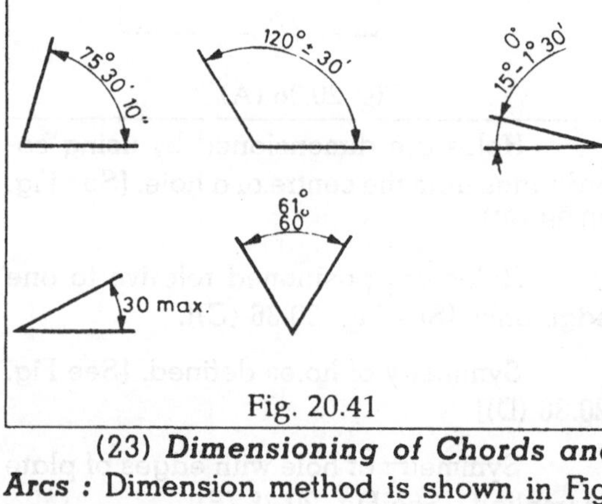

Fig. 20.41

(23) ***Dimensioning of Chords and Arcs :*** Dimension method is shown in Fig. 20.42 while a chordal dimension refers to

the straight distance, arcual distance is the exact arc length, both being equal to the length of the corresponding dimension line. In the case of arcual dimensions, the dimension shall be identified by (⌒).

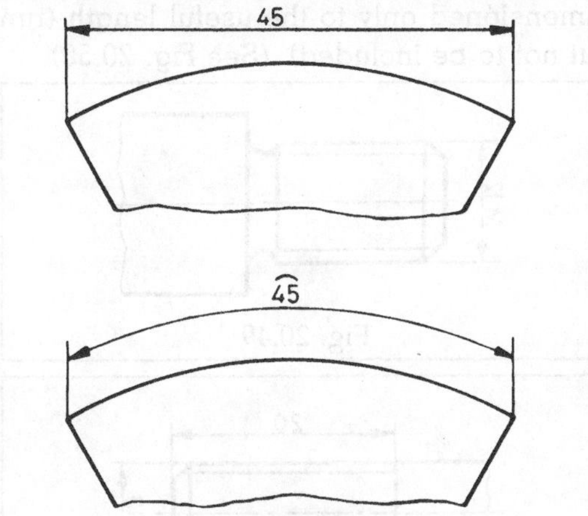

Fig. 20.42

(24) ***Dimensioning Bent Parts*** : The method of dimensioning bent parts is shown in Fig. 20.43, Fig. 20.44 and Fig. 20.45. Bends in tubes and pipes should preferably be dimensioned to the centre line. (See Fig. 20.46)

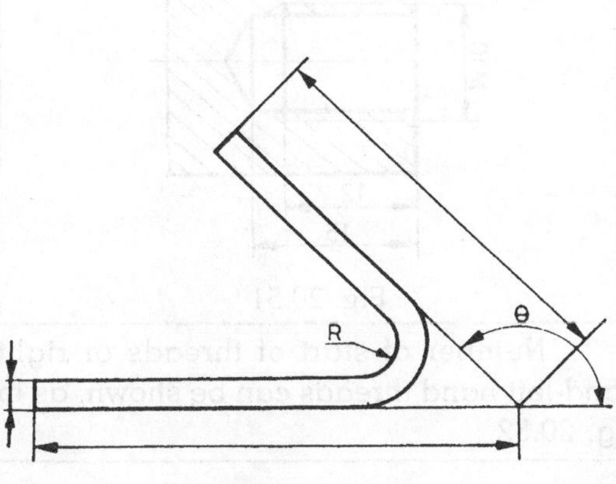

Fig. 20.43

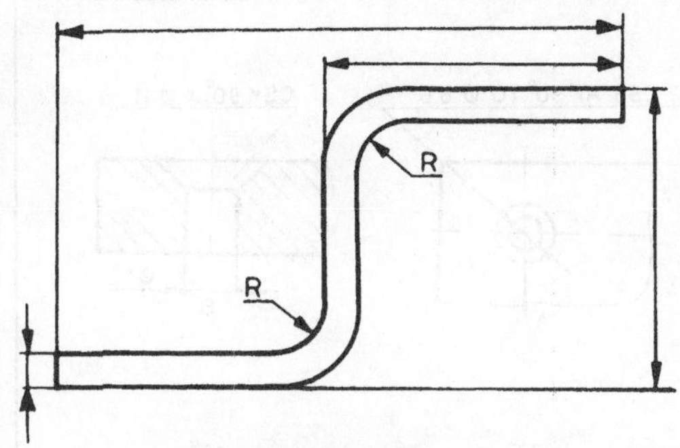

Fig. 20.44

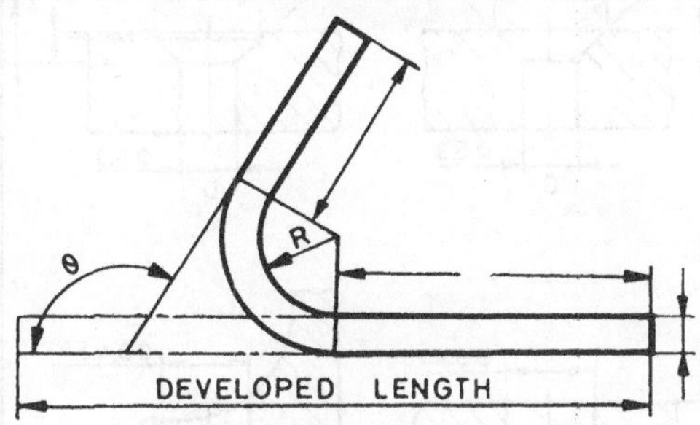

Fig. 20.45

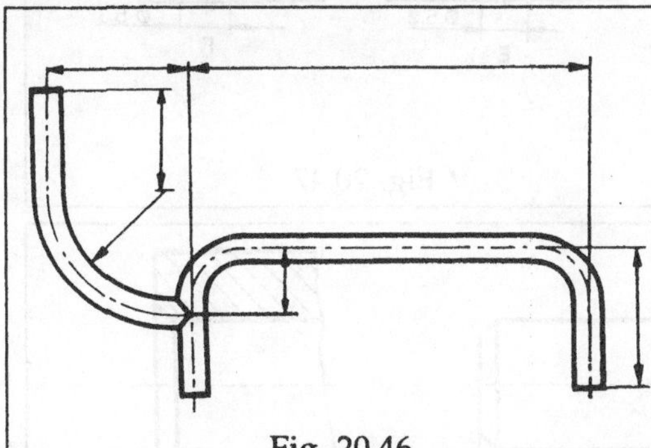

Fig. 20.46

(25) ***Dimensioning of Countersinks, Counter bores and Spot Faces*** : Some typical examples for the method of dimensioning these features are shown in Fig. 20.47 A, B, C, D, E and F.

(26) ***Dimensioning of Chamfers*** : Methods of dimensioning of chamfers are shown in Fig. 20.48 (A), (B), (C) and (D). While dimensioning 45° chamfer angles, the (usual) methods, shown in Fig. 20.48 (A) and (B) are to be followed.

While dimensioning chamfer angle other than 45°, methods, shown in Fig. 20.48

(C) and (D) are to be followed.

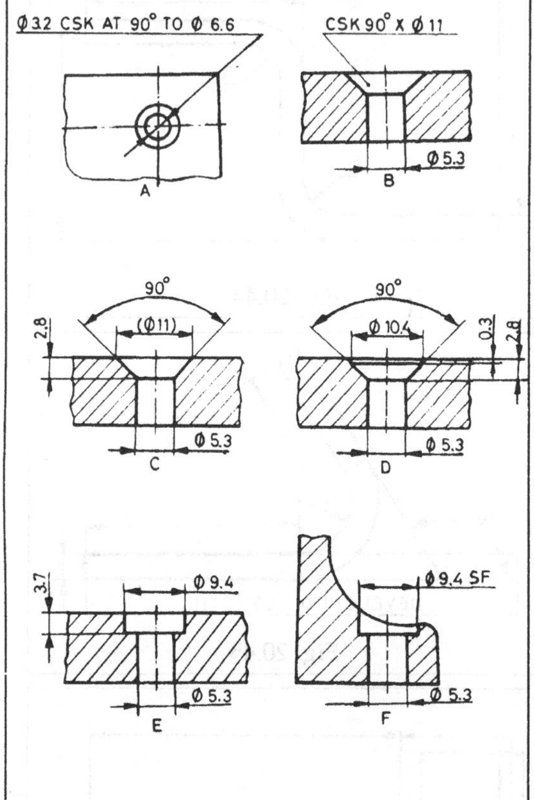

Fig. 20.47

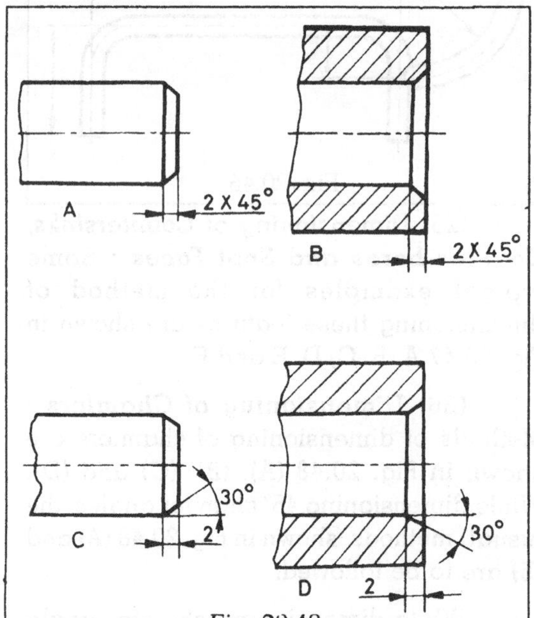

Fig. 20.48

(27) ***Methods of Dimensioning Screw Threads*** : Screw threads are dimensioned by indicating the corresponding designation for the size of the threads. Fig. 20.49 shows the method of dimensioning external threads. Length of threads are to be dimensioned only to the useful length (run out not to be included). (See Fig. 20.50)

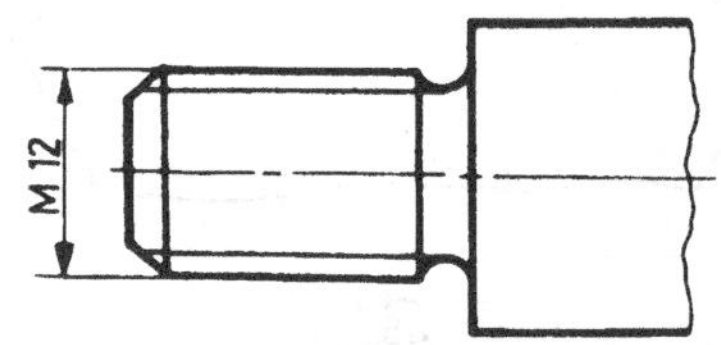

Fig. 20.49

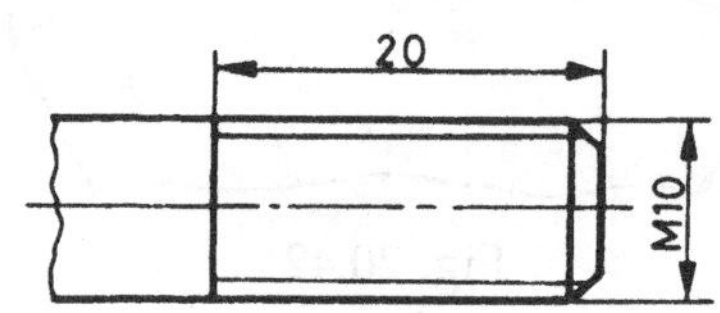

Fig. 20.50

While dimensioning internal threads, it is also necessary to dimension the length of drilled hole before tapping. (See Fig. 20.51)

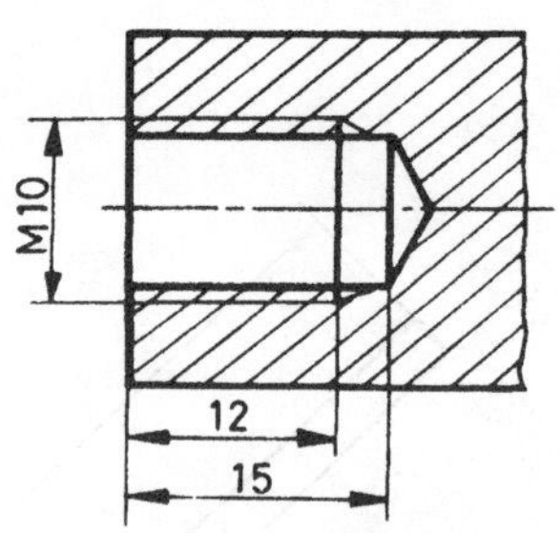

Fig. 20.51

Number of start of threads or right hand-left hand threads can be shown, as in Fig. 20.52.

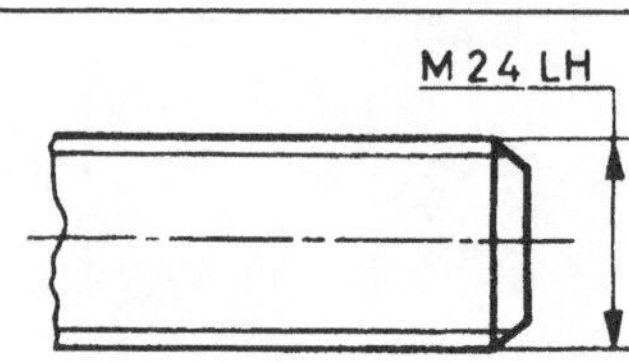

Fig. 20.52 (A)

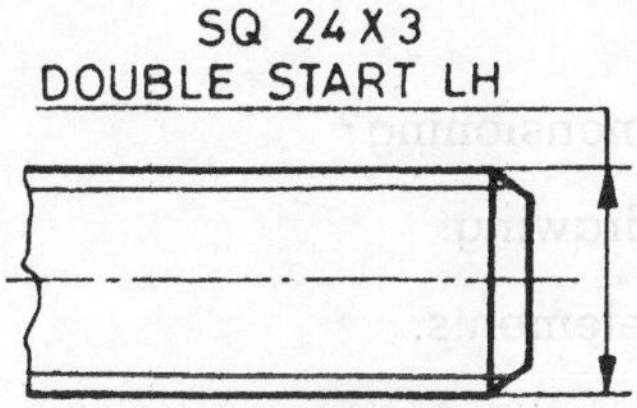

Fig. 20.52 (B)

For example, square threads of 24 mm nominal diameter, 3 mm pitch, double start with left hand threads shall be designated as under :

SQ 24 x 3 DOUBLE START LH.

(28) ***Dimensioning Keyways*** : Keyways for parallel and tapered keys in cylindrical shafts and hubs shall be dimensioned as shown in Figs. 20.53, 20.54, 20.55.

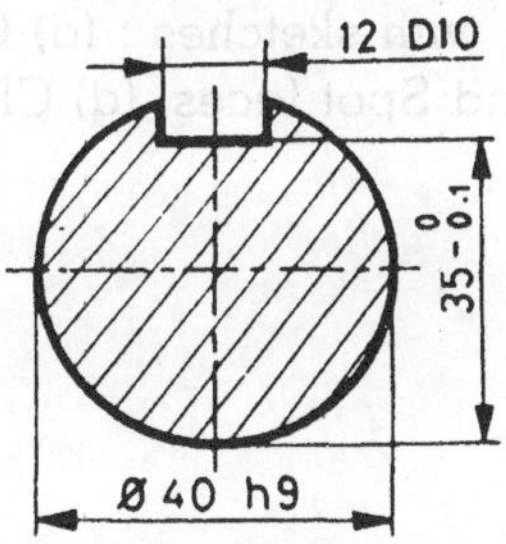

Fig. 20.53

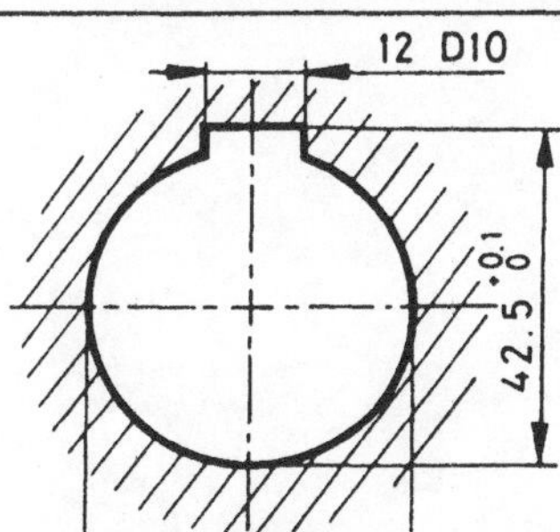

Fig. 20.54

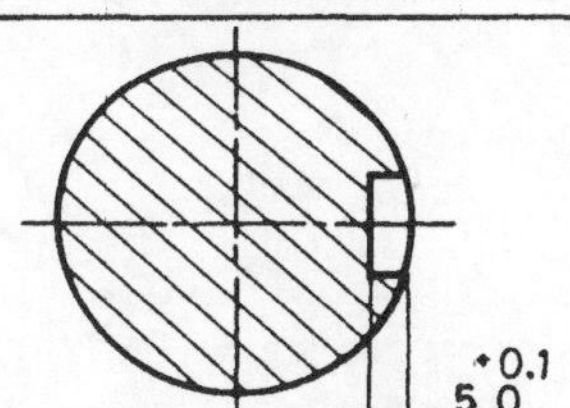

Fig. 20.55

(29) ***Dimensioning of Tapered Features*** : The following dimensions may be used in suitable combinations depending on technological requirements, to define the size and form of a tapered feature.

(a) Diameter (or width) at each end of the tapered feature, see Fig. 20.56 or 20.57.

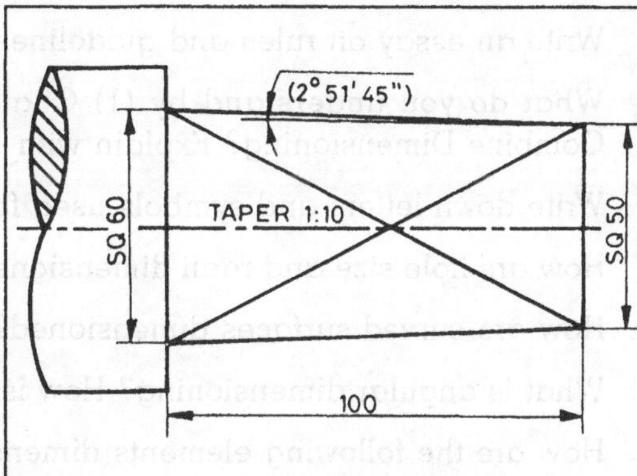

Fig. 20.56

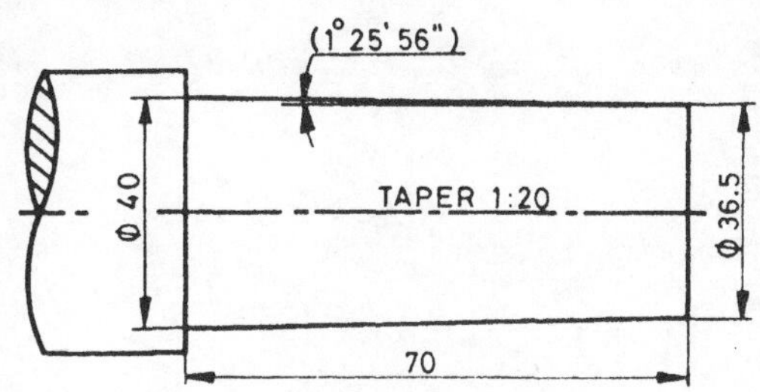

Fig. 20.57

(b) Length of tapered feature.

(c) The half angle of taper is indicated in bracket.

(d) Taper is placed parallel to centre line.

Exercise

1. Why is dimensioning necessary? What is perfect dimensioning?
2. Enlist dimensioning elements and show them on a drawing.
3. Discuss in details various aspects of dimensioning elements.
4. Show clearly the difference between size and location dimensions.
5. Explain clearly with illustrations two systems of placing of dimensions.
6. Write an essay on rules and guidelines for dimensioning, giving illustrations.
7. What do you understand by (1) Chain Dimensioning (2) Parallel Dimensioning (3) Combine Dimensioning? Explain with sketches.
8. Write down letters and symbols used for identifying shape, features of elements.
9. How are hole size and radii dimensioned in drawing from smallest to be biggest sizes?
10. How are curved surfaces dimensioned?
11. What is angular dimensioning? How is it done?
12. How are the following elements dimensioned? Explain with sketches : (a) Chords and Arcs, (b) Bent Parts, (c) Countersinks, Counter bores and Spot faces, (d) Chamfers, (e) Screw threads, (f) Key ways, (g) Taper Features.

CHAPTER 21
FASTENERS

1. INTRODUCTION

Fasteners are used to hold together or to assemble together seperate parts. Fasteners are of two types :

1. **Permanent Fasteners**
2. **Temporary Fasteners**

Seperate parts are permanently fastened together by welding or riveting and are temporarily fastened by threaded fasteners. In case of threaded fasteners,seperate parts can be assembled and disassambled as and when needed. Other temporary fastening devices are pins,cotters,keys, etc.

2. SCREW THREAD TERMINOLOGY

2.(a) Screw Thread :

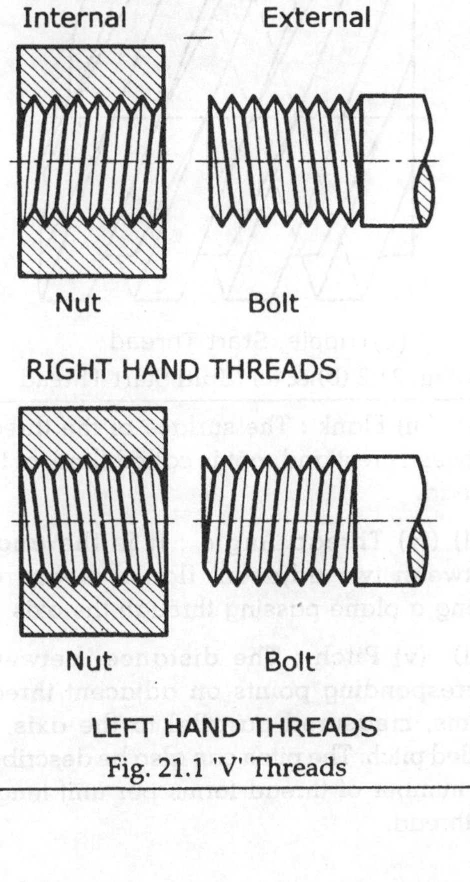

Fig. 21.1 'V' Threads

A ridge of uniform section in the form of a helix on the outside or inside surface of a cylinder or a cone is known as screw thread.

2.(b) Straight Screw Thread :

The thread formed on the surface of a cylinder is called straight thread.

2.(c) Taper Screw Thread : (Not shown)

The thread formed on the surface of a cone is called taper thread.

2.(d) External Thread : (See Fig. 21.1)

The thread formed on the external surface of the cylinder or a cone is known as external thread.

2.(e) Internal Thread : (See Fig. 21.1)

The thread formed on the internal surface of the cylinder or a cone is known as internal thread.

2.(f) Right Hand Thread : (See Fig. 21.1)

A nut screws on a bolt,when turned clockwise,then the thread is called right hand thread. Unless otherwise stated,threads are always right hand.

2.(g) Left Hand Thread : (See Fig. 21.1)

A nut screws off a bolt,when turned clockwise,then the thread is called as left hand thread .

2.(h) V-Thread or Triangular Thread : (See Fig. 21.1)

When a ridge of uniform section of the thread is triangular in shape then the thread is named as V-thread or triangular thread.

2.(i) Square Thread : (See Fig. 21.2)

When a ridge of uniform section of thread is square in shape then the thread is named as square thread.

2.(j) Single Start Thread : [See Fig. 21.2(a)]

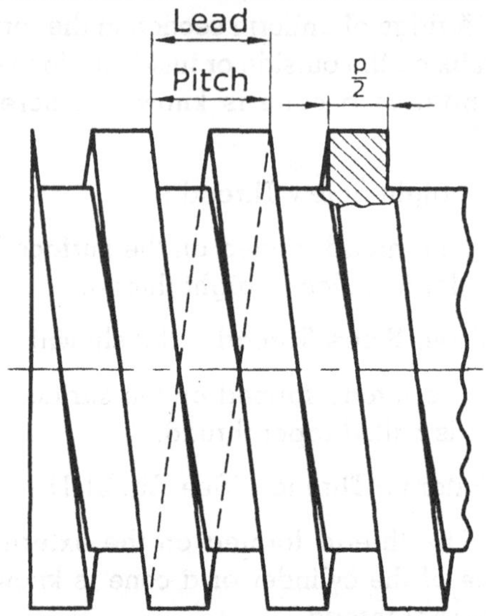

Fig. 21.2 (a) Single Start Thread

When only one thread (helix) is wrapped round the cylinder then the thread is known as a single start thread.

2.(k) Multistart Thread : [See Fig. 21.2(b) and (c)].

When more than one thread (helix) are wrapped parallel round the cylinder then the thread is known as multistart thread. In case of multistart thread-

Lead = n x Pitch

where n = number of start.

In Fig. 21.2(b) and (c) double and triple start threads are shown.

2.(l) Elements of A Thread : (See Fig. 21.3)

Various elements of a screw "V" thread are shown in Fig. 21.3. Each element is defined below :

2.(l) (i) Crest : It is the outer-most edge in the external thread and inner-most edge in the internal thread.

2.(l) (ii) Root : It is the inner-most portion of an external thread and outer-most portion of an internal thread.

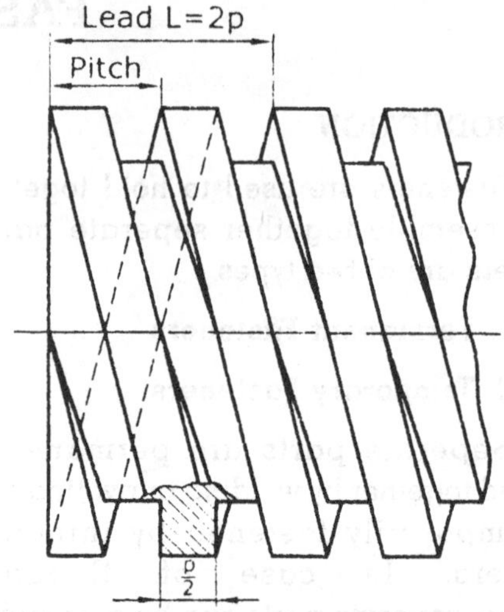

(b)Double Start Thread

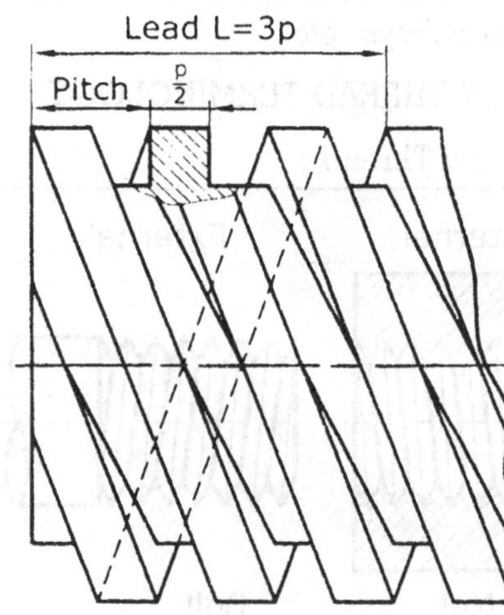

(c)Tripple Start Thread

Fig. 21.2 (b) & (c) Multi Start Thread

2.(l) (iii) Flank : The surface of the thread between crest and root is called flank of the thread.

2.(l) (iv) Thread Angle : It is the angle between two adjacent flanks measured along a plane passing through the axis.

2.(l) (v) Pitch : The distance, between corresponding points on adjacent thread forms, measured parallel to the axis, is called pitch. The pitch can also be described as number of thread forms per unit length of thread.

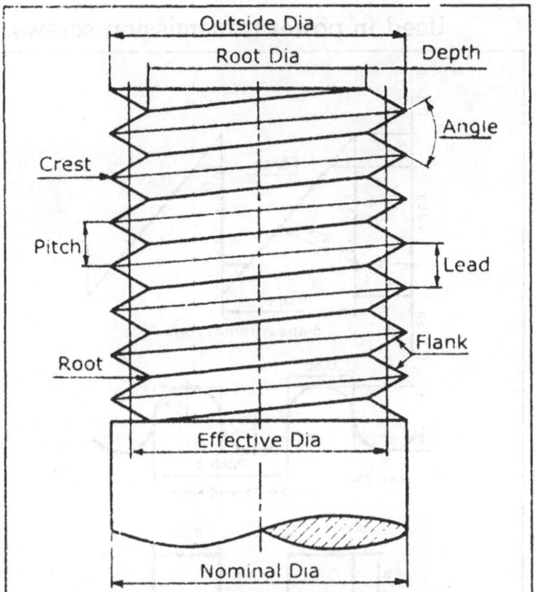

Fig. 21.3 Elements of Thread

2.(l) (vi) Lead : The distance between corresponding points on the same (helix) thread form,measured parallel to the axis, is called lead.

In other words, it is the distance by which threaded part move axially with respect to a fixed mating part when given one complete revolution.

2.(l) (vii) Major or Nominal Diameter :

It is the largest diameter of screw thread.

2.(l) (viii) Minor or Root or Core Diameter :

It is the smallest diameter of a screw thread.

2.(l) (ix) Pitch or Effective Diameter :

It is the diameter of an imaginary cylinder equal to the length of a line, perpendicular to the axis, measured between the points, where it cuts the flanks on to opposite sides of the axis of the thread.

2.(l) (x) Depth of A Thread :

Radial distance, measured at right angles to the axis between the root and the crest is known as the depth of a thread.

2.(l) (xi) Slope of A Thread :

It is equal to half of a lead.

3. FORMS OF SCREW THREADS :

(See Fig. 21.4)

Various forms of screw threads are shown in Fig. 21.4

3.(1) Whitworth Thread : (B.S.W.)

It is the standard form adopted in U.K.

3.(2) British Association Thread : (B. A. Thread)

Used for small size screws.

3.(3) Sellers or American Standard Thread :

Used as standard thread form in America.

3.(4) Internal Unified Thread Form :

International organisation for standardization adopt this thread as the basic profile.

3.(5) External Unified Thread Form :

It is slightly different from the internal type.

3.(6) Acme Thread Form :

It is the modified square thread. It is easier to manufacture. Acme threads are thicker at the root and hence are stronger threads. Used in half nuts, used for frequent engagement and disengagement.

3.(7) Buttress Thread Form :

It is a combination of square and "V" threads. Vertical flanks take axial force. Used in Carpenters' bench vice.

3.(8) Knuckle Thread Form :

Square threads are modified by rounding off the corners of the square thread. It is used in the circumstances of heavy wear and rough usage.

3.(9) Square Thread Form :

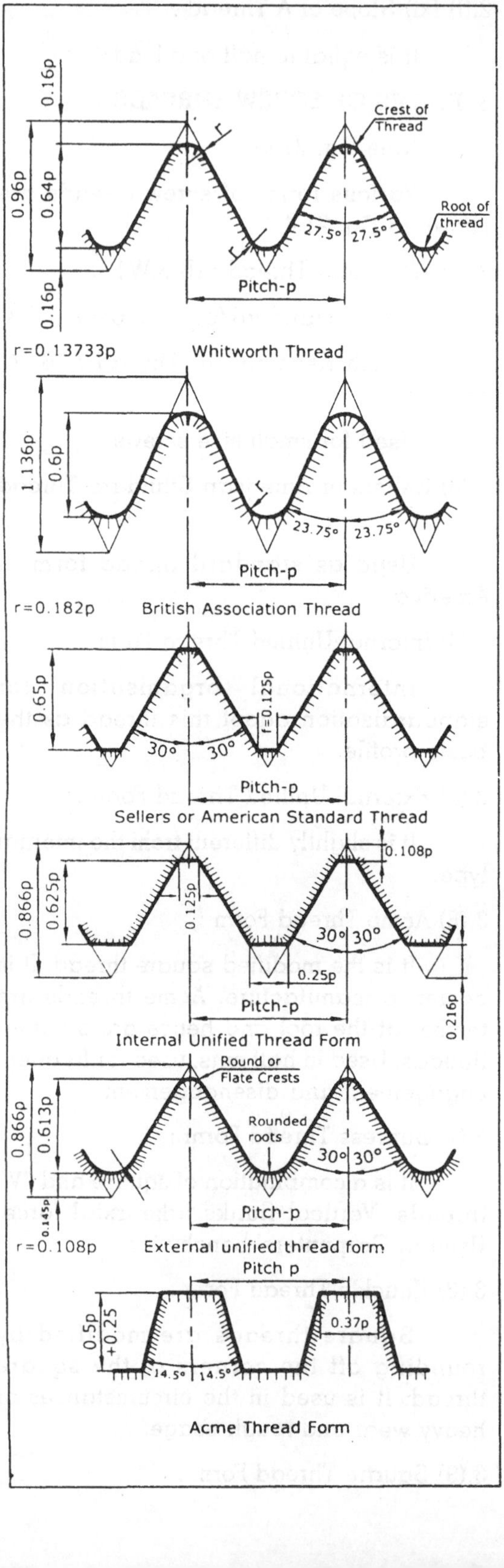

Used in power transmission screws.

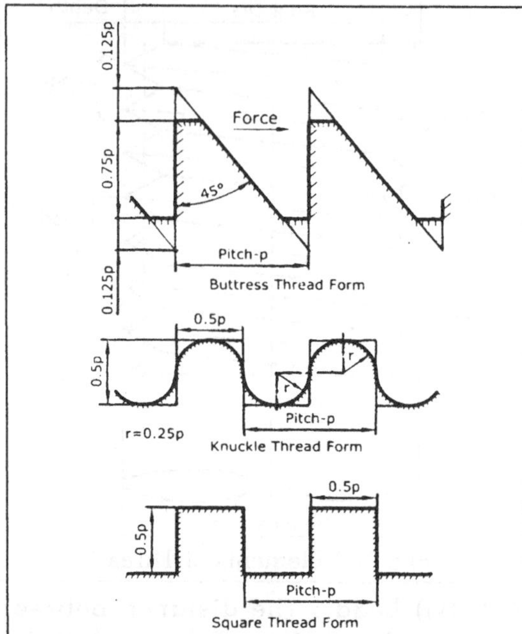

Fig. 21.4 Forms of Screw Threads

4. HEXAGONAL BOLT-NUT-WASHER

(See Fig. 21.5)

It is one kind of screw pair used for fastening parts temporarily. Bolt has external threads while nut has internal threads. Spanner is used to tighten nut on bolt. Mostly hexagonal/square bolt-nut pair is used along with a washer.

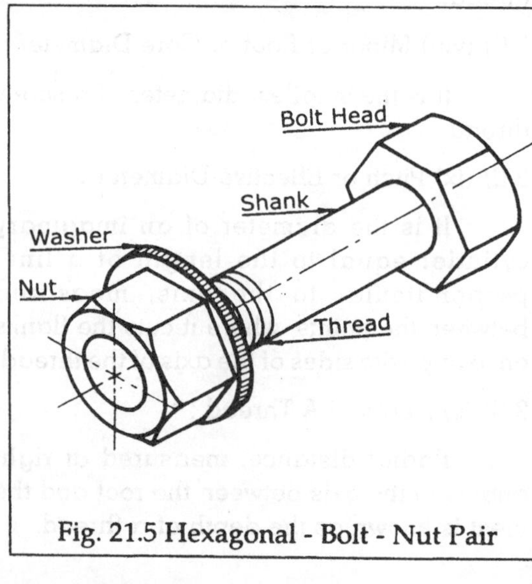

Fig. 21.5 Hexagonal - Bolt - Nut Pair

5. **TYPES OF NUTS :** (See Fig. 21.6)

Different varieties of nuts are used for different purposes. Six different varieties of nuts as given below, are shown in Fig. 21.6.

(a) Flanged nut	**(d) Capstan nut**
(b) Cap nut	**(e) Ring nut**
(c) Dome nut	**(f) Wing nut**

5.(a) Flanged Nut : See Fig. 21.6(a)

It is hexagonal nut with washer as an integral part.

5.(b) Cap Nut : See Fig. 21.6 (b)

It is a hexagonal nut with a cylindrical cap at the top side to protect bolt end from corrosion. It also prevents leakage through bolt-nut pair.

5.(c) Dome Nut : See Fig. 21.6(c)

It is a hexagonal nut with a spherical dome at the top side. By providing smooth spherical dome harmful sharp edges are avoided.

5.(d) Capstan Nut : See Fig. 21.6(d)

It is a cylindrical nut with equispaced six holes drilled on the curved surface as shown. Tommy bar tool is used for turning the nut on the bolt.

5.(e) Ring Nut :See Fig. 21.6(e)

It is a cylindrical ring nut with equispaced slots provided on the circumference as shown. C-spanner tool is used for turning the nut.

5.(f) Wing Nut : See Fig. 21.6(f)

This type of nut is used where it is required to be operated frequently and that too by hand i.e. by thumb and fingers. No special tool is required.

Drawings in Fig. 21.6(d) (e) and (f) are given in third angle projection system.

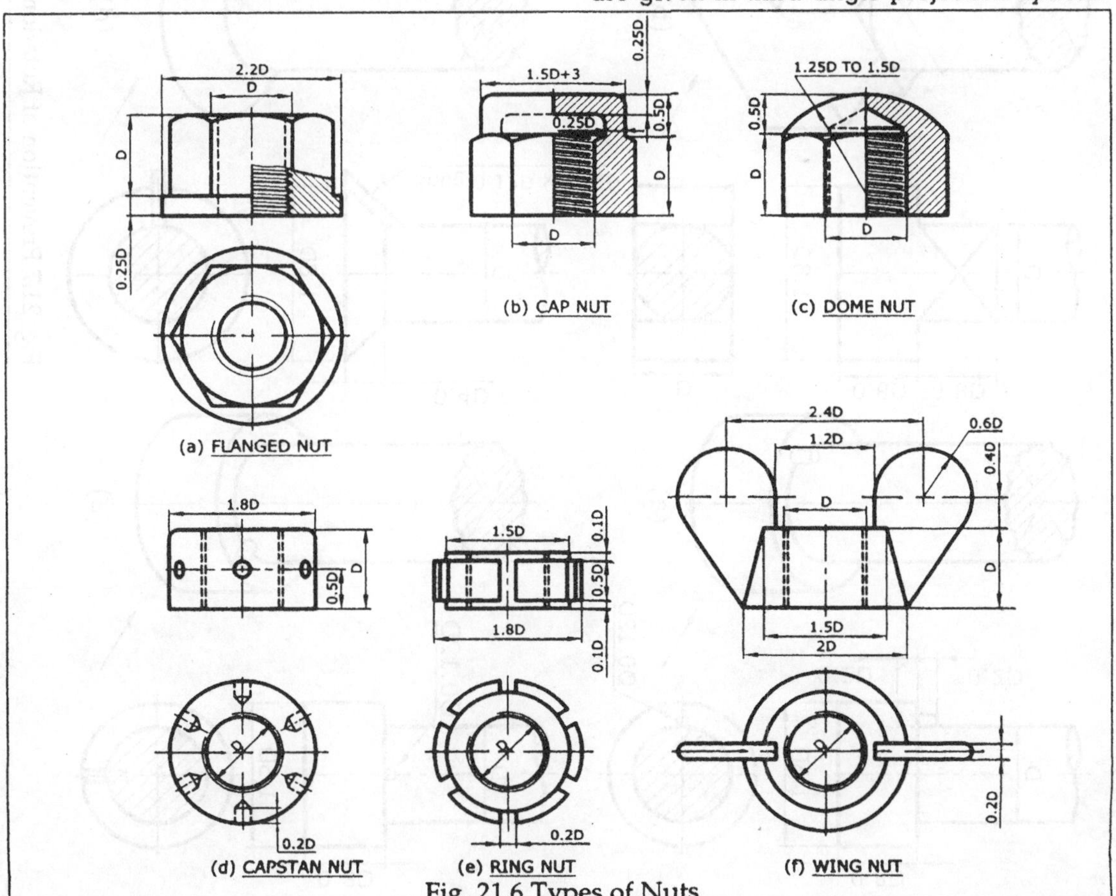

Fig. 21.6 Types of Nuts

6. PREVENTION OF ROTATION OF A BOLT (See Fig. 21.7)

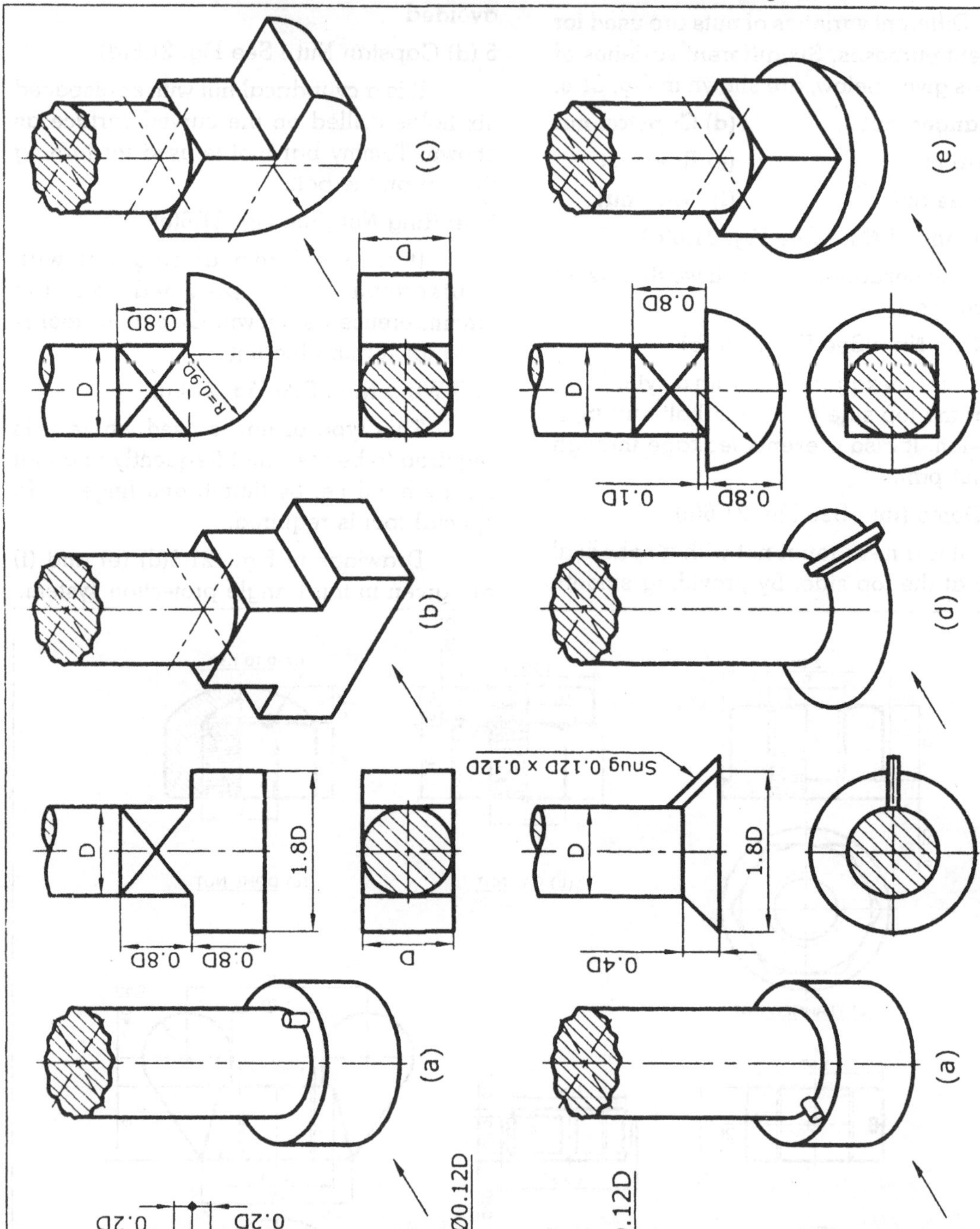

Fig. 21.7 Prevention of Rotation of Bolts

While tightening a nut on a bolt, the bolt rotates along with the nut and hence the nut cannot be tightened.

There are two ways of preventing bolt rotation along with rotation of nut.

1. By holding the bolt head by means of a spanner.
2. By providing some rotation prevention technique.

When it is not convenient or not possible to hold a bolt head by means of a spanner, rotation prevention technique, using pin or square neck or snug, is used, as shown in Fig. 21.7.

Six different types of bolts, as given below, are used to prevent the rotation of the bolt itself.

(a) Cheese headed bolt - 2 types - Fig. 21.7(a)

(b) T-headed bolt - Fig. 21.7 (b)

(c) Hook bolt - Fig. 21.7 (c)

(d) Countersunk headed bolt with snug - Fig. 21.7(d)

(e) Cup headed bolt with square neck - Fig. 21.7(e)

Figures of different types of bolts are self explanatory and hence detail discussion is not given. A pin, a square neck or a snug prevent a bolt from rotation.

7. LOCKING ARRANGEMENTS OF NUTS

See Fig. 21.8

Due to the vibration, nut has tendancy to screw off the bolt. This tendancy sometimes may cause serious accident, perticularly in vehicles. To prevent nut from getting loose, some methods are employed.

Some of the important methods, known as locking devices, are listed below and shown in Fig. 21.8.

(a) Lock nut or Check nut - see Fig. 21.8(a)

(b) Locking by split pin - See Fig. 21.8(b)

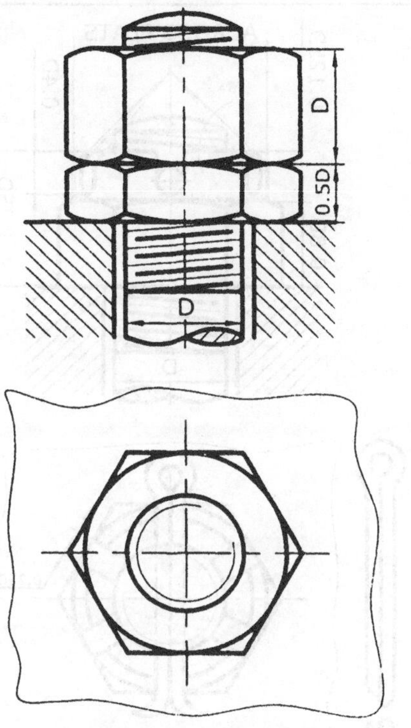

Fig. 21.8 (a) Lock Nut

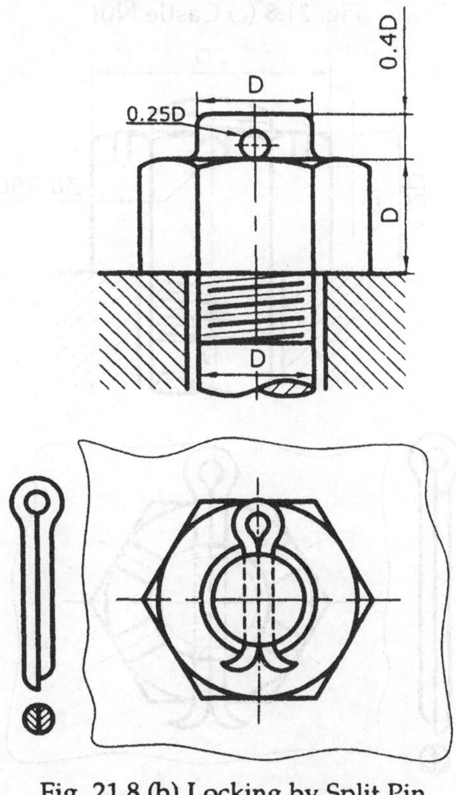

Fig. 21.8 (b) Locking by Split Pin

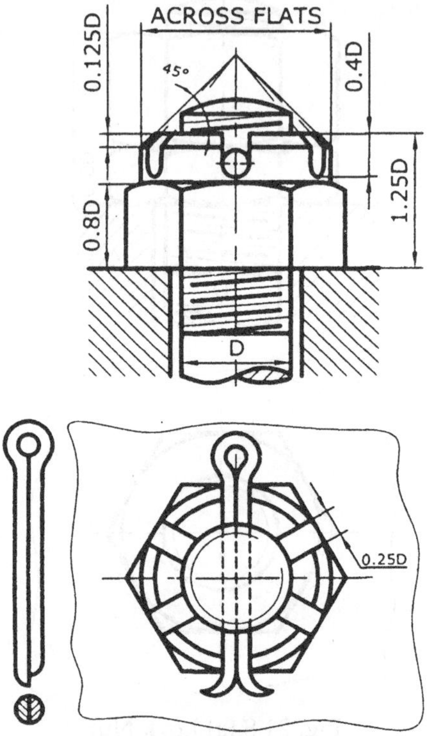

Fig. 21.8 (c) Castle Nut

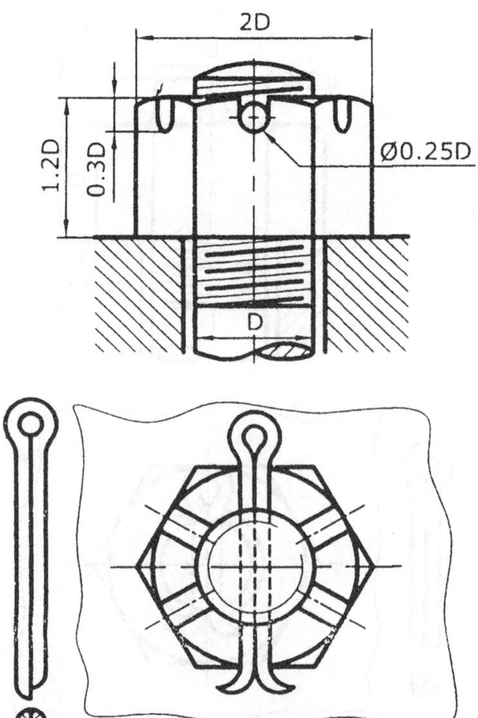

Fig. 21.8 (d)j Slotted Nut

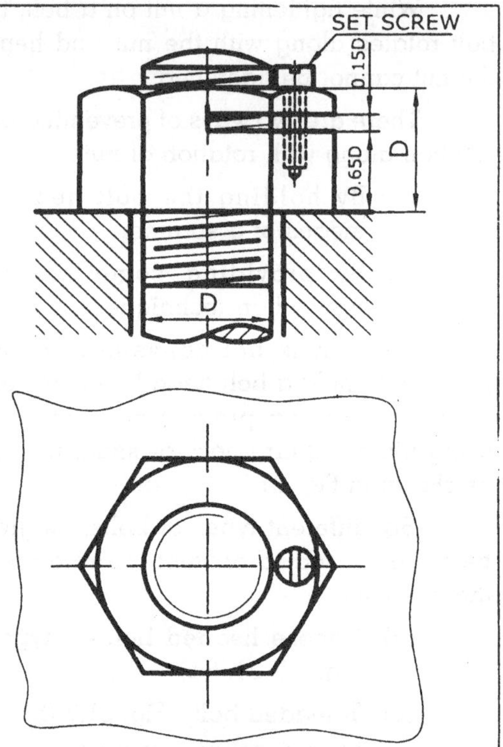

Fig. 21.8 (e) Sawn Nut

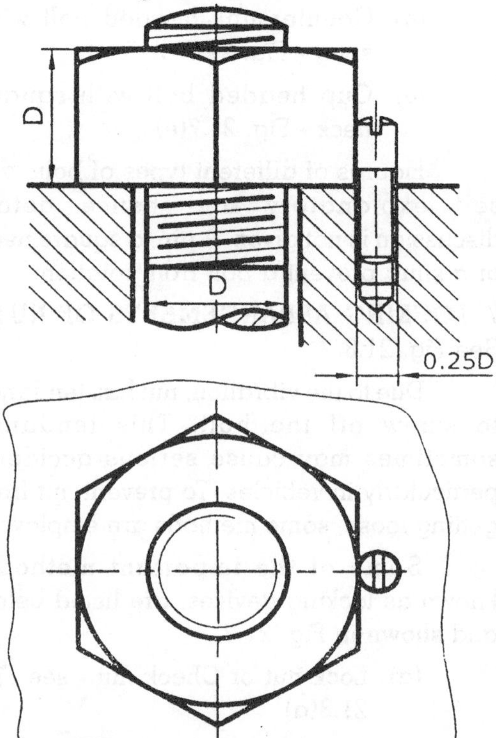

Fig. 21.8 (f) Locking by A Pin

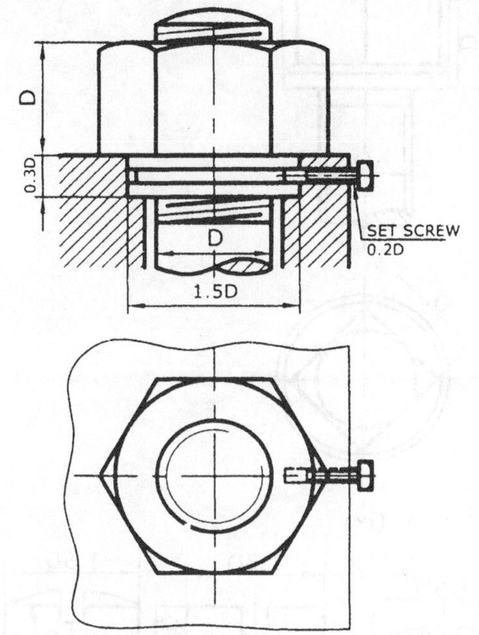

Fig. 21.8 (g) Penn, Ring or Grooved Nut

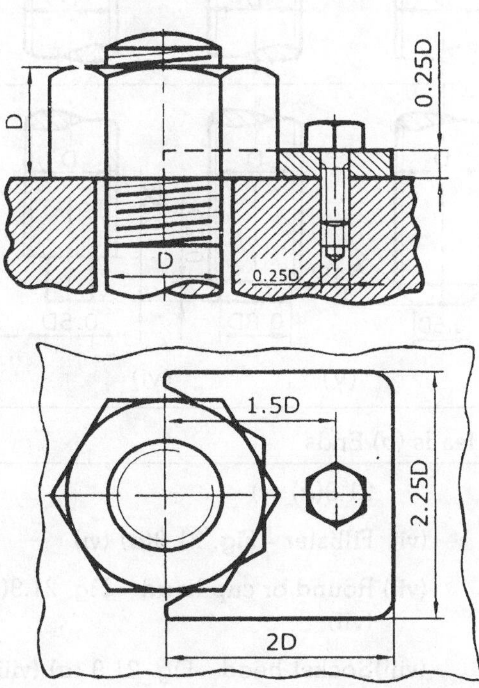

Fig. 21.8 (h) Locking Plate

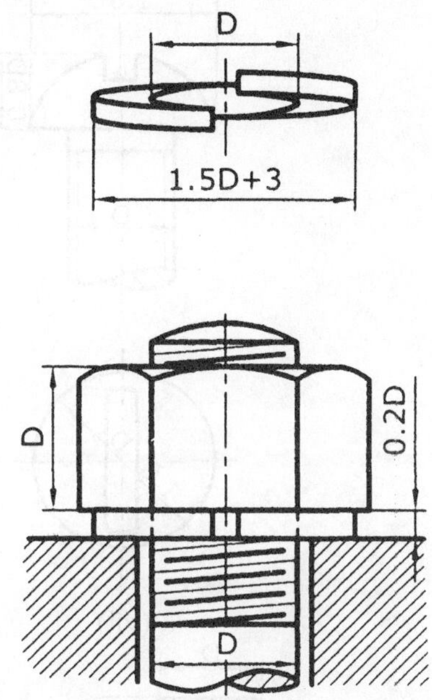

Fig. 21.8 (i) Spring Washer

(c) Castle nut - See Fig. 21.8 (c)

(d) Slotted nut - see Fig. 21.8 (d)

(e) Sawn nut - See Fig. 21.8 (e)

(f) Locking by a pin - See Fig. 21.8 (f)

(g) Penn, ring or grooved nut - See Fig. 21.8 (g)

(h) Nut with a locking plate - See Fig. 21.8 (h)

(i) Nut with a spring washer - see Fig. 21.8 (i)

Figures are self explanatory and hence detail discussion is not given. Figures given are in first angle projection system.

8. SET-SCREWS : See Figs. 21.9(a) & (b)

It is used when relative movement between two adjoining pieces is required to be prevented. It is screwed in tapped hole of one piece and it presses on the other adjoining piece.

Set screws have heads of different varieties as given below: See Fig. 21.9(a).

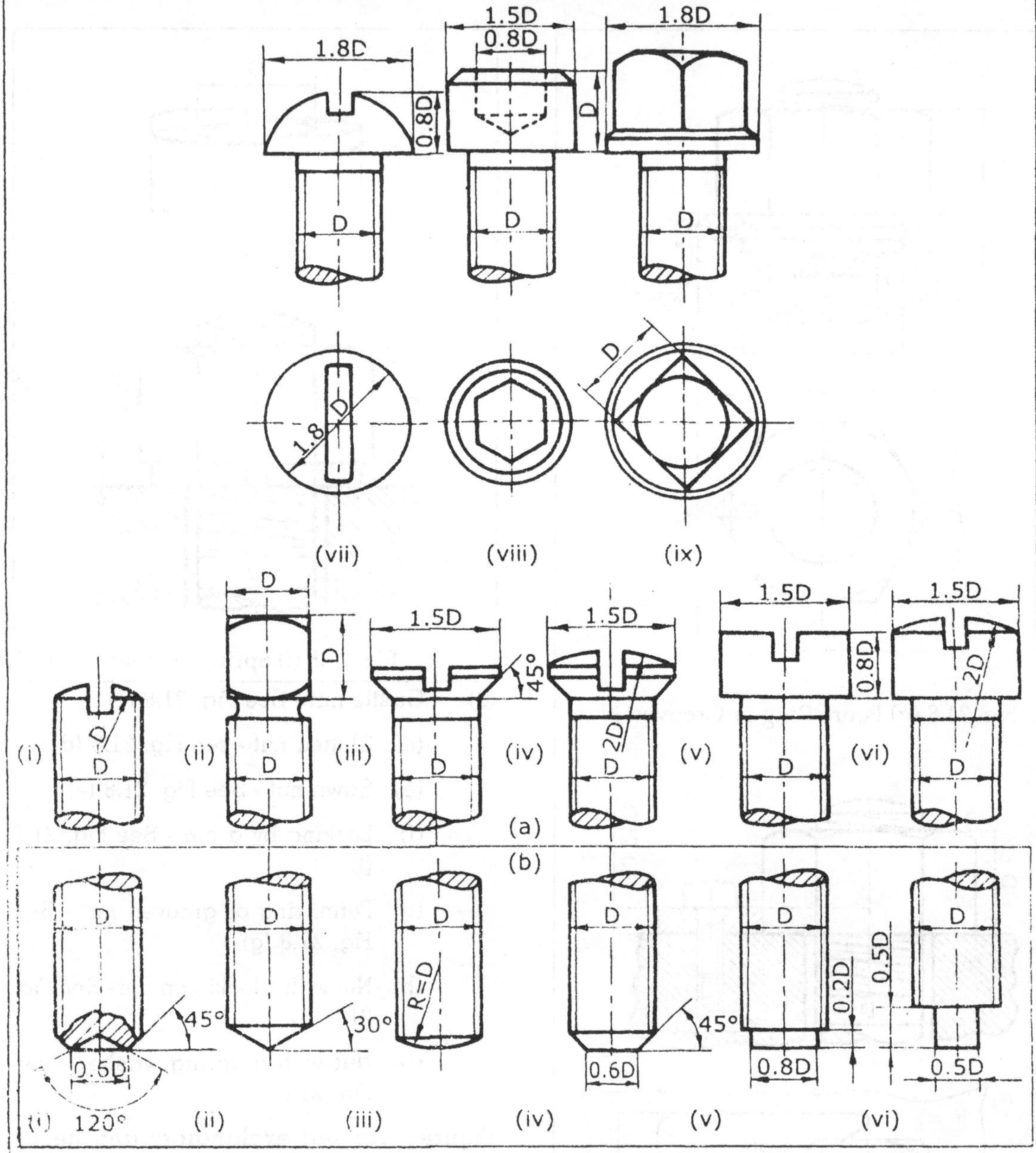

Fig. 21.9 Set Screws (a) Heads (b) Ends

(i) Headless grub screw - Fig 21.9(a) (i)

(ii) Square head - Fig. 21.9(a) (ii)

(iii) Countersunk head - Fig. 21.9(a) (iii)

(iv) Rounded counter sunk head - Fig. 21.9(a) (iv)

(v) Cylindrical or cheese head - Fig. 21.9(a) (v)

(vi) Fillister - Fig. 21.9(a) (vi)

(vii) Round or cup head - Fig. 21.9(a) (vii)

(viii) Socket head - Fig. 21.9 (a) (viii)

(ix) Collar screw with square head - Fig. 21.9 (a) (ix)

Over and above these heads, square and hexagonal heads are also used. Similarly, set screws have ends of different varieties as given below. See Fig. 21.9(b)

(i) Cup end - Fig. 21.9(b) (i)

(ii) Conical end - Fig. 21.9 (b) (ii)

(iii) Oval end - Fig 21.9 (b) (iii)

(iv) Flat end - Fig 21.9(b) (iv)

(v) Half dog end - Fig. 21.9(b) (v)

(vi) Full dog end - Fig. 21.9(b) (vi)

Fig. 21.9(a) and Fig. 21.9(b) are self explanatory and hence detail discussion is not given here. Screw drivers, spanners, wrenches, etc. are used to operate heads of set screws.

9. **STUD OR STUD BOLT** : See Fig. 21.10

It is a cylindrical piece threaded at both ends. see Fig.21.10(a). One end of the stud is for insertion in a tapped hole and the other threaded end of the stud receives nut or nuts.

Fig. 21.10(b) illustrates the use of a stud. It is used for holding cylinder heads of automobile engines. The stud replaces bolt when it is not convenient to accomodate a bolt head. Stud with square neck, as shown in Fig.21.10(c), can be tightened by engaging square neck in a spanner.

A stud with collar as shown in Fig. 21.10 (A) is used in stuffing box assembly .

10. (A) **FOUNDATION BOLTS** & (B) **LIFTING EYE BOLTS** : See Fig.21.11 (a), (b), (c), (d), (e) & (f).

Foundation bolts are used for putting machines to their foundations i.e. fixing them on floor.

Foundation bolts are suspended from holes provided in the base plate of the machine, inside the holes digged earlier in the floor. After levelling the machines cavities of floor surrounding foundation bolts are filled up by a mixture of water, cement, sand, grit, etc. as shown in the figures. Materials like lead, sulphur, etc. can also be used in place of cement.

There are many varieties of foundation bolts. Some of them are as under:

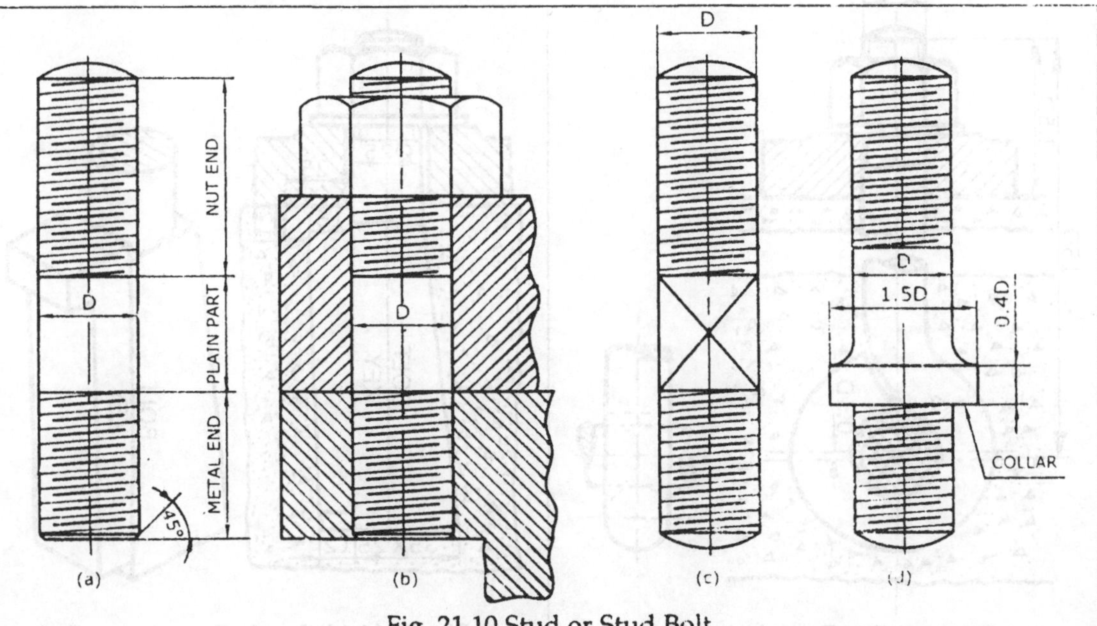

Fig. 21.10 Stud or Stud Bolt

(a) Bent or twisted foundation bolt - Fig. 21.11(a)

(b) Eye foundation bolt - Fig. 21.11(b)

(c) Rag foundation bolt - Fig. 21.11(c)

(d) Lewis foundation bolt Fig. 21.11(d)

(e) Cotter foundation bolt - Fig. 21.11(e)

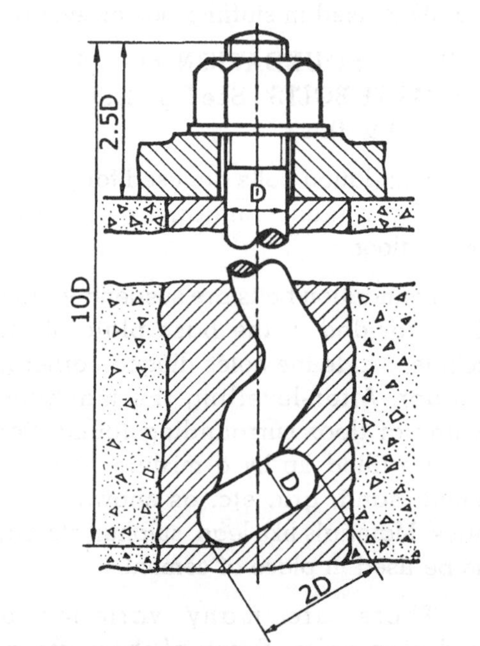

Fig. 21.11 (a) Bent or Twisted Foundation Bolt

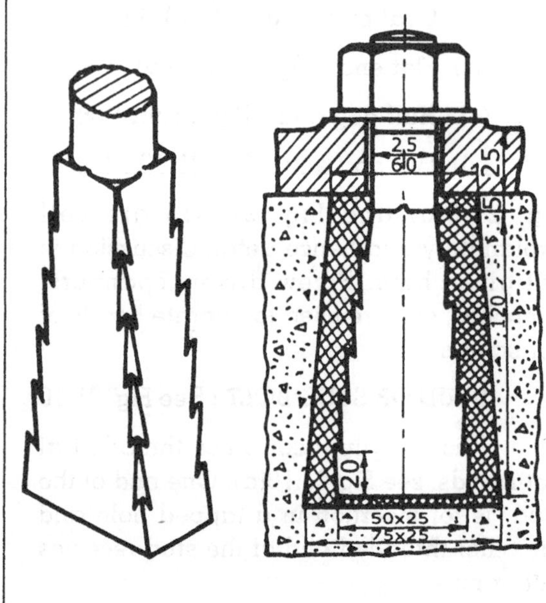

Fig. 21.11 (c) Rag Foundation Bolt

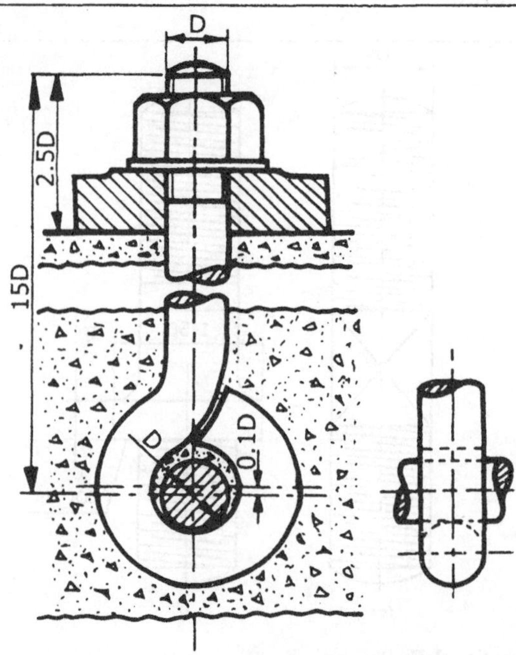

Fig. 21.11 (b) Eye Foundation Bolt

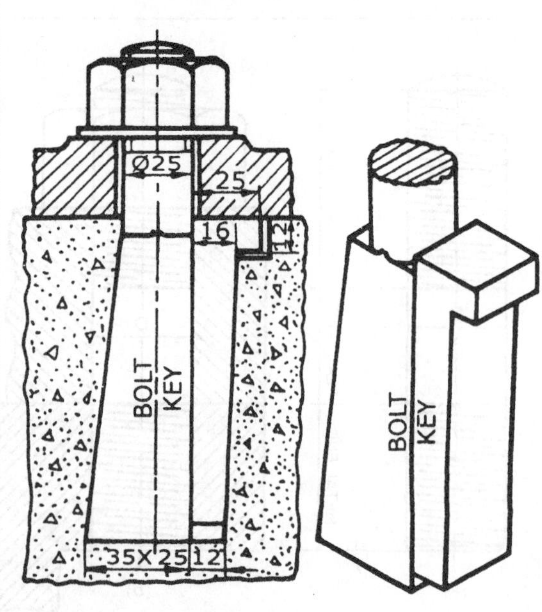

Fig. 21.11 (d) Lewis Foundation Bolt

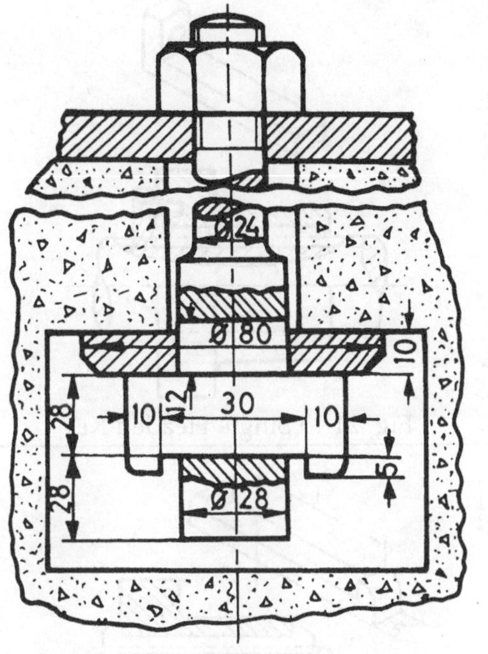

Fig. 21.11 (e) Cotter Foundation Bolt

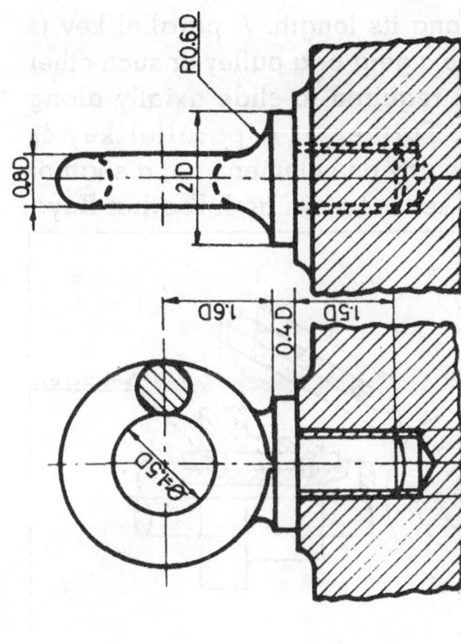

Fig. 21.11 (f) Lifting Eye Bolt

Figures of foundation bolts are self explanatory and hence a detailed discussion is not given.

10 (B) Lifting Eye Bolt : This bolt is fitted in the tapped hole, provided on the top side approximately in the centre of a heavy machine. Normally the location of this lifting eye bolt is vertically above the C. G. of the machine. Machine is lifted by inserting crane hook in the eye of this bolt. See Fig. 21.11

11. **KEYS :**

See Figs. 21.12 to 21.21

A key is a square, rectangular or circular cross-sectional piece of steel of length L, when inserted in a keyway kept between shaft and hub, prevents relative rotational motion between a shaft and hub. See Fig. 21.12 to understand how key, shaft and hub assembly is done.

Keys may be divided into the following two main groups and three sub-groups:

1. Parallel Keys	A. Sunk Keys
2. Taper keys	B. Saddle Keys
	C. Feather Keys

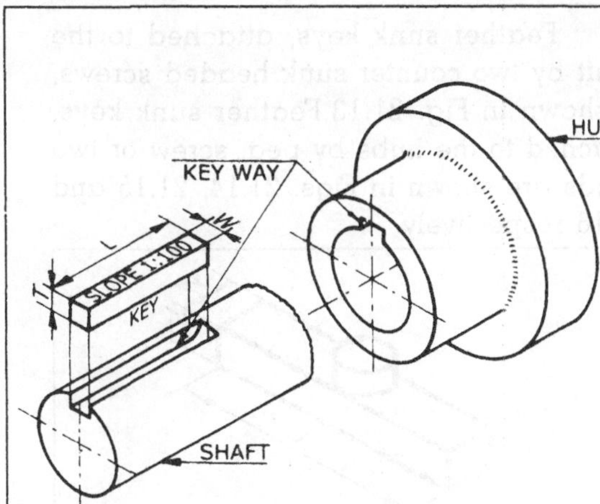

Fig. 21.12 Assembly of Shaft - Hub - Key

11.1 **Parallel Keys :**

Parallel key is uniform in cross-section along its length. A parallel key is used where a gear or a pulley or such other element is required to slide axially along the shaft. Further, if a parallel key is attached to any one element, i.e. a shaft or a hub, the key is known as **A Feather Key**.

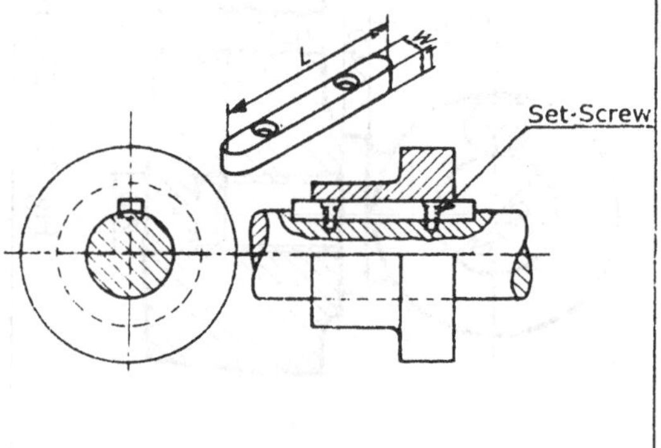

Fig. 21.13 Parallel Sunk Key

Feather sunk keys, attached to the shaft by two counter sunk headed screws, is shown in Fig. 21.13. Feather sunk keys, attached to the hubs by peg, screw or two heads are shown in Figs. 21.14, 21.15 and 21.16 respectively.

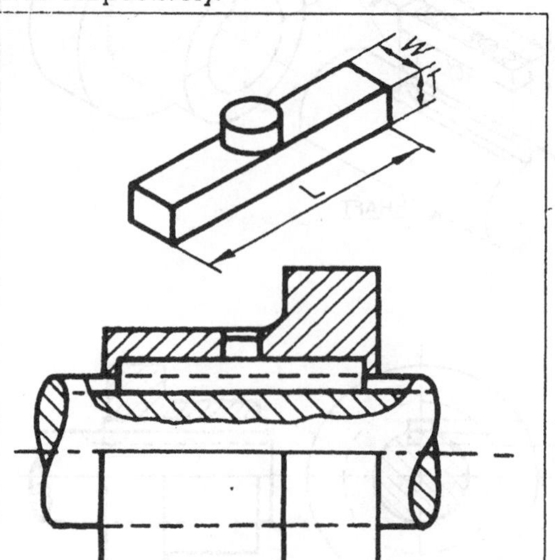

Fig. 21.14 Peg Key

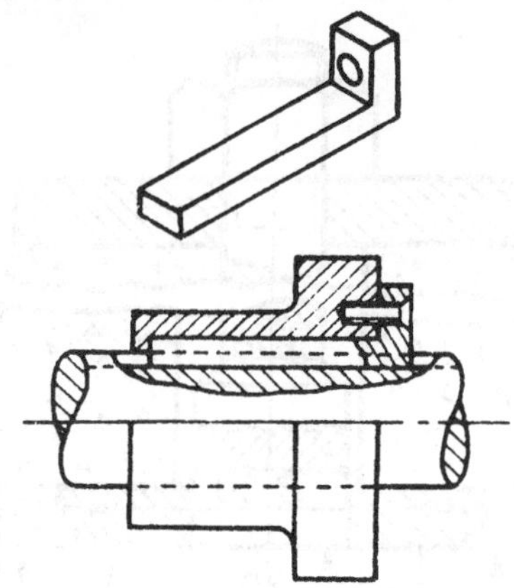

Fig. 21.15 Single Headed Key

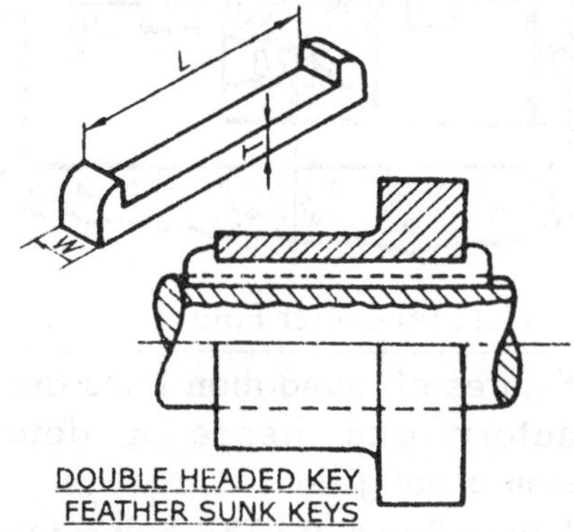

Fig. 21.16 Double Headed Key

TAPER KEYS :

Taper keys are uniform in width and tapered along thickness. Taper keys are used when along with relative rotational motion relative axial movement is also to be prevented. Taper keys are of two main varieties.

1. Taper Sunk Keys
2. Taper Saddle Keys

Gib headed key shown in Fig. 21.17 is a tapered sunk key. Hollow saddle key shown in Fig. 21.18 and flat saddle key shown in Fig. 21.19 are taper saddle keys.

A key that engages a key slot, in both the shaft and the hub, is known as a sunk key.

In case of saddle keys, key way is not provided in the shaft and key does not engage in the shaft.

Feather keys are keys attached to any one element i.e. either shaft or hub.

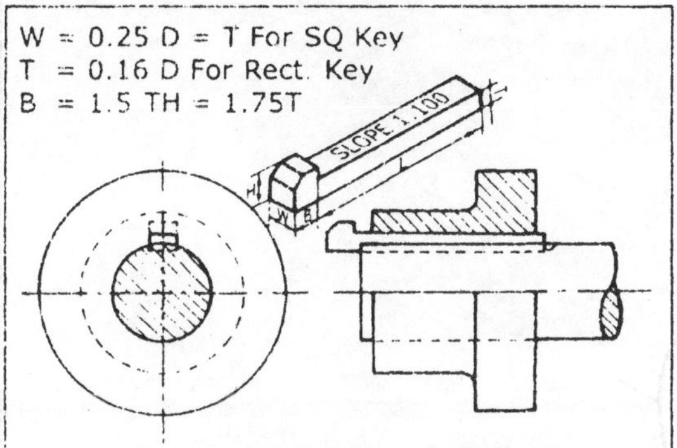

Fig. 21.17 Gib Headed Key

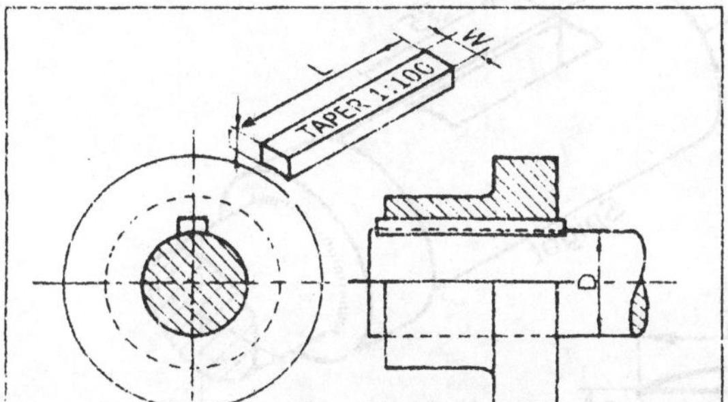

Fig. 21.18

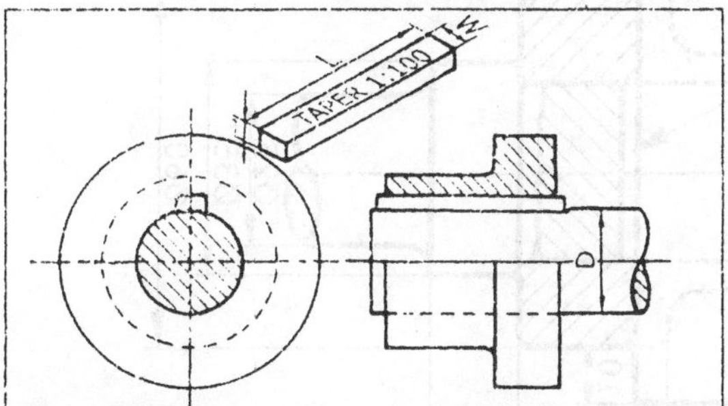

Fig. 21.19 Flat Sanddle Key Hollow Saddle Key

11.3 Woodruff Key: See Fig.21.20

It is one kind of sunk key. It is an adjustable sunk key. It is particularly used on tapered shaft. Its more depth in shaft tends to weaken the shaft. It is extensively used in automobile work and machine tools.

11.4 Pin Key/Round Key: See Fig. 21.21

It is a simple key with circular cross section. It can be made with taper or without taper. This type of key is suitable for low power transmission.

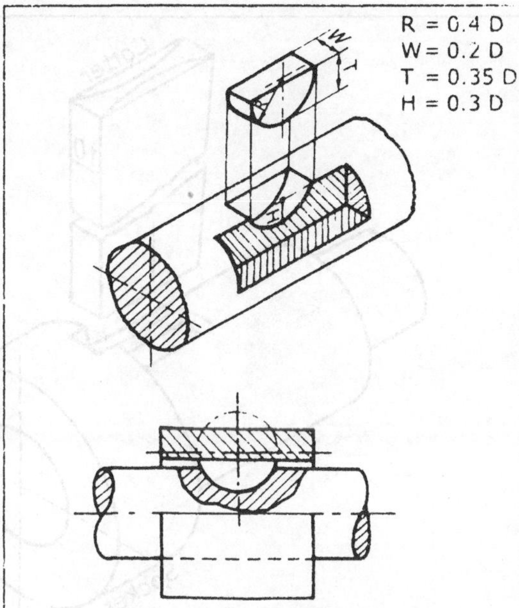

Fig. 21.20 Woodruff Key

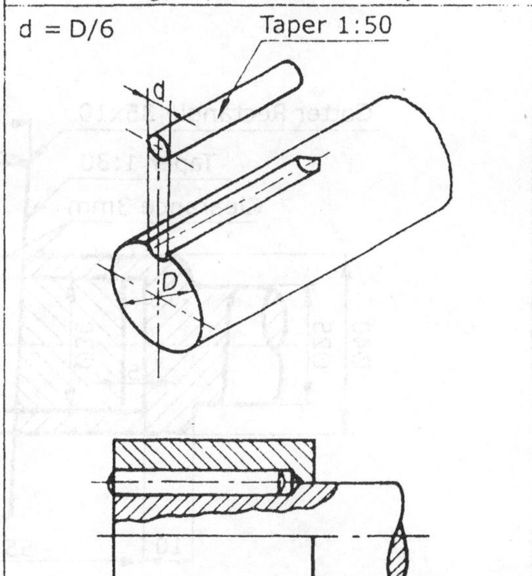

Fig. 21.21 Parallel/Taper Pin Key

Constructional features along with proportional dimensions of various types of keys are shown in Fig.21.12 to Fig. 21.21. Figures are self explanatory and hence a detailed discussion is not given.

12. **COTTER JOINT** : See Fig. 21.22

A cotter is used to connect two rods rigidly by inserting cotter at right angles to

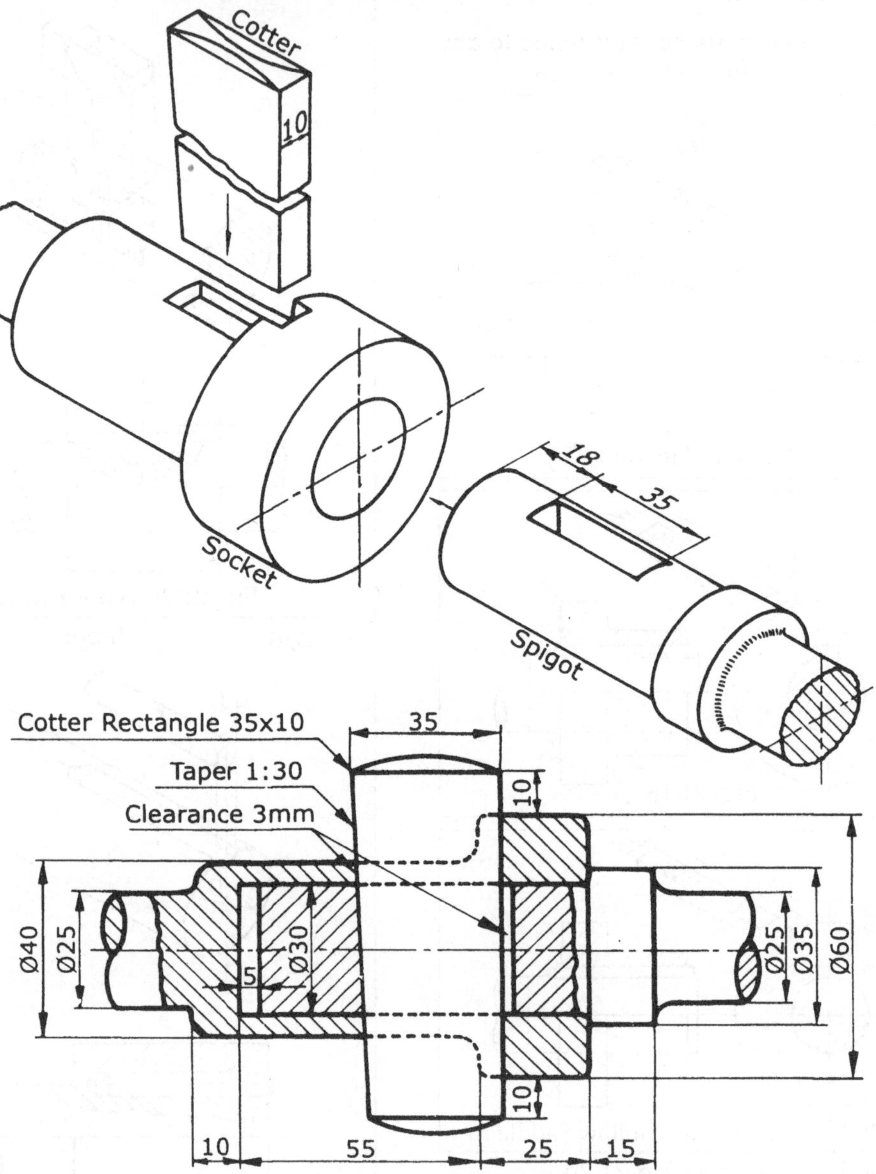

Fig. 21 22 Cotter Joint with Socket and Spigot Ends

the axes of two rods. Cotter joint is capable of transmitting tensile as well as compressive forces. Cotter is uniform in thickness and taper in width. Taper is generally given on one side only.

Socket and spigot cotter joint is shown in Fig. 21.22. One of the rods end is forged to form socket end and other rod's end is forged to get spigot end. Spigot end is inserted inside the socket end and then the cotter is inserted at right angles, as shown in figure to get a joint. Clearances shown in the figure are absolutely essential for proper functioning of the joint.

Constructional features and proportionate dimensions are clearly shown in Fig. 21.22 and hence further detail discussion is not necessary.

Sleeve cotter joint can be made by inserting two rods from opposite sides of the sleeve and then inserting two cotters at right angles.

13. KNUCKLE JOINT/PIN JOINT : See Fig. 21.23.

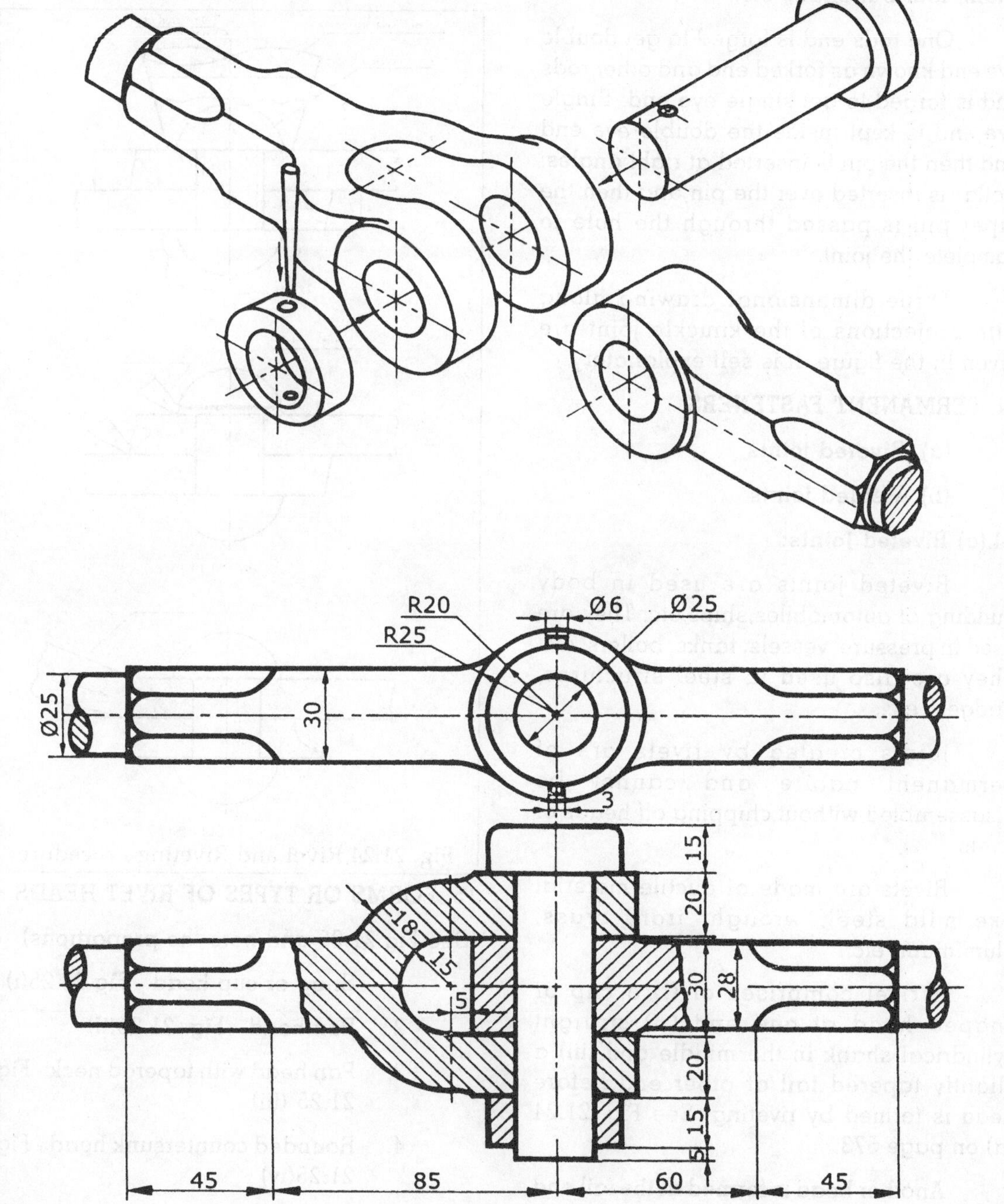

Fig. 21.23 Knuckle Joint or Pin Joint

Knuckle or pin joint is used when two rods are required to be joined permitting relative angular movement between them. It is used in link mechanism like slider crank chain, four bar chain, etc.

One rods end is forged to get double eye end known as forked end and other rods end is forged to get single eye end. Single eye end is kept inside the double eye end and then the pin is inserted at right angles. Collar is inserted over the pin and then the taper pin is passed through the hole to complete the joint.

Three dimensional drawing along with projections of the knuckle joint are given in the figure. It is self explanatory.

14. **PERMANENT FASTENERS**

(a). **Riveted joints**

(b). **Welded joints**

14.(a) Riveted Joints:

Riveted joints are used in body building of automobiles,ships,etc. They are used in pressure vessels, tanks, boilers, etc. They are also used in steel structures, bridges, etc.

Joints created by rivets are of permanent nature and cannot be disassembled without chipping off heads of rivets.

Rivets are made of ductile material like mild steel, wrought iron, brass, aluminium, etc.

A rivet comprises of (i) a cup or snaped head at one end (ii) straight cylindrical shank in the middle and (iii) a slightly tapered tail at other end before head is formed by riveting. See Fig. 21.24 (a) on page 573.

Another head is formed at the tail end by hammering on special tool having a cavity of replica of desired shape of head. See Fig. 21.24 (a)

To creat leak proof joints, the plates are firmly forced together by caulking and fullering tools. For caulking process, see Fig. 21.24 (b) and for fullering process, see Fig. 21.24 (c).

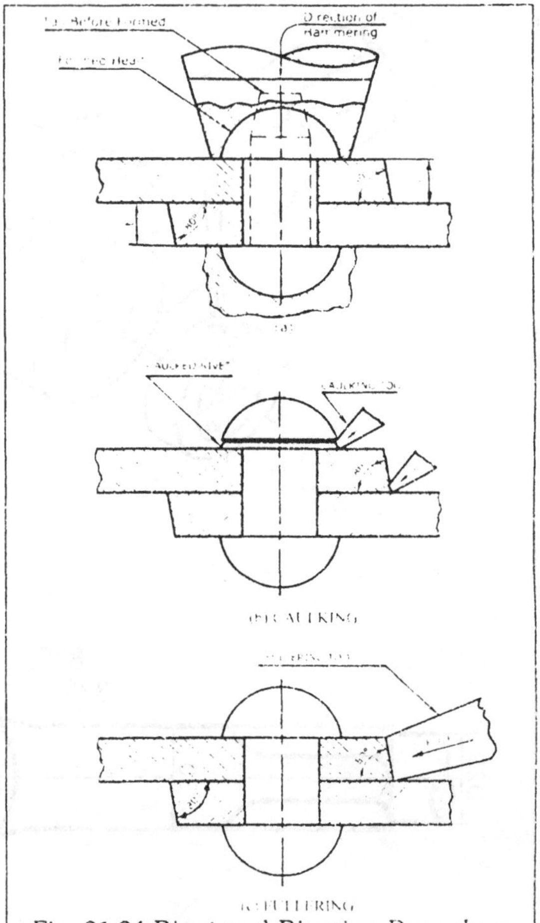

Fig. 21.24 Rivet and Riveting Procedure

15. **FORMS OR TYPES OF RIVET HEADS**

(See Fig.21.25 and note the proportions)

1. Snap or cup head - Fig.21.25(i)
2. Pan head - Fig. 21.25(ii)
3. Pan head with tapered neck - Fig. 21.25 (iii)
4. Rounded countersunk head - Fig. 21.25(iv)
5. Countersunk head - Fig. 21.25 (v)
6. Snap head 2 - Fig. 21.25(vi)

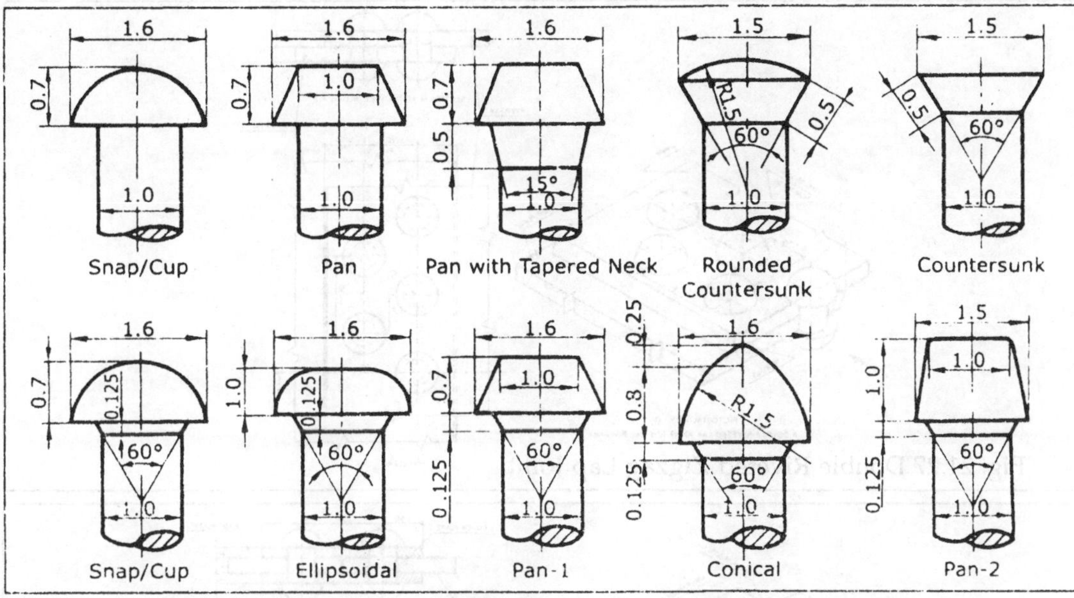

Fig. 21.25 Forms of Rivet Heads

7. Ellipsoidal head - Fig. 21.25(vii)
8. Pan head type 1 - Fig. 21.25(viii)
9. Conical head - Fig. 21.25(ix)
10. Pan head type 2 - Fig. 21.25(x)

16. TYPES OF RIVETED JOINTS

There are two main types of riveted joints :

(a). Lap joints **(b). Butt joints.**

16.(a) Lap joints:

When the two plates, to be connected are overlapped in position in the joint then the joint is called the lap joint. Depending upon the number of rows of rivets, the joints are further classified as single riveted lap joint, double riveted lap joint,triple riveted lap joint, etc. Further depending upon the arrangement of rivets in adjoining rows of rivets they are classified as chain riveting

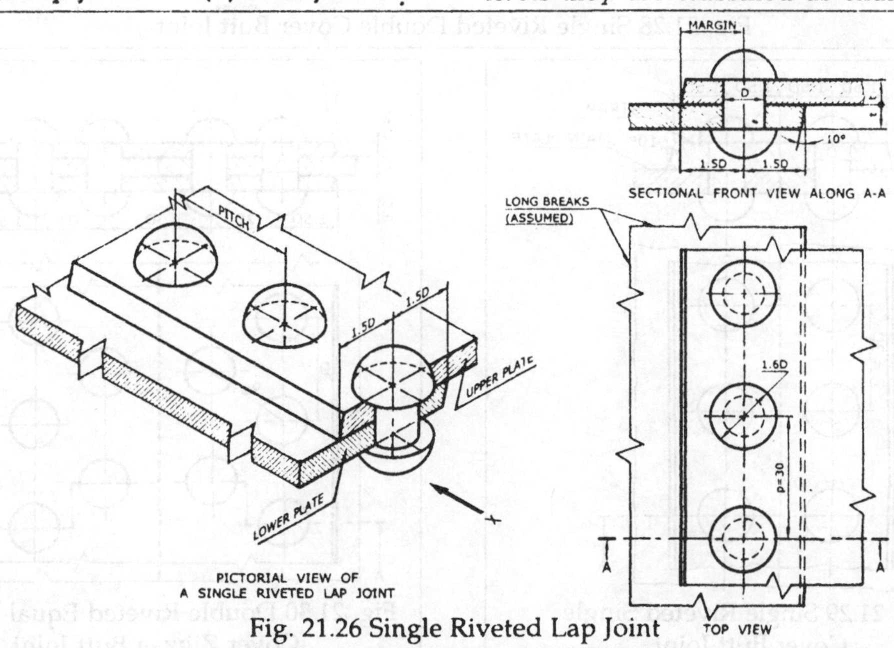

Fig. 21.26 Single Riveted Lap Joint

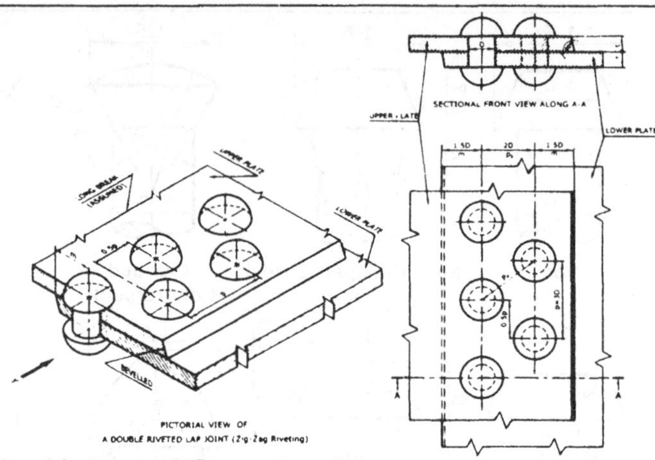

Fig. 21.27 Double Riveted Zigzag Lap Joint

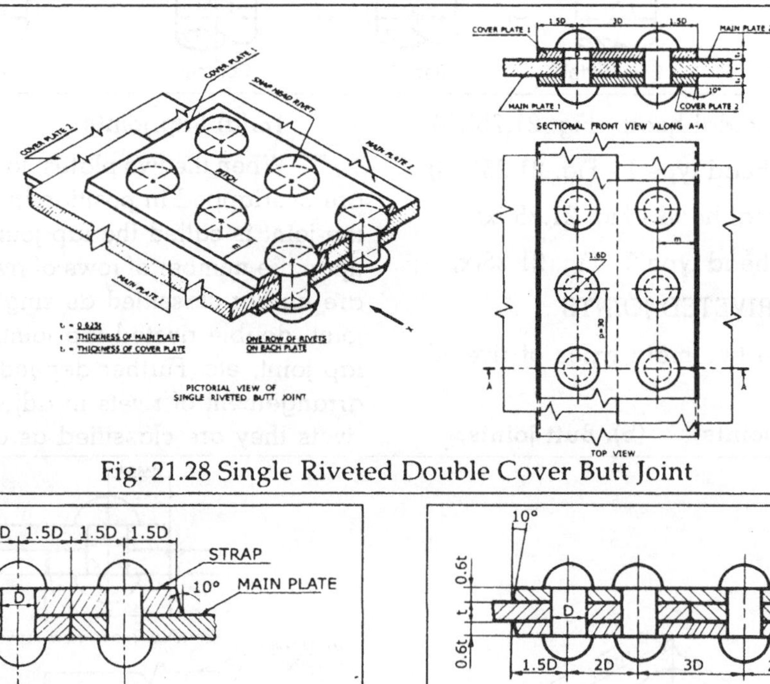

Fig. 21.28 Single Riveted Double Cover Butt Joint

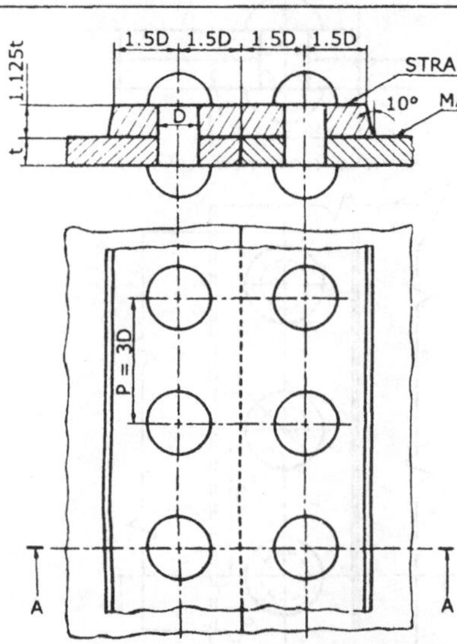

Fig. 21.29 Single Riveted Single Cover Butt Joint

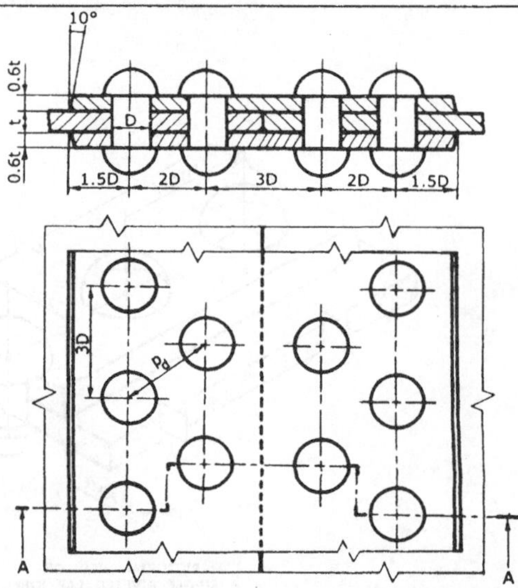

Fig. 21.30 Double Riveted Equal Double Cover Zigzag Butt Joint

and zigzag riveting.

Out of many varieties of lap joints two varieties are shown in Fig. 21.26 and Fig. 21.27.

In Fig. 21.26, a single riveted lap joint is shown and in Fig. 21.27, a double riveted zigzag lap joint is shown.

Other types of lap joints, given below, can be drawn on the similar lines.

(1) Double riveted chain type lap joint

(2) Triple riveted chain type lap joint

(3) Triple riveted zigzag type lap joint

16.(b) Butt Joints :

In butt joints the edges of the two plates, to be joined, are kept butt against each other and then the joint between them is made by single cover plate or by double cover plate.

Depending upon the number of cover plates, joint is given the name, single strap or double strap joints. If these two straps are equal then they are named as equal cover straps and if they are unequal, they are named as unequal cover straps or unequal cover plates.

Depending upon the number of rows of rivets passing through each plate, the joint is given the name single riveted, double riveted, triple riveted, etc.

Depending upon the arrangements of rivets in adjoining rows, they are named as chain riveting or zigzag riveting.

Many varieties of butt joints are listed below. Out of these varieties few are shown in Fig. 21.28 to Fig.21.30, mentioned against their names.

1. Single riveted single cover butt joint - Fig. 21.29
2. Double riveted single cover chain butt joint.
3. Double riveted single cover zigzag butt joint.
4. Triple riveted single cover chain butt joint.
5. Triple riveted single cover zigzag butt joint.
6. Single riveted double cover butt joint - Fig. 21.28
7. Double riveted equal double cover chain butt joint.
8. Double riveted equal double cover zigzag butt joint - Fig. 21.30
9. Double riveted unequal double cover chain butt joint.
10. Double riveted unequal double cover zigzag butt joint.
11. Triple riveted equal cover chain butt joint.
12. Triple riveted unequal cover chain butt joint.
13. Triple riveted equal cover zigzag butt joint.
14. Triple riveted unequal cover zigzag butt joint.

On the similar lines to three varieties already given, one can draw all other varieties easily.

14.(b) WELDED JOINTS :

Another permanent fastening device is welding. Two pieces of metal are joined by fusing them together. Welding has replaced riveting. Now a days, fabrication by welding has become very common. Welding can be done by electric arc, gas, electric resistance, etc.

Some of the forms of the welded joints as per IS:696-1960 with their sectional representations and appropriate symbols are shown in Fig. 21.31. Students should study these symbols critically and should remember them. In drawing and design office one is required to use them frequently.

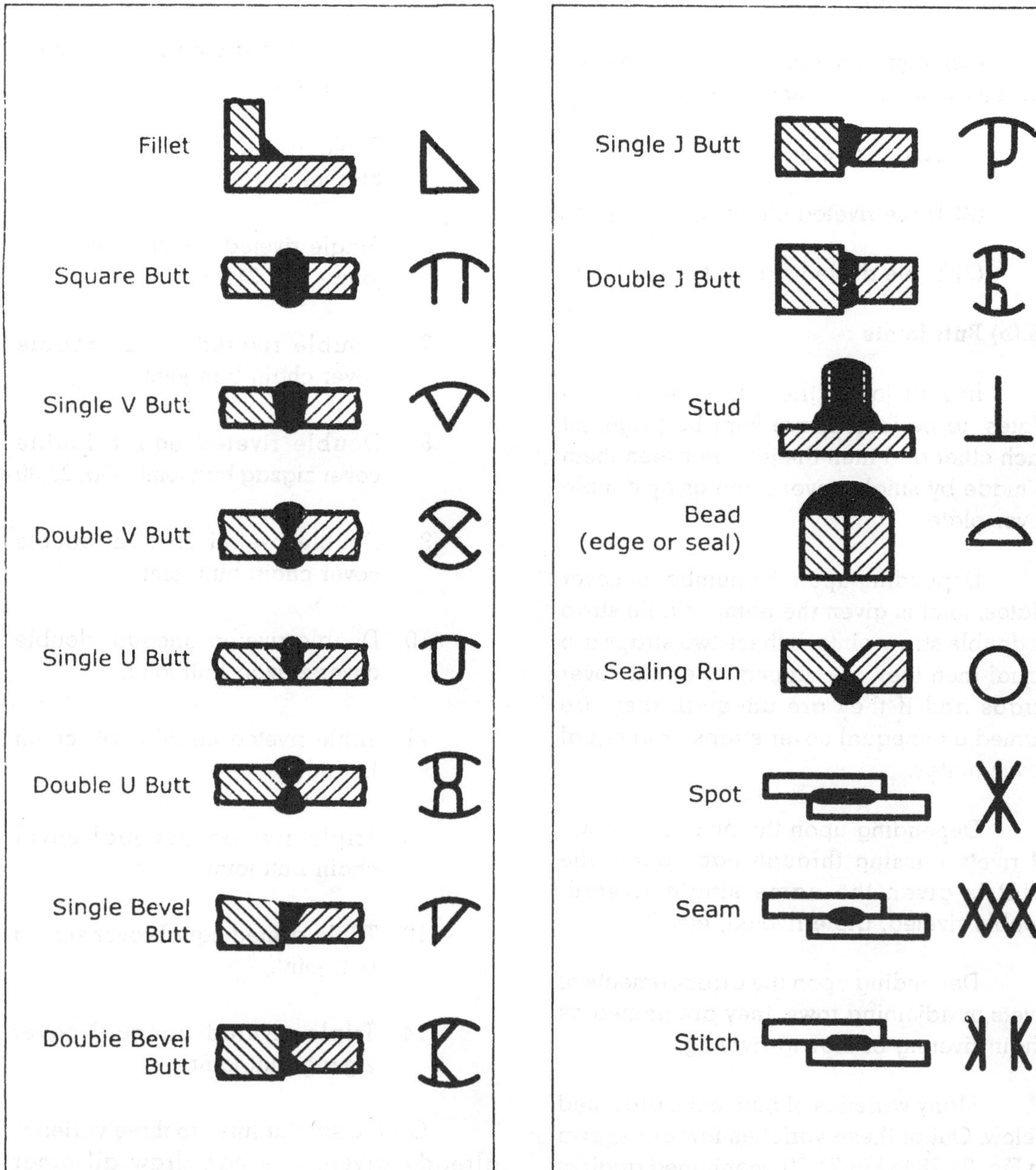

Fig. 21.31 Welded Joints with Symbols

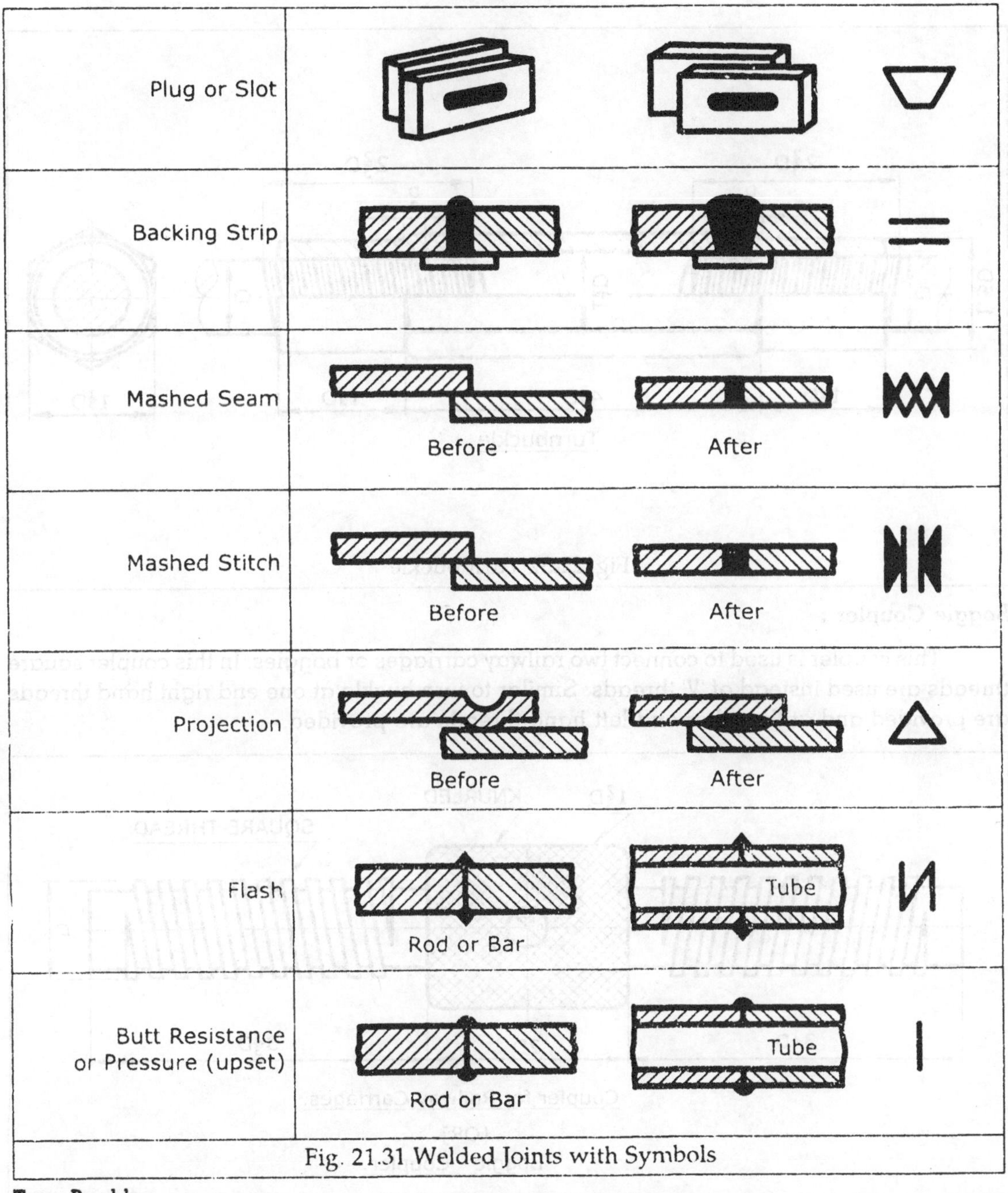

Fig. 21.31 Welded Joints with Symbols

Turn Buckle

Two rods, one having right hand 'V' threads and other having left hand 'V' threads are engaged at both ends of a hexagonal shaped [outer side] nut. Central portion of the nut is kept hollow to accomodate length of the rods. At the centre of the nut inside diameter is more than diameter of the rod and this portion is without threads. When nut is turned by spanner, both the rods come nearer or move away as per the direction of rotation of nut.

Turn buckles are used to tighten guy wires/ropes of telephone poles, electrical poles etc.

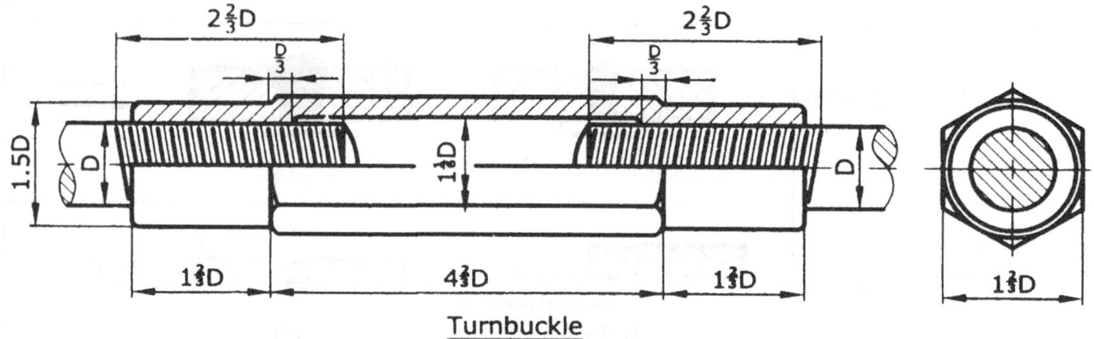

Fig. 21.32 Turn Buckle

Boggie Coupler :

This coupler is used to connect two railway carriages or boggies. In this coupler square threads are used instead of 'V' threads. Similar to turn buckle at one end right hand threads are provided and at the other end left hand threads are provided.

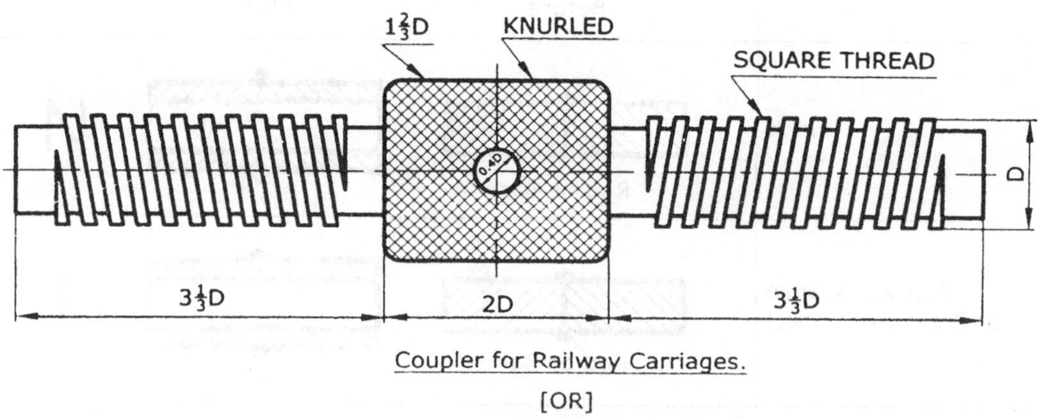

Fig. 21.33 Boggie Coupler

Cotter Joint With Sleeve : (OR) Double Cotter Joint :

In this joint two ends of the two rods are inserted from the opposite sides in a hollow cylindrical sleeve as shown. Two cotters are driven tightly in the slots provided in the sleeve as well as in the rods as shown.

Outer edges of the cotter are perpendicular to the axis of the joint, while inner edges are tapered. Bars subjected to axial forces are joined with this joint.

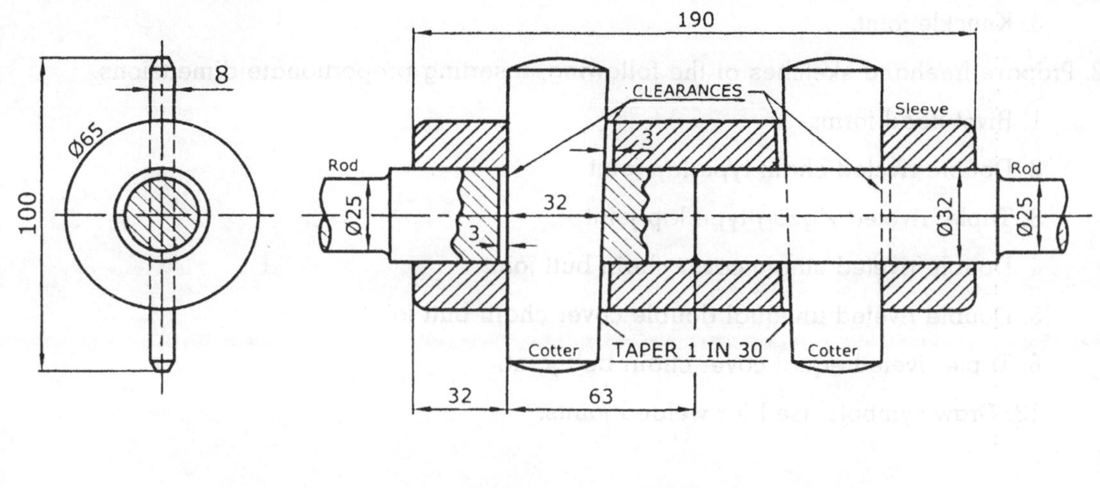

Fig. 21.34 Cotter Joint With Sleeve OR Double Cotter Joint

EXERCISE

1. What are fasteners? Explain the difference between permanent and temporary fasteners.

2. Show on sketch the different terms used for screw thread and define each term.

3. Explain the difference between single start and multistart threads with the help of a sketch. Explain lead and pitch of thread.

4. Draw freehand proportionate sketches of the following forms of screw threads.

1. Acme thread 2. Buttress thread3. Whitworth thread. 4. Unified internal and external threads 5. Sellers thread.

5. Enlist the various types of nuts you know and draw their freehand sketches inserting proportionate dimensions.

6. What is prevention of rotation of bolt? Draw different types of such bolts.

7. What do you understand by locking arrangements of nuts? Enlist and draw sketches of various locking arrangements of nuts, inserting proportionate dimensions.

8. Explain with sketch the use of set-screw. Draw different varieties of heads and ends of set screws.

9. When is stud preferred to bolt? Draw different varieties of studs showing their uses.

10. What are keys? Classify them. Draw freehand sketches of the following keys.

1. Round key 2. Woodruff key 3. Feather key 4. Gib headed sunk key.

11. What is the functional difference between the cotter joint and the knuckle joint? Draw the following freehand sketches inserting necessary proportionate dimensions.

1. Socket and spigot cotter joint
2. Sleeve cotter joint.
3. Knuckle joint.

12. Prepare freehand sketches of the following, inserting proportionate dimensions.
    1. Rivet head forms
    2. Double riveted chain type lap joint
    3. Triple, riveted zigzag type lap joint
    4. Double riveted single cover chain butt joint
    5. Double riveted unequal double cover chain butt joint
    6. Triple riveted equal cover chain butt joint.

13. Draw symbols used for welded joints.

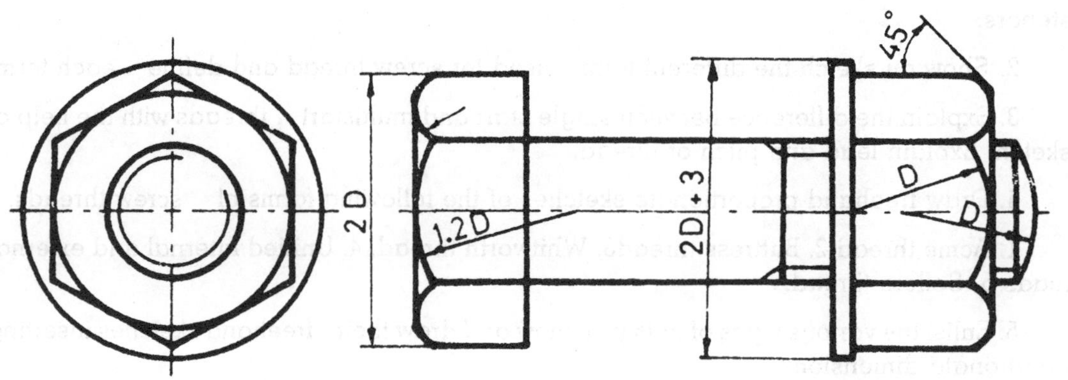

Fig. 21.35 Bolt - Nut - Washer Pair with Proportions

CHAPTER - 22
SHAFT COUPLINGS

1. INTRODUCTION:

Shaft couplings are used to connect two shafts for the purpose of transmission of motion and power. e.g. an electric motor is connected to a centrifugal pump by coupling so as to drive it.

Couplings are classified as under:

(i) Rigid or Fast Couplings.
(ii) Flexible Couplings
(iii) Loose or Disengaging Couplings.
(iv) Non-aligned Couplings.

2. RIGID OR FAST COUPLINGS : (Fig.22.1 to 22.6)

Rigid couplings are used to connect only co-axial shafts. This coupling provides rigid connection between two shafts and does not allow any relative motion between shafts.

Rigid couplings are further sub-classified as under:

(a) Muff Couplings -	(i) Box or Muff Couplings	Fig.22.1
	(ii) Half Lap Muff Couplings	Fig.22.2
	(iii) Split Muff Couplings	Fig.22.3
(b) Flanged Couplings	(i) Flanged Couplings	Fig.22.4
	(ii) Protected Type Flanged Couplings	Fig.22.5
	(iii) Solid or Forged Flanged Couplings	Fig.22.6

2.(a) (i) BOX OR MUFF COUPLINGS: Fig 22.1

It consists of cast iron cylindrical sleeve called "Muff". Two shafts are butt against each other in the middle of the muff, as shown. A key is driven in the keyway between the muff and the shaft. A sunk key is inserted on both the shafts.

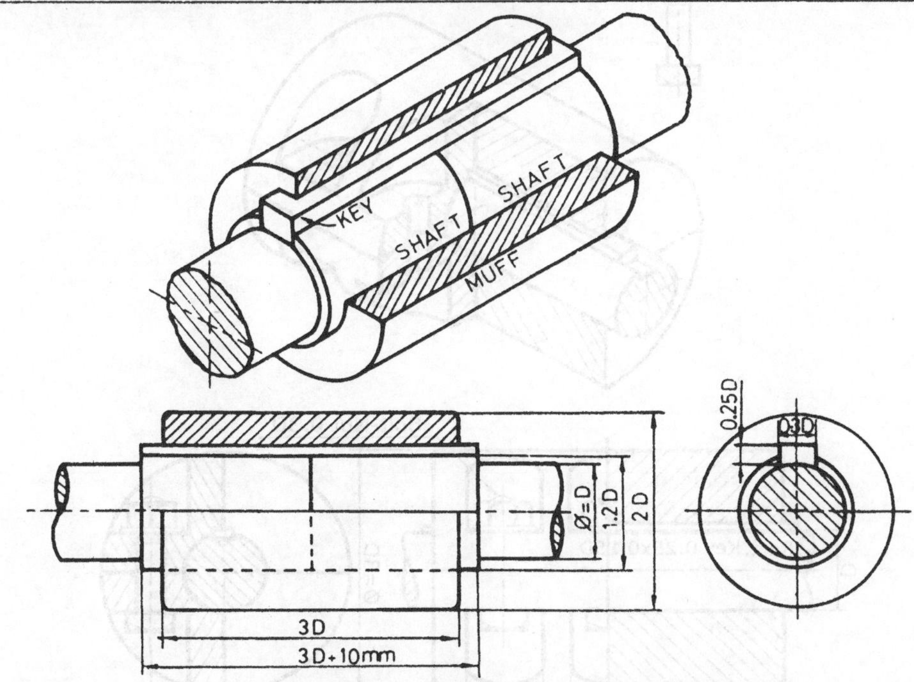

Fig. 22.1 Box or Muff or Sleeve Coupling

2.(a) (ii) HALF LAP MUFF COUPLINGS : (Fig. 22.2)

Two shafts are in overlapped position inside the muff instead of butt position. The taper provided in the overlap prevents the axial separation of shafts. Hollow saddle key is used. Overlap portion helps in power transmission.

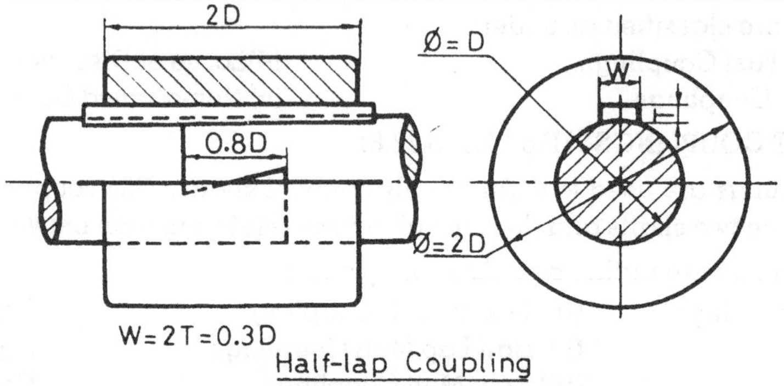

Fig. 22.2 Half - lap Muff Coupling

2.(a) (iii) SPLIT MUFF COUPLINGS : (Fig. 22.3)

Muff is made in two equal semi-cylindrical halves. They are assembled together by means of 2 or 3 bolts, nuts and washer pairs, as shown. Assembly of two split muffs with shafts and key is conveniently done without disturbing the set-up. This is the main advantage of split muff couplings.

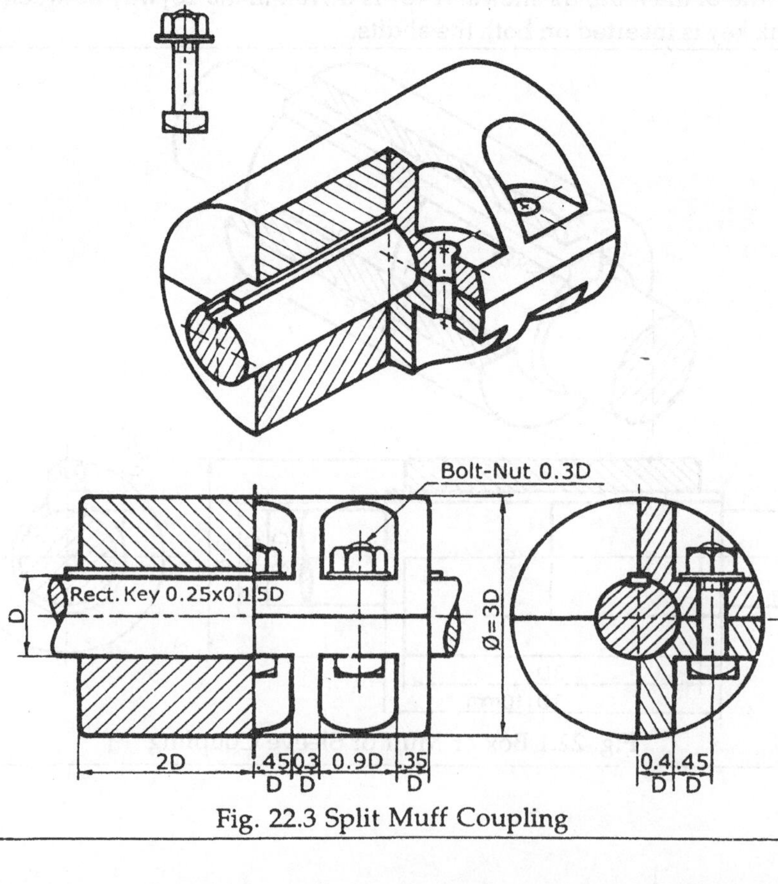

Fig. 22.3 Split Muff Coupling

2.(b)(i) FLANGED COUPLINGS (UNPROTECTED) : (Fig. 22.4)

This is very widely used coupling. Two similar cast iron flanges are keyed to the two ends of shafts. Taper sunk keys are commonly used. For the purpose of alignment, cylindrical portion of one flange is fitted into a corresponding recess of another flange. Bolt-Nut pairs, 4 to 6 in numbers, are used to connect two flanges.

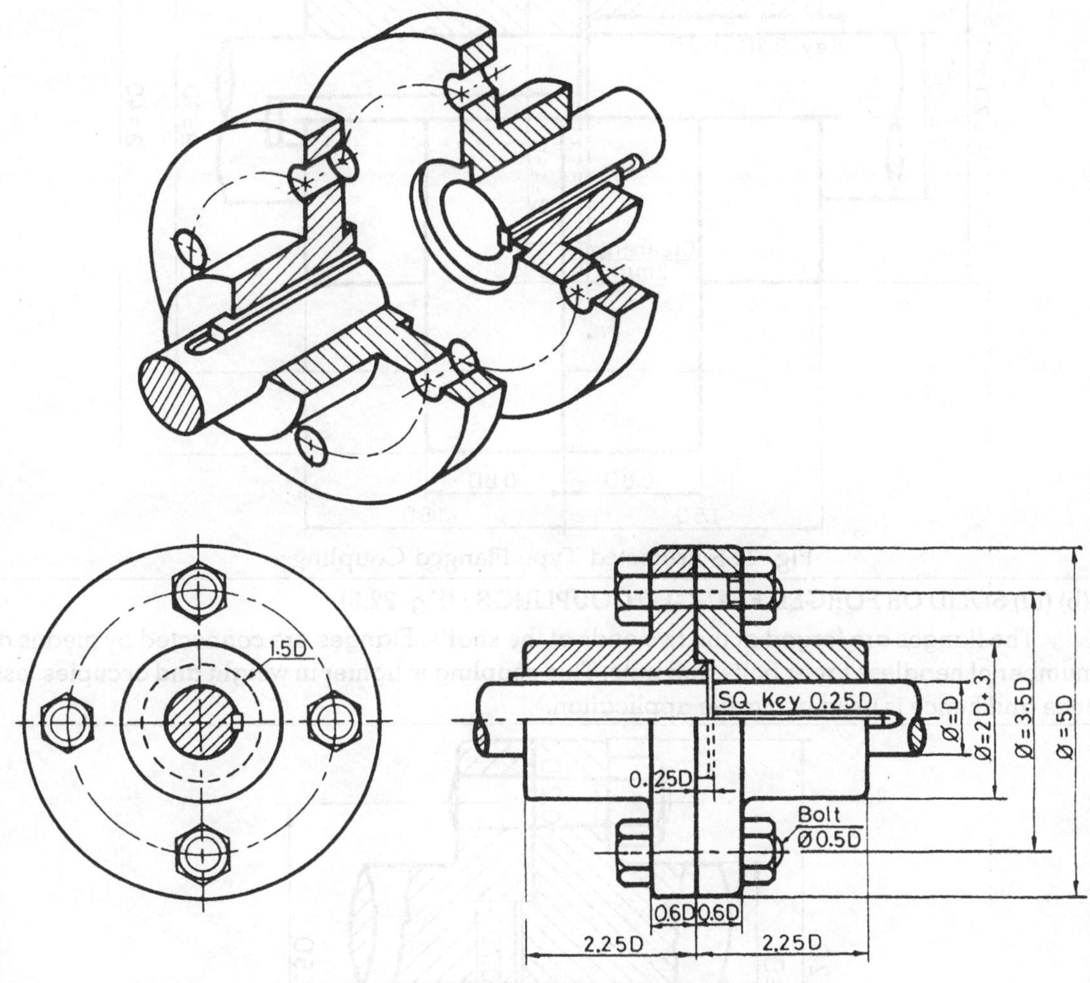

Fig. 22.4 Flanged Coupling (Unprotected)

2.(b)(ii) PROTECTED TYPE FLANGED COUPLINGS : (Fig. 22.5)

In this coupling both the flanges are provided with annular projection just to cover bolt-head, on one side and nut on the other side. This arrangement prevents bolts and nuts from catching the clothes of workmen and thus ensures safety. All other things are similar to the flanged coupling.

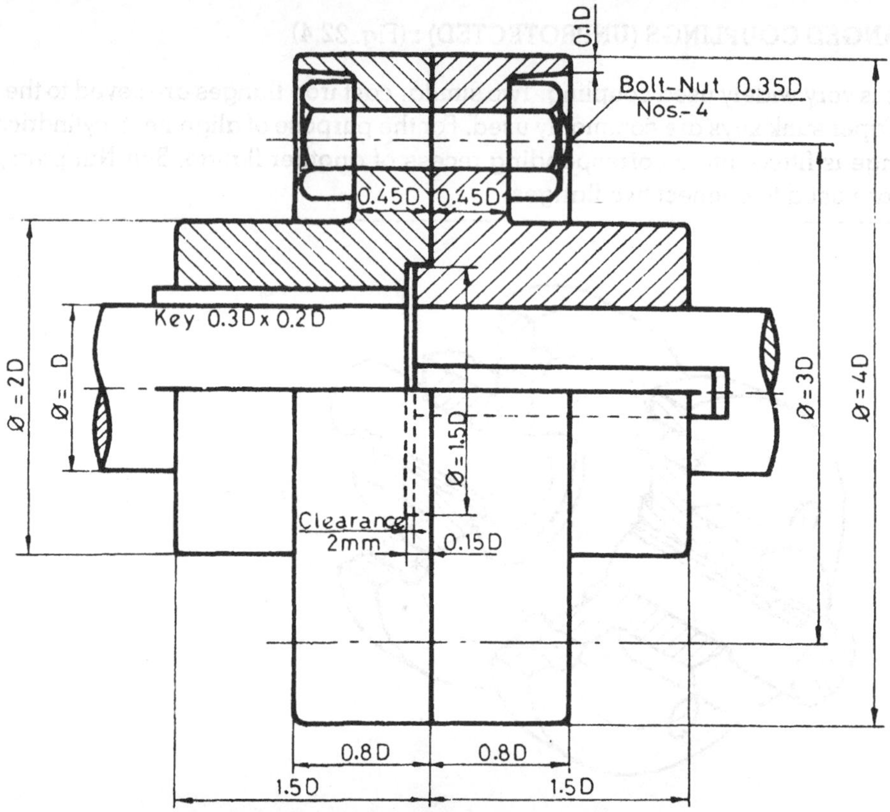

Fig. 22.5 Protected Type Flanged Coupling

2.(b) (iii) SOLID OR FORGED FLANGED COUPLINGS : (Fig. 22.6)

The flanges are forged at the two ends of the shafts. Flanges are connected by means of a number of headless taper bolts and nuts. This coupling is lighter in weight and occupies less space and hence is used in marine application.

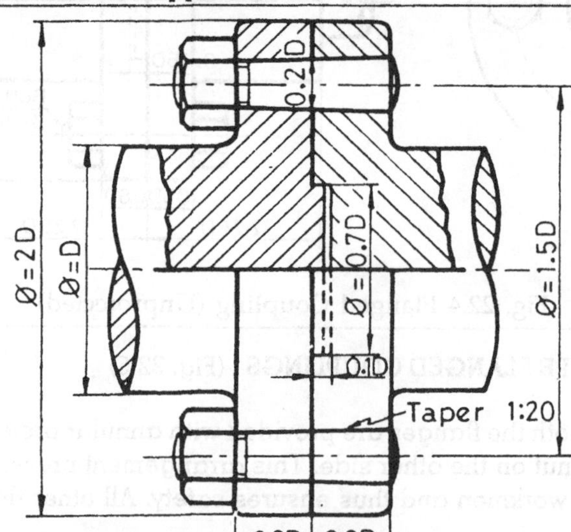

Fig. 22.6 Solid or Forged Flanged Coupling

3. PIN-TYPE FLEXIBLE FLANGED COUPLINGS : (Fig. 22.7)

A flexible coupling allows variation in the alignment of shafts and permits limited amount of relative motion. Rubber or leather washers, as shown in the figure, are used to absorb shocks. These washers also act as insulators.

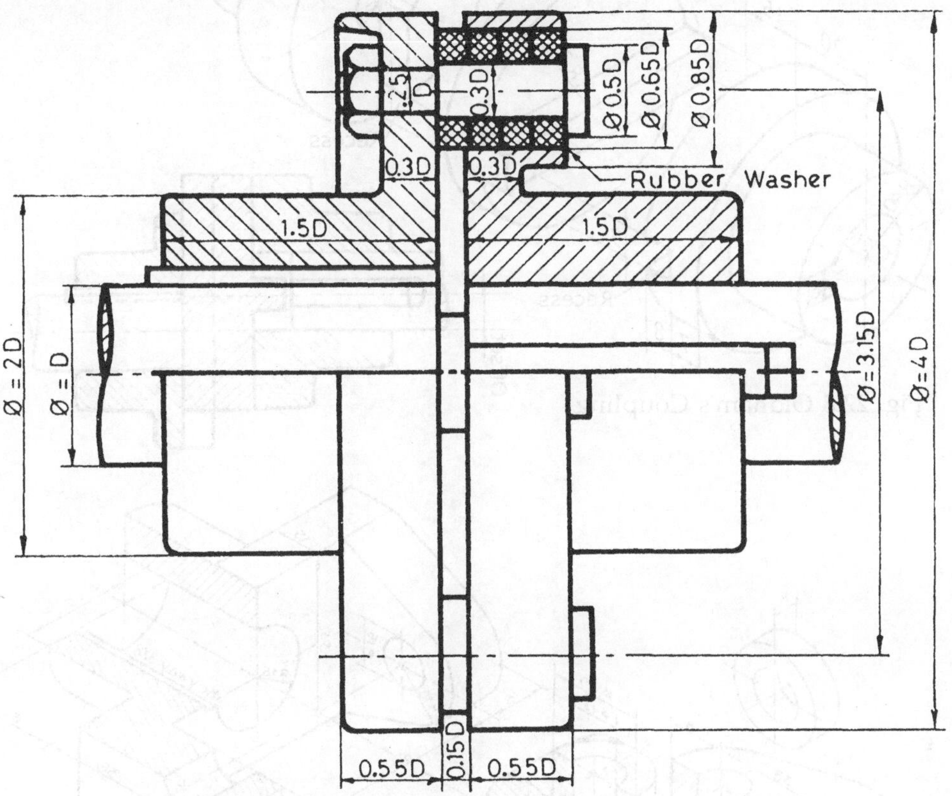

Fig. 22.7 Pin - Type Flexible Flanged Coupling

4. LOOSE OR DISENGAGING COUPLINGS :

In this coupling shafts can be engaged and disengaged at will. Following are the couplings of this type :

(i) Claw Couplings

(ii) Friction Clutches

(iii) Dog Clutches, etc.

Illustrations of the above couplings are not given.

5. NON-ALIGNED COUPLINGS (OLDHAM'S COUPLINGS) : (Fig. 22.8)

This type of coupling is used for connecting two shafts whose axes are not co-axial but are parallel. Oldham's coupling is such a coupling. The two mutually right-angled tongues, provided on two opposite sides of central circular disc, slide inside corresponding recesses provided on two flanges, keyed on two shafts. The figure is self explanatory.

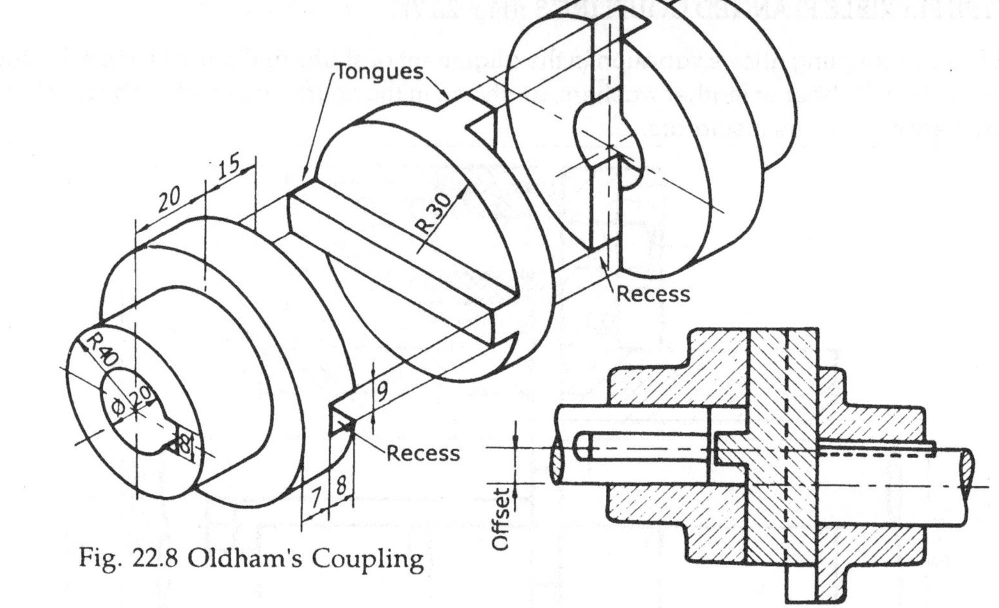

Fig. 22.8 Oldham's Coupling

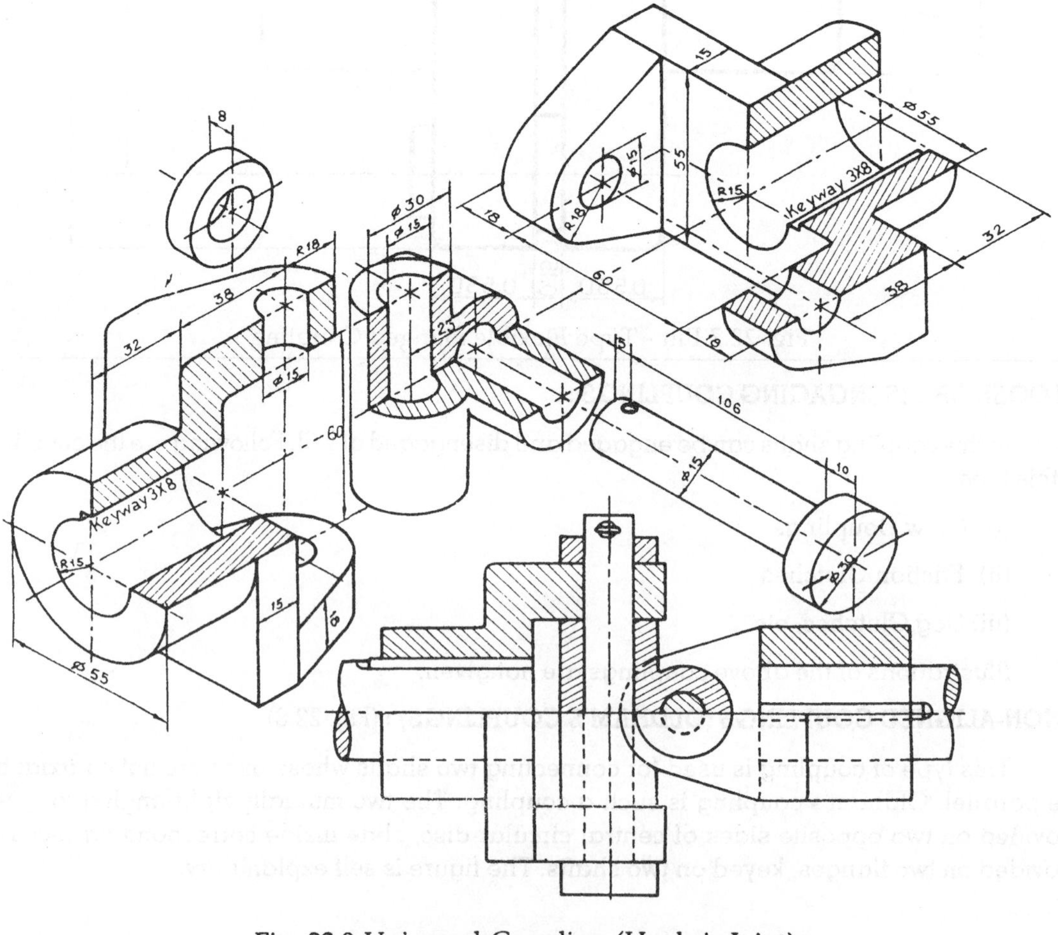

Fig. 22.9 Universal Coupling (Hooke's Joint)

6. UNIVERSAL COUPLINGS (HOOKE'S JOINT) : (Fig.22.9)

This coupling is used to connect two shafts whose axes intersect at an angle. Variation in angle between shafts is permitted even during rotation. This type of couplings are used in automobiles and machine tools. See figure on page 185.

EXERCISE

1. Classify couplings?
2. List different types of couplings you know, stating special features and use of each one.
3. What is the difference between rigid and flexible couplings? Where are they used?
4. Name and sketch the coupling used for coupling parallel shafts, having smaller distance between them. Give proportionate dimensions.
5. Name and sketch the coupling used for joining two shafts at some angle. Give proportionate dimensions.
6. Name and draw sketches of all types of muff couplings showing clearly the difference between them. Give proportionate dimensions.
7. Name and draw sketches of all types of flanged couplings showing clearly the difference between them. Give proportionate dimensions.

CHAPTER - 23
BEARINGS

1. INTRODUCTION :

The function of bearings is to support rotating shafts. A shaft rotates inside the bearing practically with zero friction.

The bearings are classified in two different manners.

(1)	(i) A Journal Bearing	[Fig. 23.1(a)]
	(ii) A Pivot Bearing	[Fig.23.1(b)]
	(iii) A Collar or Thrust Bearing	[Fig. 23.1 (c)]
(2)	(i) Sliding Bearings	[Fig. 23.2 to 23.6]
	(ii) Rolling or Antifriction Bearings	[Fig. 23.7]

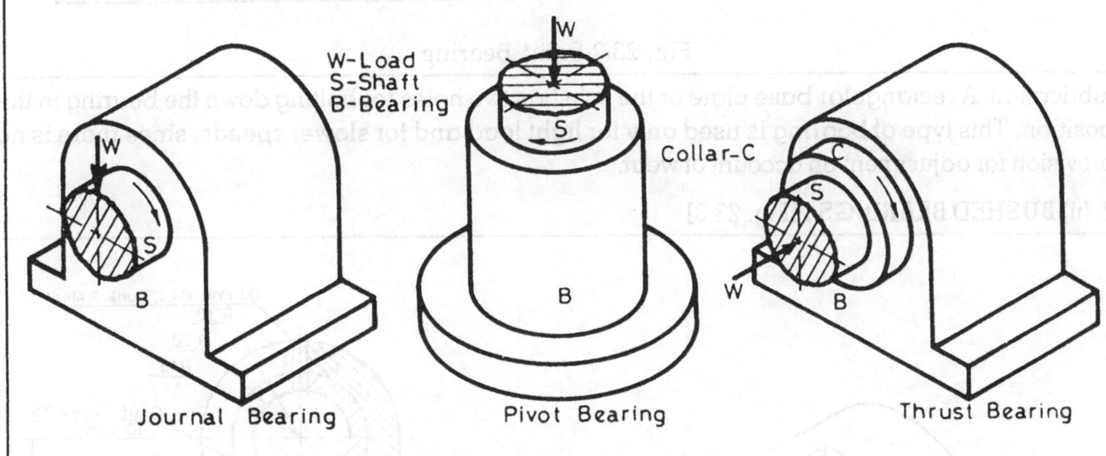

Fig. 23.1 (a) (b) (c) Types of Bearings

2. A JOURNAL BEARING : [Fig.23.1 (a)]

A bearing, in which the load on the shaft is perpendicular to the axis, is known as journal bearing. See Fig. 23.1(a). The portion of the shaft rotating within the bearing is known as journal.

Some of the journal bearings are :

(i) Solid Bearing - Fig.23.2

(ii) Bushed Bearing - Fig.23.3

(iii) Pedestal Bearing or Plummer Block Bearing - Fig. 23.4

2.(i) SOLID BEARINGS : [Fig. 23.2]

A solid bearing is the simplest kind of a journal bearing. It consists of a cast iron block with a hole to accomodate the journal (shaft). Oil hole is drilled at the top for the purpose of

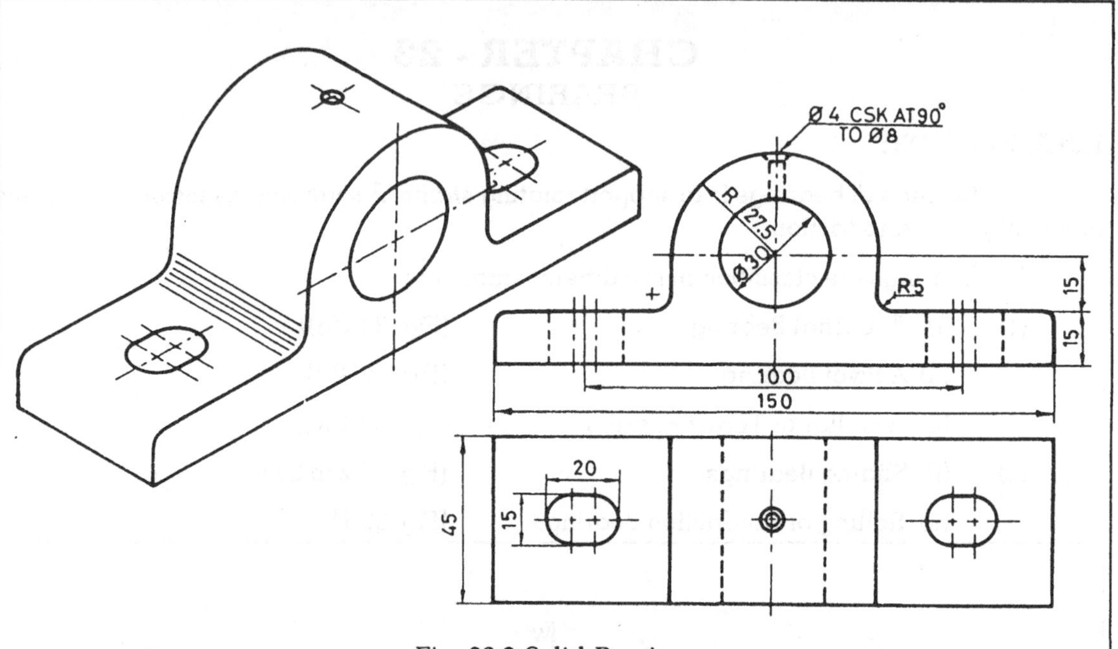

Fig. 23.2 Solid Bearing

lubrication. A rectangular base plate or the sole has two holes for bolting down the bearing in the position. This type of bearing is used only for light load and for slower speeds, since there is no provision for adjustment on account of wear.

2. (ii) **BUSHED BEARINGS :** [Fig. 23.3]

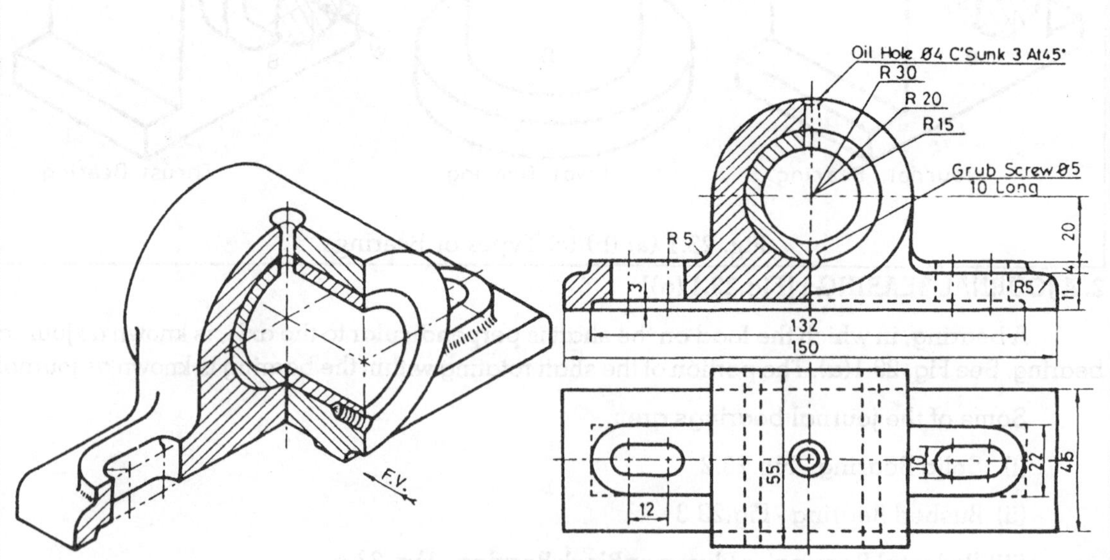

Fig. 23.3 Bushed Bearing

It is a modified form of solid bearing. Brass, phospher bronze or gun metal bush is press-fitted inside the bearing. Bush is hollow cylindrical piece forced or press-fitted inside the solid bearing. The journal or the shafts rotates in the bush. Wear and tear takes place only on bushes and hence periodically they are replaced. Grub-screw or a dowel-pin prevents rotation of bush along with rotation of shaft.

In the above two bearings, insertion of the shaft in the bearing is endwise only.

2.(iii) PEDESTAL BEARINGS OR PLUMMER BLOCK BEARINGS : [Fig. 23.4]

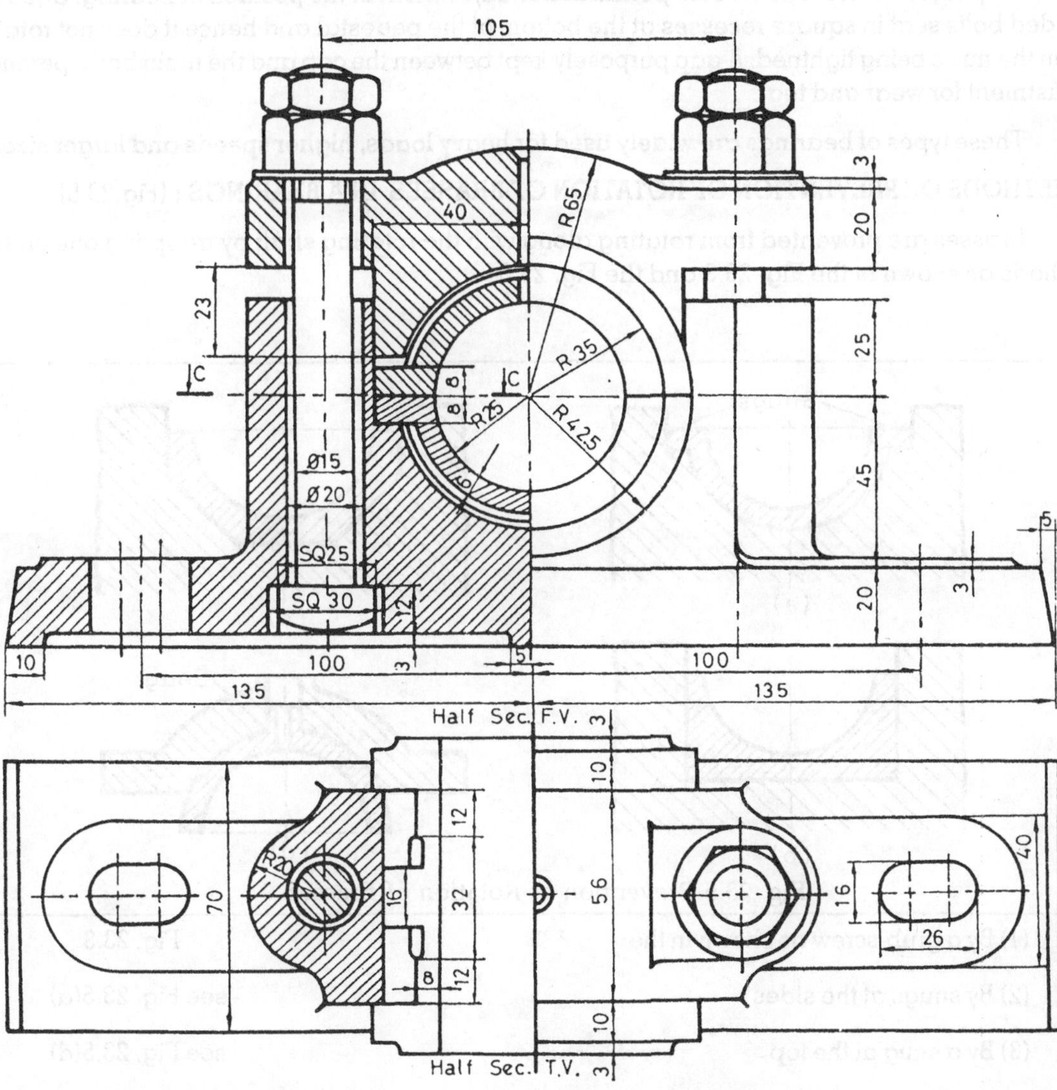

Fig. 23.4 Pedestal or Plummer Block Bearing

It consists of mainly,

(1) A cast iron pedestal with a sole,

(2) Gunmetal or phospher bronze brasses in two halves.

(3) A cast iron cap,

(4) Two pairs of square headed bolt, nut and washer made of mild steel.

Provision of separate cap and brasses in two halves facilitates easy placing and removal of the shaft in and from the bearing. Adjustment for wear can also be easily done in this bearing.

Brasses are prevented from rotating along with shaft by any one of the methods discussed in the later part of this chapter. Collars at either end on brasses prevent axial movement of brasses inside the bearing.

Elliptical bolt holes in the sole permit minor adjustment in the position of bearing. Square headed bolts seat in square recesses at the bottom of the pedestal and hence it does not rotate when the nut is being tightned. A gap purposely kept between the cap and the main body, permits adjustment for wear and tear.

These types of bearings are widely used for heavy loads, higher speeds and larger sizes.

3. **METHODS OF PREVENTION OF ROTATION OF BRASSES IN A BEARINGS :** [Fig.23.5]

Brasses are prevented from rotating along with the rotating shaft by adopting one of the methods as shown in the Fig. 23.3 and the Fig. 23.5

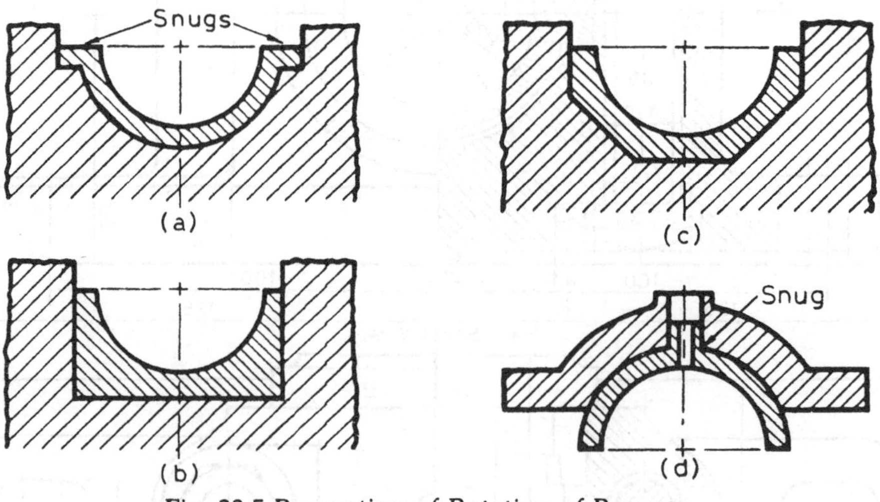

Fig. 23.5 Prevention of Rotation of Brasses

(1) By a grub-screw as shown in the Fig. 23.3

(2) By snugs at the sides - see Fig. 23.5(a)

(3) By a snug at the top - see Fig. 23.5(d).

(4) By making outer shape of one of the brasses rectangular - see Fig. 23.5(b)

(5) By making outer shape of one of the brasses octagonal - see Fig. 23.5(c)

Outer rectangular or octagonal shape of the brasses may be kept for entire length or may be kept at the ends only.

Brasses are made thicker where they are likely to wear more. Thicker brasses, at the place of wear, take care of wear and tear and hence they are used for longer duration.

4. PIVOT BEARINGS : [Fig.23.1(b) and Fig.23.6)

These bearings are used for vertical shafts with axial loads. The bottom surface of the shaft end rests on the surface of the bearing. See Fig. 23.1(b) and Fig 23.6.

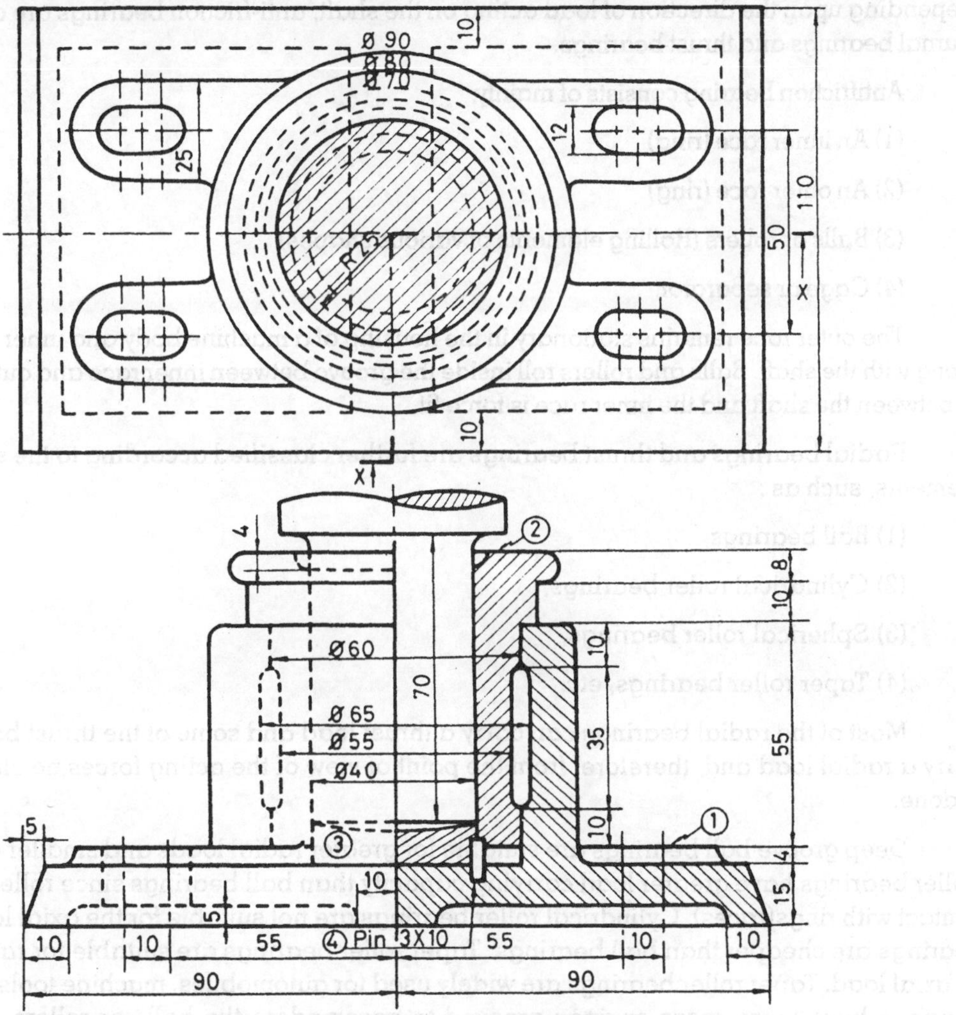

Fig. 23.6 Pivot or Foot-Step Bearing

It consists of mainly,

(1) A cast iron block with sole

(2) A gun metal bush with a collar at the top.

(3) Concave steel disc placed at the bottom in the bush.

(4) M.S. pin half in the disc, half in the bush and half in the housing located off centre.

Concave steel disc takes axial thrust and the bush takes the radial load. M.S. pin prevents, the disc and the bush both from rotating along with the rotating shaft. The collar of the bush serves as an oil cup. Oil is used for the purpose of lubrication. Soles in most of the bearings are recessed at the bottom to reduce machining.

5. **ANTI-FRICTION BEARINGS :** (Fig. 23.7)

So far we have studied sliding bearings. Friction in sliding bearings is much more than rolling bearings. **Anti-friction bearings** have rolling friction. Ball and roller bearings are Anti-friction bearings. Depending upon the direction of load acting on the shaft, anti-friction bearings are classified as journal bearings and thrust bearings.

Antifriction bearing consists of mainly,

(1) An inner race (ring)

(2) An outer race (ring)

(3) Balls or rollers (Rolling elements of various shapes)

(4) Cage or separator.

The outer race remains stationary in the housing of a machine body and inner race rotates along with the shaft. Balls and rollers roll inside the groovo between inner race and outer race. The fit between the shaft and the inner race is force fit.

Radial bearings and thrust bearings are further classified according to the shape of the elements, such as :

(1) Ball bearings,

(2) Cylindrical roller bearings,

(3) Spherical roller bearings,

(4) Taper roller bearings, etc.

Most of the radial bearings can carry a thrust load and some of the thrust bearings can carry a radial load and, therefore, from the point of view of the acting forces no clear division is done.

Deep groove ball bearings are suitable for greater radial loads and smaller axial loads. Roller bearings have greater load carrying capacity than ball bearings since rollers have line contact with rings(races). Cylindrical roller bearings are not suitable for the axial loads. Roller bearings are cheaper than ball bearings. Taper roller bearings are suitable for radial as well as axial load. Taper roller bearings are widely used for automobiles, machine tools, etc. Thrust bearings have loose races or rings grooved to accomodate the balls or rollers. Thrust ball bearings are used for the axial loads. If the thrust is likely to occur in both the directions double direction ball thrust bearing is used.

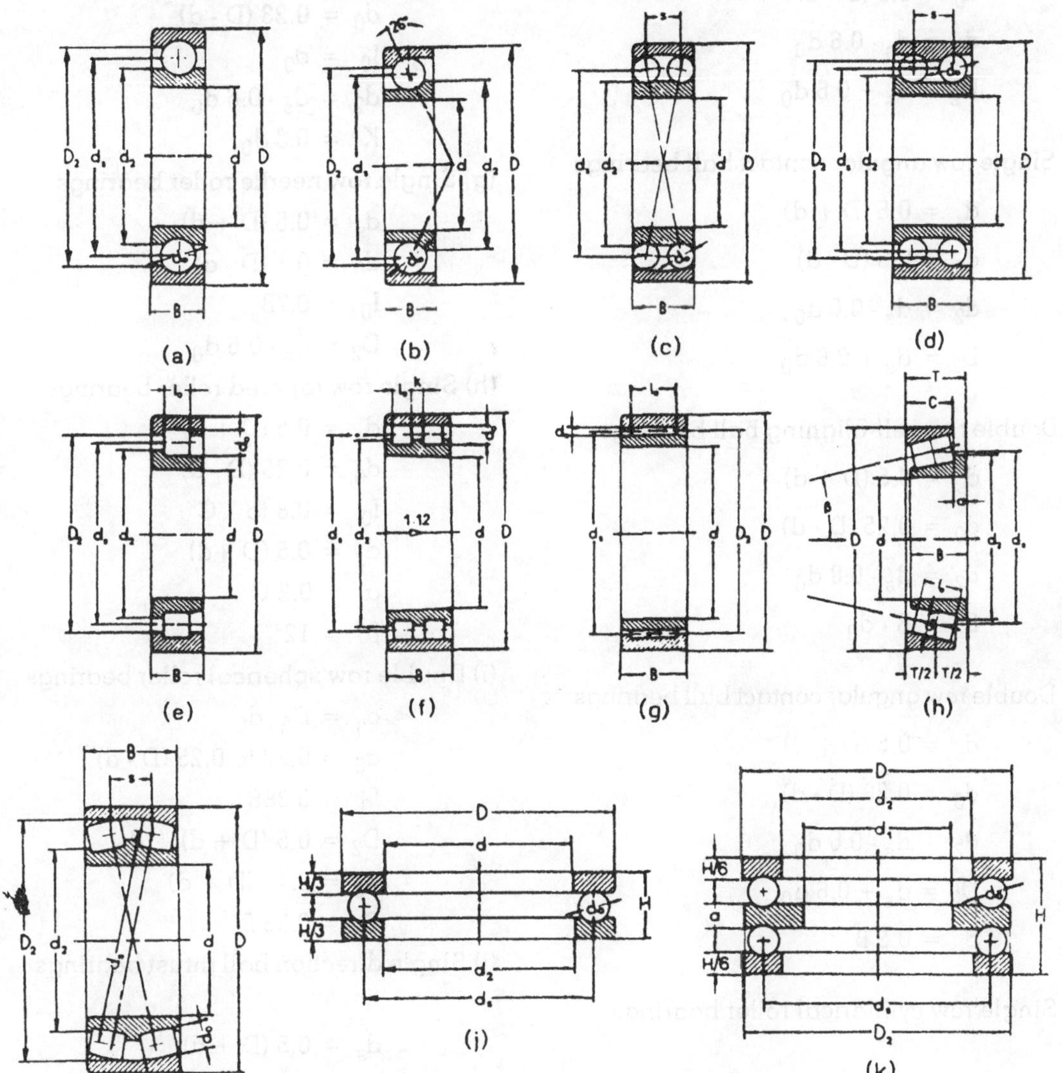

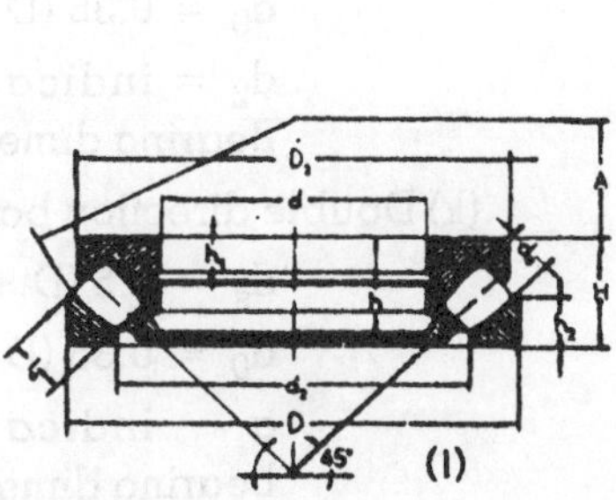

(l) Spherical roller thrust bearings

d_0 = 0.37 to 0.42 H

l_0 = 0.50 to 0.52 H

Other dimensions from table.

Fig. 23.7 Antifriction Bearing

(a) Single row deep groove ball bearings

$d_s = 0.5 (D + d)$

$d_0 = 0.3 (D - d)$

$d_2 = d_s - 0.6 d_0$

$D_2 = d_s + 0.6 d_0$

(b) Single row angular contact ball bearings

$d_s = 0.5 (D + d)$

$d_0 = 0.3 (D - d)$

$d_2 = d_s - 0.6 d_0$

$D_2 = d_s + 0.6 d_0$

(c) Double row self-aligning ball bearings

$d_s = 0.5 (D + d)$

$d_0 = 0.25 (D - d)$

$d_2 = d_s - 0.8 d_0$

$S = B - d_0$

(d) Double row angular contact ball bearings

$d_s = 0.5 (D + d)$

$d_0 = 0.28 (D - d)$

$d_2 = d_s - 0.6 d_0$

$D_2 = d_s + 0.6 d_0$

$S = 0.5 B$

(e) Single row cylindrical roller bearings

$d_s = 0.5 (D + d)$

$d_0 = 0.25 (D - d)$

$I_0 = 1 \text{ to } 1.5 d_0$

$d_2 = d_s - 0.6 d_0$

$D_2 = d_s + 0.6 d_0$

(f) Double row cylindrical roller bearings with tapered bore

$d_s = 0.5 (D + d)$

$d_0 = 0.23 (D - d)$

$I_0 = d_0$

$d_2 = d_s - 0.6 d_0$

$K = 0.3 d_0$

(g) Single row needle roller bearings

$d_s = 0.5 (D + d)$

$d_0 = 0.1 (D - d)$

$I_0 = 0.7B$

$D_2 = d_s - 0.6 d_0$

(h) Single row tapered roller bearings

$d_s = 0.5 (D + d)$

$d_0 = 0.25 (D - d)$

$I_0 = 0.8 \text{ to } 1 C$

$d_2 = 0.5 (D + d)$

$a = 0.2 C$

$\beta = 12°$

(i) Double row spherical roller bearings

$d_s = D_2 - d_0$

$d_0 = 0.22 \text{ to } 0.25 (D - d)$

$I_0 = 0.38B$

$D_2 = 0.5 (D + d) + d_0$

$d_2 = 0.45 (D + d)$

$S = 0.53 B$

(j) Single direction ball thrust bearings

$d_s = 0.5 (D + d)$

$d_0 = 0.36 (D - d)$

d_2 = indicated in respective Bearing dimension tables

(k) Double direction ball thrust bearings

$d_s = 0.5 (D + d_2)$

$d_0 = 0.35 (D - d_2)$

a indicated in respective bearing dimension

d_1 tables

d_2

Some of the rolling bearings are as under :

(1) Single Row Deep Grooved Ball Bearings - Fig. 23.7(a)

(2) Single Row Angular Contact Ball Bearings - Fig. 23.7(b)

(3) Double Row Self-aligning Ball Bearings - Fig. 23.7(c)

(4) Double Row Angular Contact Ball Bearings - Fig. 23.7(d)

(5) Single Row Cylindrical Roller Bearings - Fig. 23.7(e)

(6) Double Row Cylindrical Roller Bearings With - Tapered Bore - Fig. 23.7(f)

(7) Single Row Needle Roller Bearings - Fig. 23.7(g)

(8) Single Row Tapered Roller Bearings - Fig. 23.7(h)

(9) Double Row Spherical Roller Bearings - Fig. 23.7(i)

(10) Single Direction Ball Thrust Bearings - Fig. 23.7(j)

(11) Double Direction Ball Thrust Bearings - Fig. 23.7(k)

(12) Spherical Roller Thrust Bearings - Fig. 23.7(l)

EXERCISE

1. Classify bearings? Draw sketches of each type.
2. What is a bush? Why it is used in bearing?
3. What are brasses ? Why are they used in bearings?
4. Explain the method of prevention of rotation of brasses in bearings with illustration.
5. Draw or sketch following bearings for a shaft of 50mm diameter.
   (i) Solid bearings
   (ii) Bushed bearings
   (iii) Plummer block bearings
   (iv) Pivot bearings
6. What are antifriction bearings? How do they differ from the sliding bearings?
7. List different types of sliding and rolling bearings stating special characteristics of each type.

Solutions to Exercises - Chapter - 15

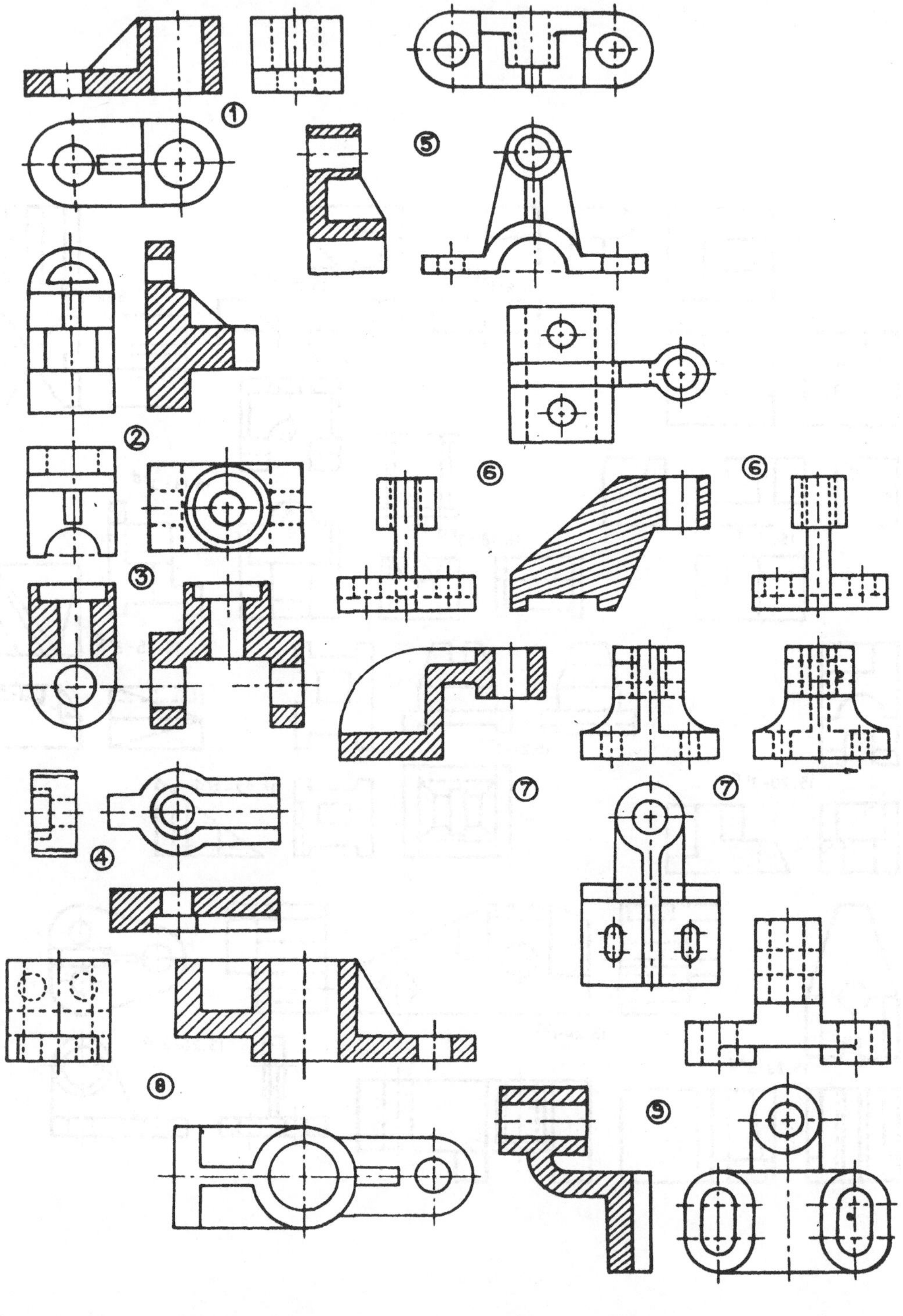

Solutions to Exercises - Chapter - 16 [May not be in sequence]

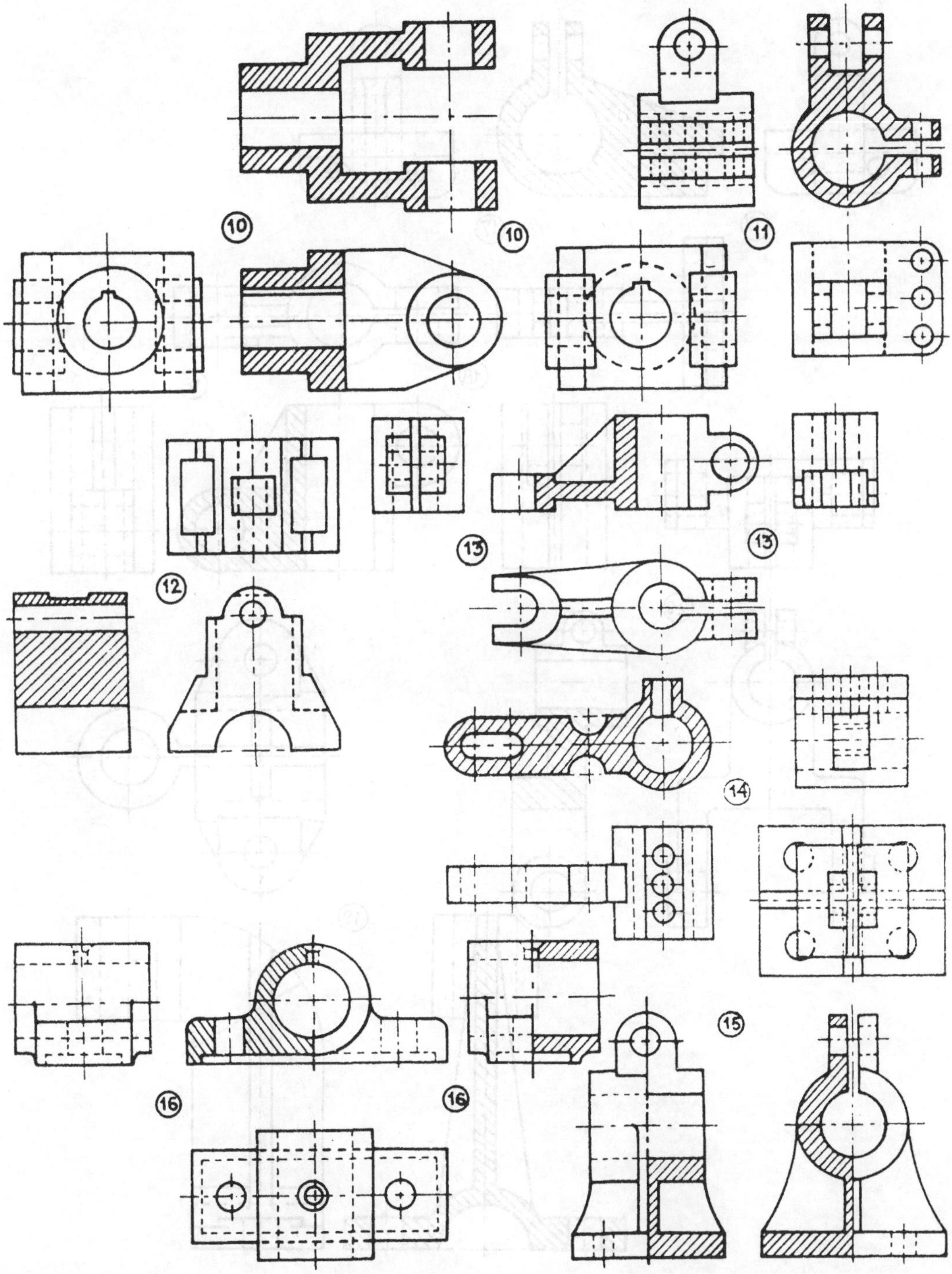

Solutions to Exercises - Chapter - 16 [May not be in sequence]

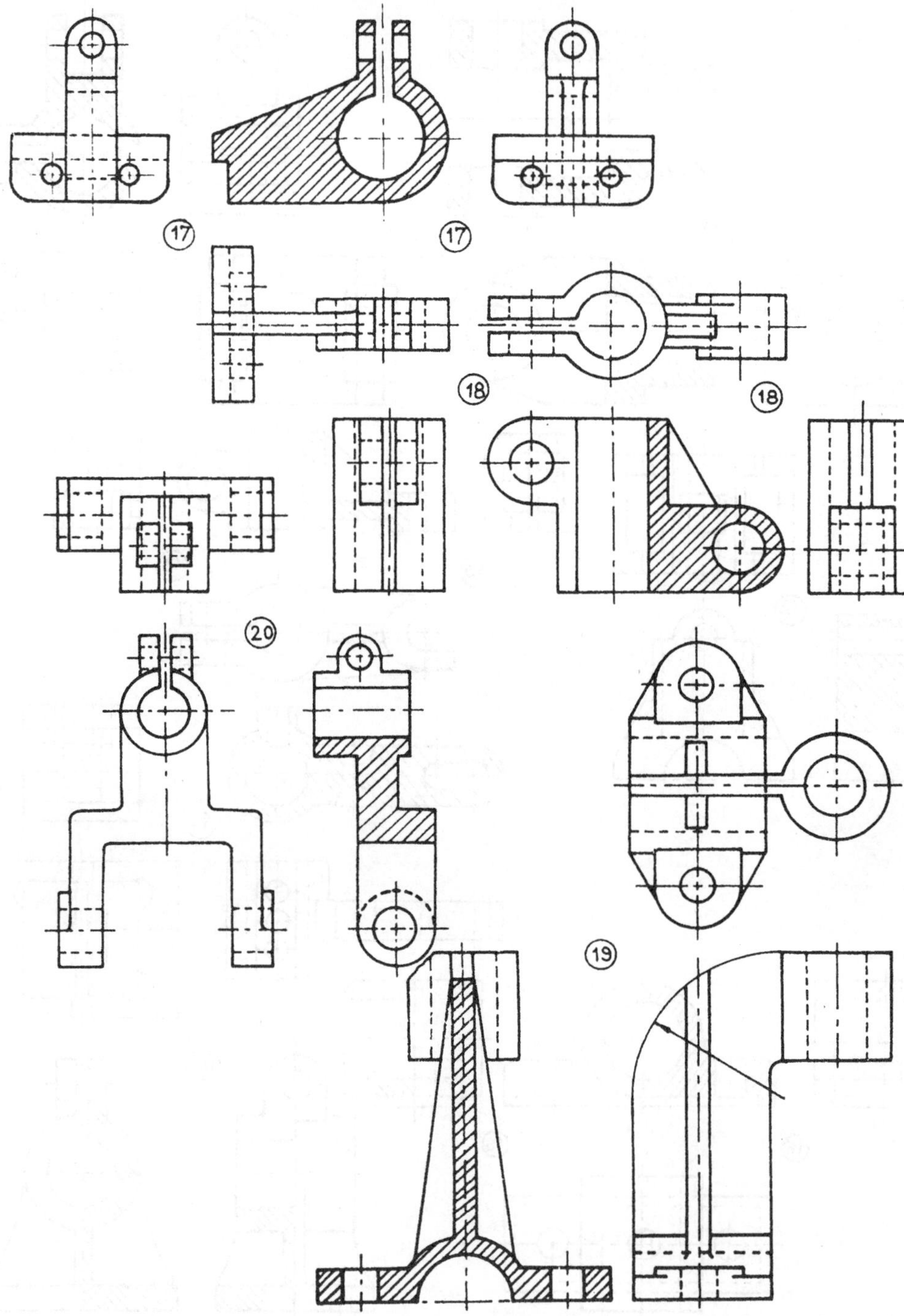

Solutions to Exercises - Chapter - 16 [May not be in sequence]

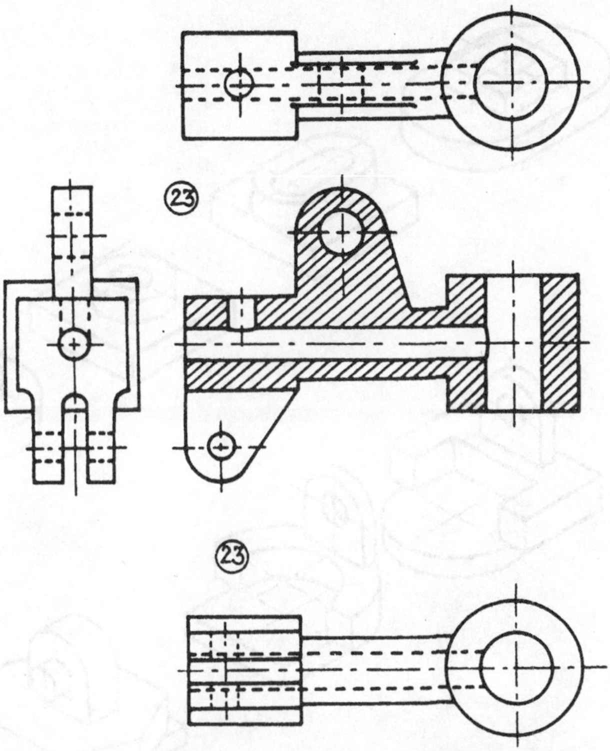

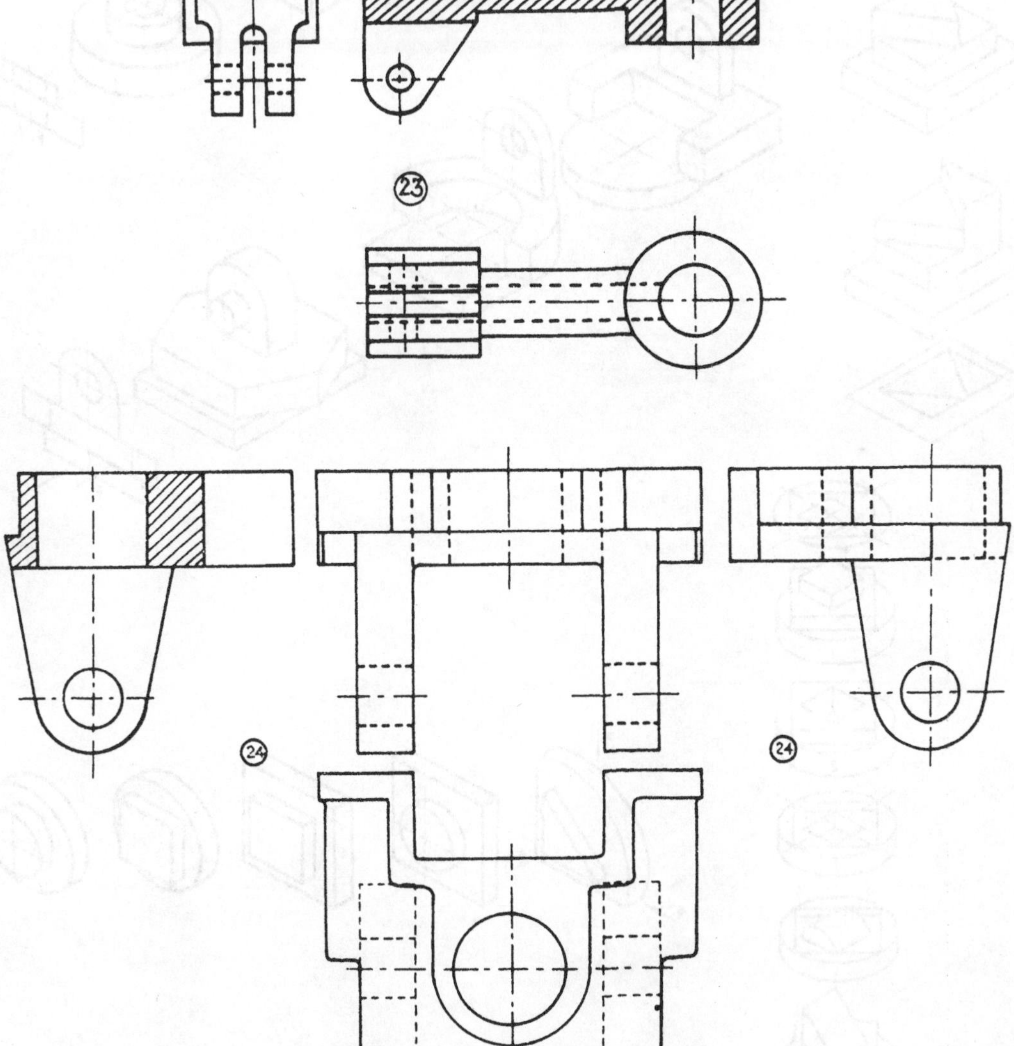

Solutions to Exercises - Chapter - 16 [May not be in sequence]

F.V.
T.V.
F.V.
T.V.
1
2
3
4
5
T.V.
S.V.
F.V.

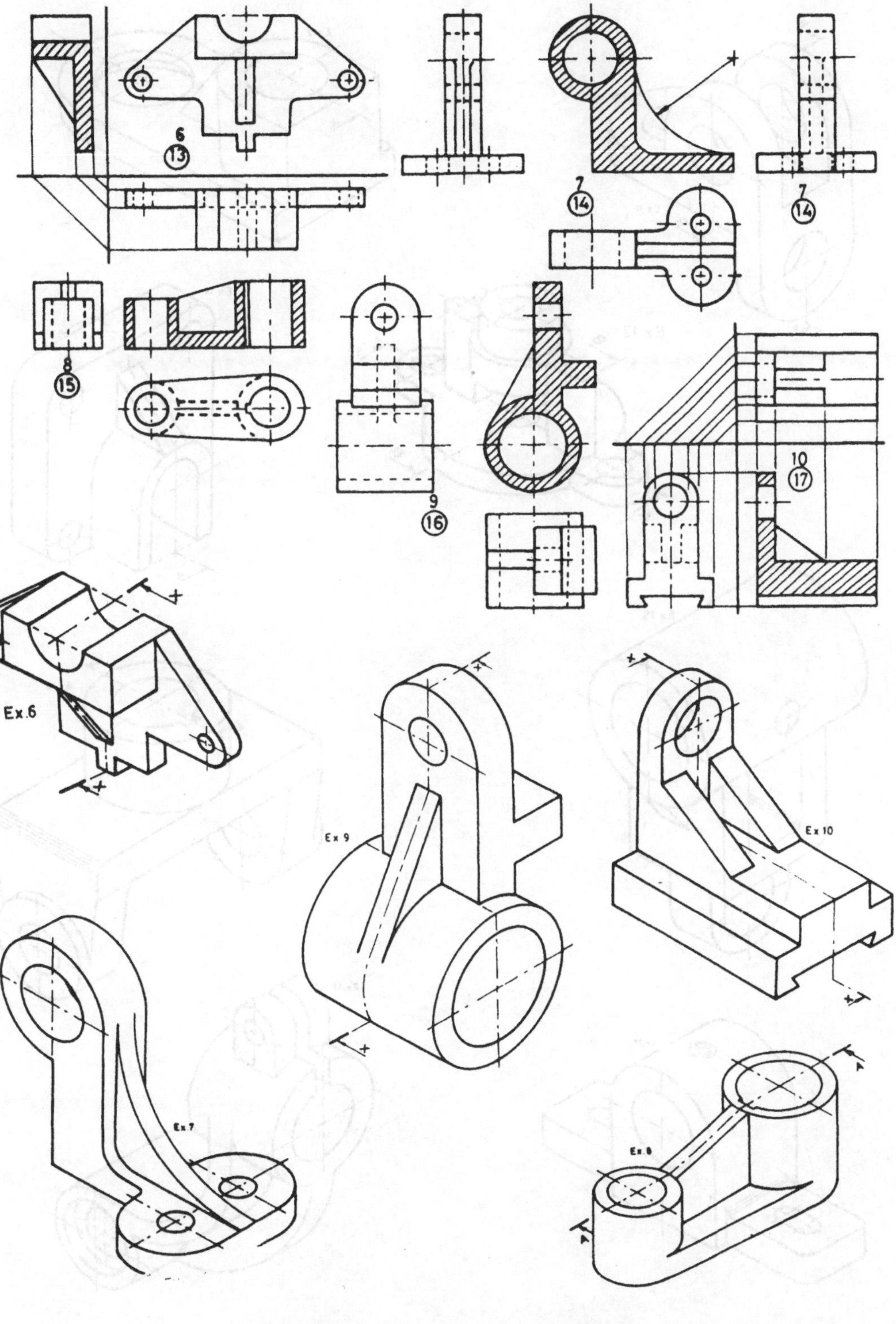
6
13
7
14
7
14
8
15
9
16
10
17
Ex.6
Ex 9
Ex 10
Ex.7
Ex.8

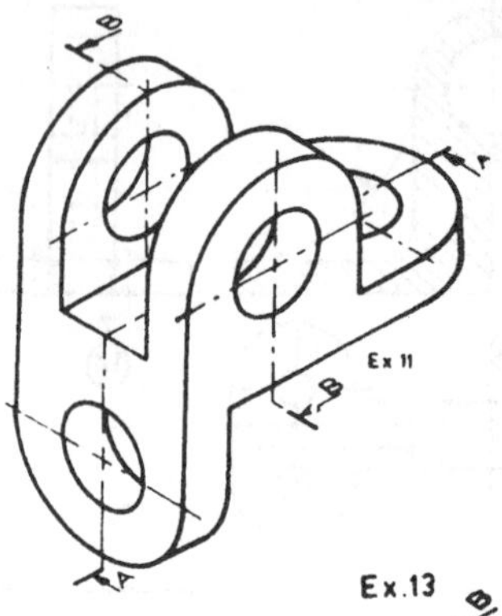

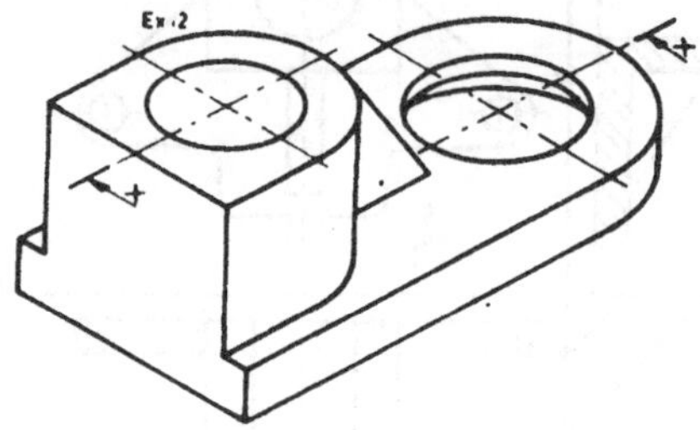

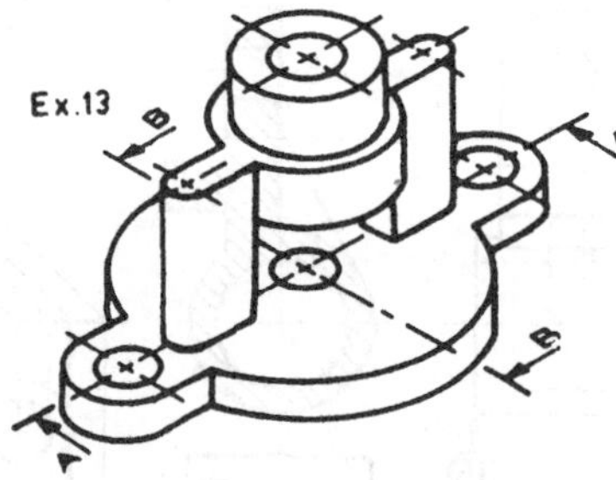

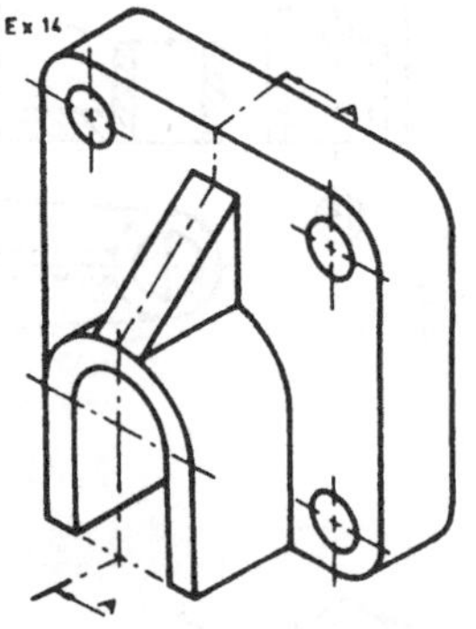

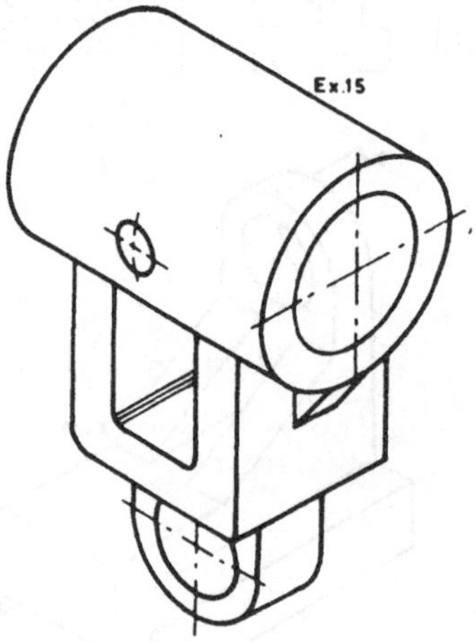

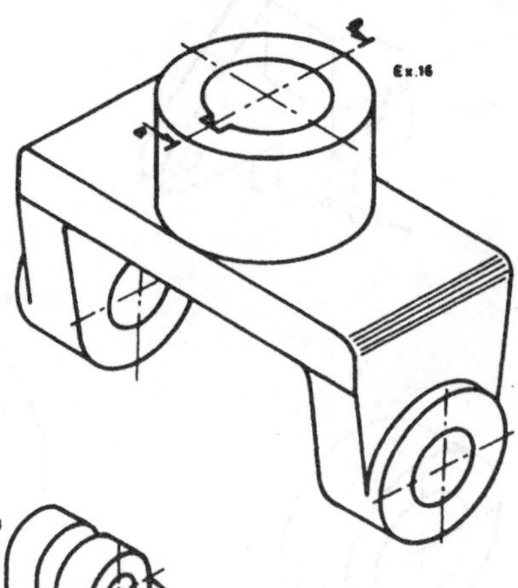

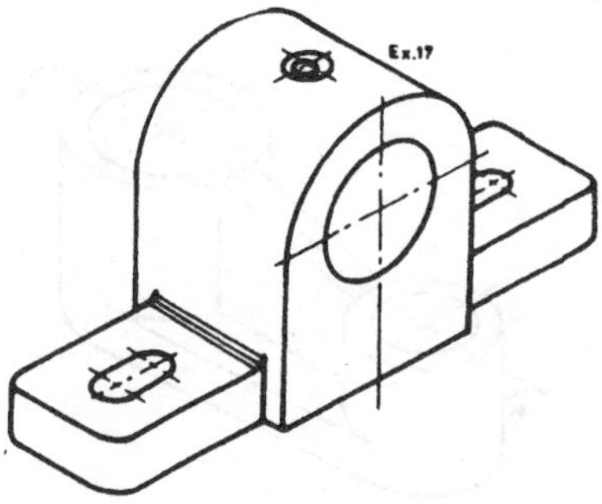

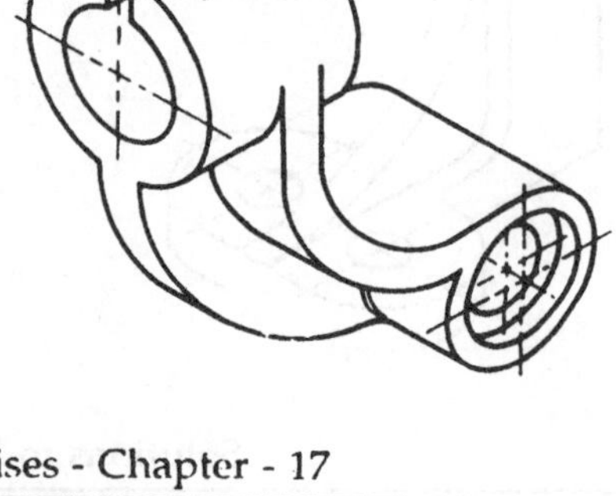

Hints for Solutions to Exercises - Chapter - 17

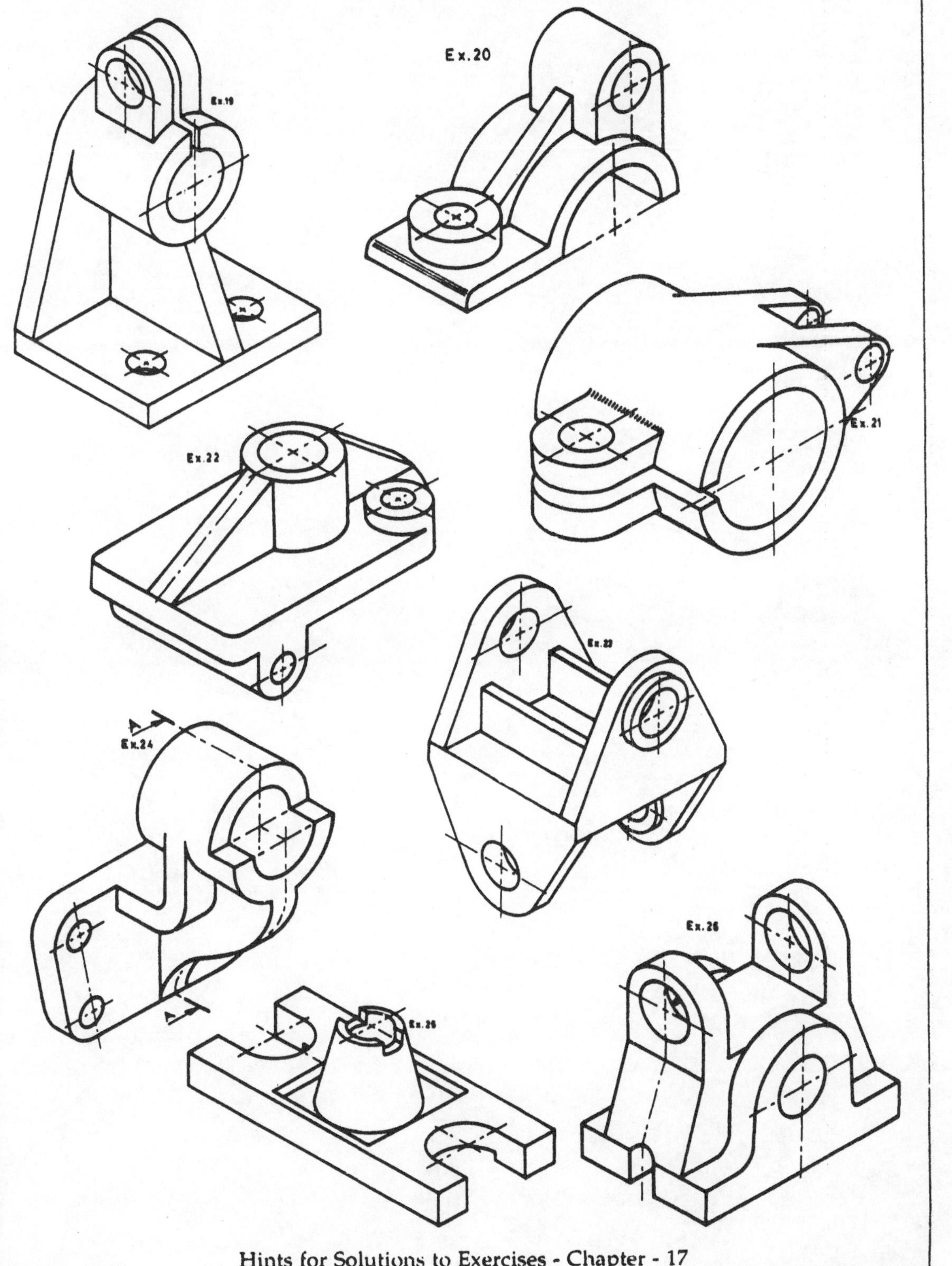

Hints for Solutions to Exercises - Chapter - 17